ENVIRONMENTAL CHANGE, CLIMATE AND HEALTH

Issues and research methods

The advent of global environmental change, with all its uncertainties and requirement for long-term prediction, brings new challenges and tasks for scientists, the public and policy makers.

A major environmental upheaval such as climate change is likely to have significant health effects. Current mainstream epidemiological research methods, in general, do not adequately address the health impacts that arise within a context in which ecological and other biophysical processes display nonlinear and feedback-dependent relationships. The agenda of research and policy advice must be extended to include the larger-framed and longer-term environmental change issues. This book identifies the nature and scope of the problem, and explores the conceptual and methodological approaches to studying these relationships, modelling their future realization, providing estimates of health impacts and communicating the attendant uncertainties.

This timely volume will be of great interest to health scientists and graduate students concerned with the health effects of global environmental change.

PIM MARTENS holds degrees in Biological and Environmental Health Sciences from Maastricht University, The Netherlands. He worked within the project Global Dynamics and Sustainable Development, launched in 1992 by the Dutch National Institute of Public Health and the Environment (RIVM). After obtaining his PhD from the Department of Mathematics, Maastricht University, he worked as assistant professor at the same Department. Since 1998, Pim Martens has been a senior researcher at the University's International Centre for Integrative Studies, where he directs the Global Assessment Centre. He is editor-in-chief of the international journal *Global Change and Human Health*. Furthermore, he was a member of the Assessment Study 'Climate, Ecosystems, Infectious Disease, and Human Health' (US National Research Council/National Academy of Sciences), and lead-author of several climate change and human health assessment reports of the Intergovernmental Panel on Climate Change (IPCC) and the World Health Organization (WHO). Pim Martens is a Fulbright New Century Scholar within the program 'Challenges of Health in a Borderless World.'

ANTHONY J. MCMICHAEL is a medical graduate from Adelaide, South Australia. He is currently Director of the National Centre for Epidemiology and Population Health, at the Australian National University, Canberra, Australia, having previously been Professor of Epidemiology at the London School of Hygiene and Tropical Medicine, UK. His research interests, over 30 years, have encompassed the social aetiology of mental health problems, the causes of occupational diseases, studies of diet and cancer, and environmental epidemiology. During the 1990s he developed a strong interest in the assessment of population health risks from global environmental change. His current research activities include studies of heatwave impacts on urban mortality patterns, of solar ultraviolet radiation and immune disorders, of dietary factors in breast cancer, and the mathematical modelling of future climate change impacts on malaria transmissibility. During 1990–1992 he chaired the Scientific Council of the International Agency for Research on Cancer. Between 1994 and 2001 he convened the review by the UN's Intergovernmental Panel on Climate Change of potential health impacts of climate change. He is a member of WHO's newly established Expert Group on Globalization and Health, and is a member of the International Scientific Panel on Population and Environment. In 2001, Cambridge University Press published his latest book, *Human Frontiers, Environments and Disease: Past Patterns, Uncertain Futures.*

ENVIRONMENTAL CHANGE, CLIMATE AND HEALTH

Issues and research methods

Edited by

P. MARTENS

Senior Researcher
International Centre for Integrative Studies,
Maastricht University, The Netherlands

A. J. McMICHAEL

Director of the National Centre for Epidemiology and
Population Health, Australian National University,
Canberra, Australia

CAMBRIDGE
UNIVERSITY PRESS

CAMBRIDGE UNIVERSITY PRESS
Cambridge, New York, Melbourne, Madrid, Cape Town, Singapore, São Paulo, Delhi

Cambridge University Press
The Edinburgh Building, Cambridge CB2 8RU, UK

Published in the United States of America by Cambridge University Press, New York

www.cambridge.org
Information on this title: www.cambridge.org/9780521114028

First published 2002
This digitally printed version 2009

A catalogue record for this publication is available from the British Library

Library of Congress Cataloguing in Publication data

Martens, Willem Jozef Meine, 1968–
Health impacts of global environmental change: concepts and methods / Pim Martens,
Anthony J. McMichael.
p.   cm.
Includes bibliographical references and index.
ISBN 0 521 78236 8
1. Environmental health.   2. Environmentally induced diseases.   3. Social medicine.
I. McMichael, A. J. (Anthony J.)   II. Title.
RA566 .M27   2002
616.9′88 – dc21   2001052884

ISBN 978-0-521-78236-4 hardback
ISBN 978-0-521-11402-8 paperback

Additional resources for this publication at www.cambridge.org/9780521114028

Contents

Contributors

Tamara Awerbuch
Department of Population and
  International Health
Harvard School of Public Health
Boston
USA

Louisa R. Beck
Ecosystem Science and Technology
  Branch
NASA Ames Research Center
Moffet Field
USA

Matthew R. Bobo
Ecosystem Science and Technology
  Branch
NASA Ames Research Center
Moffet Field
USA

Menno Bouma
Disease Control and Vector
  Biology Unit
London School of Hygiene and
  Tropical Medicine
London
UK

Diarmid H. Campbell-Lendrum
Infectious Diseases Department
London School of Hygiene and
  Tropical Medicine
London
UK

Ann G. Carmichael
History Department
Indiana University
Bloomington
USA

Manish A. Desai
Center for Occupational and
  Environmental Health
School of Public Health
University of California
Berkeley
USA

Kristie L. Ebi
Global Climate Change Research
EPRI
Palo Alto
USA

Paul R. Epstein
Center for Health and the Global
  Environment
Harvard Medical School
Boston
USA

Andy Haines
London School of Hygiene and
  Tropical Medicine
London
UK

Anthony E. Kiszewski
Department of Population and
  International Health
Harvard School of Public Health
Boston
USA

Uriel Kitron
College of Veterinary Medicine
University of Illinois
Urbana
USA

R. Sari Kovats
Department of Epidemiology and
  Population Health
London School of Hygiene and
  Tropical Medicine
London
UK

Katrin Kuhn
Department of Infectious and Tropical
  Diseases
London School of Hygiene and
  Tropical Medicine
London
UK

Richard Levins
Department of Population and
  International Health
Harvard School of Public Health
Boston
USA

Pim Martens
International Centre for Integrative
  Studies
Maastricht University
Maastricht
The Netherlands

Anthony J. McMichael
National Centre for Epidemiology and
  Population Health
Australian National University
Canberra
Australia

Bettina Menne
WHO European Centre for
  Environment and Health
Rome Division
Rome
Italy

Millicient Fleming Moran
Department of Applied Health Science
Indiana University
Bloomington
USA

Tim O'Riordan
CSERGE
School of Environmental Sciences
University of East Anglia
Norwich
UK

Terry W. Parr
Centre for Ecology and Hydrology
Merlewood
UK

Jonathan A. Patz
Department of Environmental Health
    Sciences
Johns Hopkins University
Bloomberg School of Public Health
Baltimore
USA

Dale S. Rothman
International Centre for Integrative
    Studies
Maastricht University
Maastricht
The Netherlands

Jan Rotmans
International Centre for Integrative
    Studies
Maastricht University
Maastricht
The Netherlands

Kirk R. Smith
Center for Occupational and
    Environmental Health
School of Public Health
University of California
Berkeley
USA

Paul Wilkinson
Department of Public Health and Policy
London School of Hygiene and
    Tropical Medicine
London
UK

Alistair Woodward
Department of Public Health
Wellington School of Medicine and
    Health Sciences
Wellington South
New Zealand

Foreword

Over the past two decades there has been a rapid evolution of research concepts and methods in relation to global environmental changes – their processes, impacts and the response options. The scale and complexity of these environmental problems are, in general, greater than those that individual scientists or their disciplines usually address. This is particularly so for those components of the topic that are furthest "downstream" from the pressures, or their drivers, that initiate the processes of global environmental change.

Indeed, in seeking to detect or forecast the population health impacts of global environmental changes there is an additional difficulty. Not only is the impact of research contingent on various assumptions, simplifications and projections made by scientists working "upstream" on the environmental change process per se, but the category of outcome – a change in the rate of disease or death – is one that usually has multiple contending explanations. If a glacier melts, then temperature increase is a very plausible explanation. Likewise, if birds, bees and buds exhibit their springtime behaviours a little earlier as background temperatures rise, that too is reasonably attributable to climatic change. However, if malaria ascends in the highlands of eastern Africa, regional climate change is just one contending explanation – along with changes in patterns of land use, population movement, increased urban poverty, a decline in the use of pesticides for mosquito control, or the rise of resistance to antimalarial drugs by the parasite.

There is also the problem of the time-frame. Much of the postulated health impact of global environmental change is likely to unfold over coming decades, as environmental stresses increase and life-support systems weaken. Yet, scientists generally prefer to work with empirical observations. Given that preference, and a well-honed body of scientific methods appropriate to empirical research, why try to use mathematical models to estimate how a change in global climatic conditions would affect patterns of infectious diseases, when the simple alternative is to sit back and wait for empirical evidence?

Well, that question is very much the nub of the issue. The world cannot afford to sit back and await the empirical evidence. The luxury of unhurried scientific curiosity must, here, be replaced by a more urgent attempt to estimate the dimensions of this problem – the health consequences of global environmental change – and then feed this information, with all its imperfections and assumptions, into the policy arena. Consideration of human health impacts is a crucial, even central, issue in the emerging international discourse on "sustainable development".

This, then, is a timely volume. There is an indisputable need to clarify the concepts and research procedures, and to illustrate recent and current research activities in this domain. The ongoing spectrum of health impact research entails learning from the recent past, detecting emergent health impacts and modelling future impacts. It also requires the assessment of how changes in world futures (social, economic, technological, political) will modulate these impacts, and how populations can or are likely to adapt to the change in environmental conditions.

If anything, this volume is overdue. The recognition of global environmental changes has already been a major spur to scientific development and methodological advances in many other disciplines, especially those elucidating and modelling the processes of change themselves. Accordingly, for example, our ability to model the world's climate system has increased many-fold over the past decade. In contrast, because of the abovementioned complexities that beset research into human health impacts, compounded by an apparent diffidence on the part of most epidemiologists and other population health scientists to engage in this unfamiliar domain, advances have been relatively slow to emerge in this disciplinary area. This volume will help to change that.

It is a well-rounded volume. The range of chapters includes attention to historical and social context, to differing conceptual domains of research, to questions about the assessment of population vulnerability, and to exploring and evaluating societal adaptation options. The challenge of scientific uncertainties is addressed – a challenge that looms large in research that deals with complex biophysical, ecological and social processes and which seeks to estimate future trajectories of population health risks.

Finally, this is an important volume because population health is so central to the formulation of humankind's "sustainable development" trajectory. If the life-support systems are weakened, and health is jeopardized, then we are all on the wrong track. Health scientists therefore have a major role and responsibility in informing this international discourse. The team of authors assembled in this book has had impressive and wide-ranging experience in the pioneering stages of this great scientific undertaking. Their shoulders should now be stood upon by others.

Robert T. Watson
Chair, Intergovernmental Panel on Climate Change,
Chief Scientist, and Director, Environmentally and Socially
Sustainable Development, World Bank

1

Global environmental changes: anticipating and assessing risks to health

ANTHONY J. McMICHAEL & PIM MARTENS

1.1 Introduction

The meaning of the word "environment" is elastic. Conventionally it refers to the various external factors that impinge on human health through exposures common to members of groups, communities or whole populations, and that are typically not under the control of individuals (i.e. the exposures are predominantly involuntary). Thus, "environmental exposures" are usually thought of as physical, chemical and microbiological agents that impinge on us from the immediately surrounding (ambient) environment.

The "environmental" roles of socioeconomic status in the determination of disease patterns, including aspects such as housing quality and material circumstances, have also claimed increasing attention from health researchers. This, however, requires a more inclusive definition of "environment" – one that embraces social and economic relations, the built environment and the associated patterns of living.

Note also that we typically view the environment as being "out there". It surrounds us, it impinges on us – but it is *not* us. This implied separateness reflects the great philosophical tradition that arose in seventeenth-century Europe as the foundations of modern empirical western science were being laid by Bacon, Descartes, Newton and their contemporaries. For several centuries this view helped us to manage, exploit and reshape the natural world in order to advance the material interests of industrializing and modernizing western society. In recent times, however, the magnitude of that environmental impact by human societies has increased exponentially. Consequently, in the light of the now-evident accruing environmental damage and the ongoing deterioration of many ecological systems, we must re-think our relationship to that "external world". We must recognize the essential dependency of human society and its economy upon the natural world. That dependency is manifest in the risks to human health that have arisen, or will arise, from the advent of these large-scale environmental changes – changes that are the current hallmark of the impact of the modern human species upon the ecosphere.

1.2 "Environment": the wider dimension

During the last quarter of the twentieth century we began to see evidence of a general disturbance and weakening of the world's life-supporting systems and processes (Loh *et al.*, 1998; Watson *et al.*, 1998). This unprecedented disruption of many of Earth's natural systems by humankind, at the global level (Vitousek *et al.*, 1997), reflects the combined pressure of rapidly increasing population size and a high-consumption, energy-intensive and waste-generating economy.

Global economic activity increased 20-fold during the twentieth century. Meanwhile, in absolute terms, the human population has been growing faster than ever in this past quarter-century, capping a remarkable fourfold increase from 1.6 to six billion during the twentieth century (Raleigh, 1999). The last three billion have been added in 14, 13 and, most recently, 12 years, respectively. While we remain uncertain of Earth's human "carrying capacity" (Cohen, 1995), we expect that the world population will approximate to nine billion by around 2050, and will probably stabilize at around 10–11 billion by the end of the twenty-first century.

In September 1999, the United Nations Environment Program issued an important report: *Global Environment Outlook 2000* (United Nations Environment Program, 1999). Its final chapter begins thus:

The beginning of a new millennium finds the planet Earth poised between two conflicting trends. A wasteful and invasive consumer society, coupled with continued population growth, is threatening to destroy the resources on which human life is based. At the same time, society is locked in a struggle against time to reverse these trends and introduce sustainable practices that will ensure the welfare of future generations ...

There used to be a long time horizon for undertaking major environmental policy initiatives. Now time for a rational, well-planned transition to a sustainable system is running out fast. In some areas, it has already run out: there is no doubt that it is too late to make an easy transition to sustainability for many of these issues ...

These are strong words. The report urges national governments everywhere to recognize the need for urgent, concerted and radical action. The report's assessment concurs with others, such as the detailed analysis of changes in major global ecosystems carried out by the World Wide Fund for Nature, leading to an estimation that approximately one-third of the planet's vitality, its natural resource stocks, have been depleted over the past three decades (Loh *et al.*, 1998). In Box 1.1 the main types of global environmental changes are addressed. It is of interest to review, as historical narrative, the changing profile and scale of human intervention in the environment. From that review, in Section 1.3, we can thus better understand how we have arrived at today's situation.

BOX 1.1

The main types of global environmental change

The main global environmental changes, of a kind that were not on the agenda a short quarter-century ago, are summarized below.

Climate change

During the 1990s, the prospect of human-induced global climate change became a potent symbol of these unprecedented large-scale environmental changes. Since 1975 average world temperature has increased by approximately 0.5 °C, and climate scientists now think this may be the beginning of the anticipated climate change due to human-induced greenhouse-gas accumulation in the lower atmosphere (Intergovernmental Panel on Climate Change, 2001). Weather patterns in many regions have displayed increasing instability, and this may be a foretaste of the increasing climatic variability predicted by many climate change modellers.

Stratospheric ozone depletion

Meanwhile, higher in the atmosphere, a separate problem exists. Depletion of stratospheric ozone by human-made industrial gases such as chlorofluorocarbons (CFCs) has been documented over several decades. Terrestrial levels of ultraviolet irradiation are estimated to have increased by around 5–10 % at mid-to-high latitudes since 1980. This problem is now projected to peak by around 2010–2020. Simulation models estimate that European and North American populations will experience an approximate 10 % excess incidence of skin cancer in the mid-twenty-first century (Martens *et al.*, 1996; Slaper *et al.*, 1996). These changes in the lower and middle atmospheres provide the most unambiguous signal yet that the enormous aggregate impact of humankind has begun to overload the biosphere. The capacity of the atmosphere to act as a "sink" for our gaseous wastes has been manifestly exceeded.

Loss of biodiversity

The loss of biodiversity is another major global environmental change. As the human demand for space, materials and food increases, so populations and species of plants and animals around the world are being extinguished at an accelerating rate – apparently much faster than the five great natural extinctions that have occurred in the past half-billion years since vertebrate life evolved. The problem is not simply the loss of valued items from nature's catalogue. It is, more seriously, the destabilization and weakening of whole ecosystems and the consequent loss of their products and their recycling, cleansing and restorative services. That is, we are losing, prior to their discovery, many of nature's chemicals and genes – of the kind that have already conferred enormous medical and health-improvement benefits. Myers (1997) estimates that five-sixths of tropical vegetative nature's medicinal goods have yet to be recruited for human benefit. Meanwhile, "invasive" species are spreading into new non-natural environments via intensified human food production, commerce and mobility. These changes in regional species composition have myriad consequences for human health. Just one example: the choking spread of the water hyacinth in eastern Africa's Lake Victoria, introduced from Brazil as a decorative plant, has provided a microenvironment for the proliferation of diarrhoeal disease bacteria and the water snails that transmit schistosomiasis (Epstein, 1999).

Nitrogen loading

Since the commercialization of nitrogenous fertilizers in the 1940s, there has been a remarkable, sixfold, increase in the human "fixation" of biologically activated nitrogen (Vitousek *et al.*, 1997). Humankind now produces more activated nitrogen than does the biosphere at large. The recent United Nations Environment Program Report (1999) suggests that disruption of the biosphere's nitrogen cycle may soon turn out to be as serious a problem as the better-known disruption of the world's carbon cycle. This increased nitrogen loading is affecting the acidity and nutrient balances of the world's soils and waterways. This, in turn, is affecting plant biochemistry, the pattern of plant pests and pathogens, and the species composition of ecosystems. Via the sequence of eutrophication of waterways, leading to algal blooms and oxygen depletion, nitrogen loading is beginning to sterilize coastal waters, such as Chesapeake Bay in Maryland, the Baltic Sea, and the Gulf of Mexico.

Terrestrial and marine food-producing systems

Meanwhile, the ever-increasing demands of agricultural and livestock production are adding further stresses to the world's arable lands and pastures. We enter the twenty-first century with an estimated one-third of the world's previously productive land significantly damaged by erosion, compaction, salination, waterlogging or chemical destruction of organic content, and with about half of that damaged land showing reduced productivity (United Nations Environment Program, 1999). Similar pressures on the world's ocean fisheries have left most of them seriously depleted. These changes compromise the capacity of the world to continue to provide, sustainably, sufficient food for humankind.

Freshwater supplies

In all continents, freshwater aquifers are being depleted of their "fossil water". Agricultural and industrial demand now often greatly exceed the rate of natural recharge. Water shortages are likely to cause tensions and conflict over coming decades (Homer-Dixon, 1994; Gleick, 2000). For example, Ethiopia and the Sudan, upstream of Nile-dependent Egypt, increasingly need the Nile's water for their own crop irrigation. Approximately 40 % of the world's population, living in 80 countries, now faces some level of water shortage. India has seen its per-person supply of freshwater drop from 5500 cubic metres per year in the 1950s to around 1800 cubic metres now, hovering just above the official scarcity threshold. By 2050 India's supply will be around 1400 cm per person – and, further, the slight drying due to global climate change that is projected by climate modelling would exacerbate this further (Cassen & Visaria, 1999).

Persistent organic pollutants

Many long-lived and biologically active chemicals have become widely distributed across the globe (Watson *et al.*, 1998). Lead and other heavy metals are present at increasing concentrations in remote environments. More worrying, various semi-volatile organic chemical pollutants (such as polychlorinated biphenyls) are disseminated towards the poles via a remarkable sequential "distillation" process through the cells of the lower atmosphere (Tenenbaum, 1998). Consequently, their concentrations are increasing in polar mammals and fish and in the traditional human groups that eat them. Their immunosuppressive effect has been demonstrated in seals, other marine mammals and rodents (Vos *et al.*, 2000). Current epidemiological studies in the Faroe Islands and elsewhere may soon tell us if humans are similarly affected.

1.3 Six phases of human ecology over the past 100 millennia

The story of human health and disease in relation to environmental conditions has deep roots in human prehistory and history. The profile of contemporary western diseases would have been as unrecognizable to your average Palaeolithic hunter-gatherer, early agrarian or nineteenth-century urban citizen as would their day-to-day procession of diseases be to our eyes. Over the past 100 millennia, humans have undergone an accelerating succession of environmental and cultural changes: dispersal, tool-making, patterns of social cohesion, agriculture, urbanism, sea-faring, and, latterly, industrialization. Six main phases can be identified, each ushering in new patterns of disease and death. Because these phases provide the backdrop to much of what follows in later chapters it may help to outline them here.

1.3.1 Hunter-gatherers

For most hunter-gatherers, the primary causes of death were physical trauma, infection or, less often, starvation. As with other animals, human life expectancy was that of young adulthood – only a successful or lucky minority completed a full reproductive lifespan. Fossil bones suggest an average lifespan of around 25 years.

The bones yield some evidence of trauma and malnutrition. The types of infections would have been limited to those compatible with small mobile human populations, probably including bacterial infections of skin, ears, nose and throat, various parasitic intestinal worms, and incidental infection with the malaria parasite and the African sleeping sickness trypanosome – both of which diseases were circulating in wild animals.

1.3.2 Agriculture, settlements and cities

Two important new influences on health emerged with agriculture, animal domestication and settlement: chronic nutritional deficiencies occurred and various "crowd infections" began to appear in urbanizing populations. Agrarian dependence on a restricted range of staple foods, with reduced meat intake, led to nutritional deficiencies. Early agrarians were distinctly shorter than their immediate hunter-gatherer predecessors. Agriculture, while greatly increasing local environmental carrying capacity, does not eliminate famines: they have persisted throughout most of history. Meanwhile, new contagious infections such as influenza, dysentery, smallpox and measles arose as mutated versions of long-established infections in newly domesticated animals or rodent pests. As villages became towns, and towns became cities, the magnificence and might of urban life unfolded, along with the crowding, oppression and squalor. Great civilizations came and went, often largely in response to the exhaustion of local agricultural systems or surface water supplies – as seemed to be the case, for example, with the 2000-year success story that once was Mesopotamia. Infectious disease epidemics occurred, sometimes in response to, and sometimes as a precursor of, great social and political upheavals.

1.3.3 Commerce, conquest and microbial confluence

Much later, as trade routes opened up, and as conquering armies spread their reach, so infectious diseases spread more widely. Smallpox and measles, unknown in Greece, reached Rome because of trade with the Middle East and Asia during the middle years of the Roman empire. The bubonic plague first arrived in cataclysmic fashion in the Roman Empire in the sixth century AD and in China shortly after. Bubonic plague (the Black Death) returned to Europe, again from the east, in the mid-fourteenth century, immediately following a devastating outbreak in China. The Spanish conquistadors in the early 1500s took measles, smallpox and other acute infectious diseases to the Americas, where, inadvertently, they proved to be terrible weapons of microbiological and psychological warfare. Relative to the genetically selected and immunologically battle-hardened Eurasian populations, Amerindians, Australian aboriginals and Pacific island populations were immunologically naive and were consequently devastated by these infections. The dissemination of many infectious diseases continues today, as poverty persists, as human mobility and trade increase, and as Third World populations urbanize.

1.3.4 Industrialization

The advent of mechanized agriculture in the eighteenth and nineteenth centuries, along with sea-freight and refrigeration, increased the food supplies to western countries. Europe's population expanded and spilled over to the Americas, southern Africa and Australasia. Industrialization and imperialism brought material wealth and social modernization to Europe. In the latter decades of the nineteenth century, improvements occurred in sanitation, housing, food safety, personal hygiene and literacy. These, in turn, led to control of infection. Later, immunization and antibiotics consolidated a new era of human supremacy over infectious diseases. Industrialization, meanwhile, also intensified the contamination of local environments with chemical pollutants. From early in the twentieth century, occupational exposures to hazardous chemicals and to ionizing radiation became more frequent.

1.3.5 Modern times: urban consumerism

Since World War II, human lifestyles in western countries have changed radically. Changes in food choices, dietary habits, smoking behaviour, alcohol consumption and physical inactivity have caused increases in various chronic noncommunicable diseases (and decreases in some others). Changes in sexual, contraceptive and reproductive behaviours have also greatly influenced patterns of infectious and noninfectious diseases – including human immunodeficiency virus and acquired immunodeficiency syndrome (HIV/AIDS), other sexually transmitted diseases, breast and ovarian cancers and cardiovascular diseases. Meanwhile, the introduction of life-saving public health and medical technology to Third World countries has reduced the childhood death toll from infectious diseases. Because this mortality decline has so far only been partially offset by a subsequent fall in fertility, rapid population increases have occurred in many of those countries in recent decades, creating additional demographic and resource pressures.

1.3.6 An increasingly full world: the advent of global environmental change

Today, the aggregate impact of the human population size and economic activity on various of the world's biophysical systems has begun to exceed the regenerative and repair capacities of those systems. Such overload has never before occurred globally; this is a historical "first". *Homo sapiens* now accounts for approximately

40 % of the total terrestrial photosynthetic product (actual or potential): by growing plants for food, by clearing land and forest, by degrading land (both arable and pastoral), and by building or paving over the land (Vitousek *et al.*, 1997).

This unfamiliar, historically unprecedented, situation of humankind overloading Earth's carrying capacity presents a special challenge to science. How can we best estimate the likely consequences for human health (or other outcomes) of the plausible future scenarios of environmental change (see Box 1.1 for an overview of the main global environmental changes affecting human health)? This question warrants careful consideration. It poses a number of challenges, some of them unfamiliar, to population health scientists. However, let us first review the recent history of evolving priorities in the topic area of "environment and health."

1.4 Environment and health: recent developments

At the 1972 United Nations Conference on the Human Environment, in Stockholm, concern was focused on the increasing release of chemical contaminants into local environments, the prospects of depletion of certain strategic materials, and some aspects of the modern urban environment. There were environmental hazards resulting from western industrial intensification, the rapid, programmed and often profligate industrialization in Soviet bloc countries, and the poorly controlled and increasingly debt-driven industrial and agricultural growth in newly-independent Third World countries. In consequence, the world experienced various serious episodes of air pollution (e.g. London in 1952), organic mercury poisoning (Minamata in 1956), heavy metal accumulation (especially lead and cadmium), pesticide toxicity and scares from environmental ionizing radiation exposures.

Today, similar toxicological environmental problems persist widely around the world. Since 1972, we have had Bhopal, Seveso, Chernobyl, and in 1999 the fatal reactor accident at Tokaimura in Japan. Air pollution is an increasing, often dramatic, problem in many large cities in the developing world.

Meanwhile, a further, unfamiliar, set of large-scale environmental problems has begun to emerge. Indeed, by the 1992 United Nations Conference on Environment and Development, in Rio de Janeiro, they were moving centre-stage. The World Commission on Environment and Development had, in the late 1980s, put "sustainable development" on the world's agenda. There was nascent recognition that we were beginning to live beyond Earth's means, that limits had been breached, and that the continuing increase in the weight of human numbers and economic activity therefore posed a new and serious problem – including risks to human health. Life-support systems were coming under threat at a global level.

These global environmental changes are a manifestation of a larger pattern of change in the scale and intensity of human affairs. Global climate change is one of the most widely discussed of these global environmental changes. In 1996, the United Nation's Intergovernmental Panel on Climate Change (IPCC) concluded that human-made changes in the global atmosphere were probably already beginning to change world climate (IPCC, 1996). During 1997 and 1998, global temperatures reached their highest levels since record keeping began in the mid-nineteenth century, and 1999 was also well above the century's average temperature. Overall, ten of the 12 hottest years of the twentieth century occurred after 1988. Around the world, during the late 1990s and turn of the century, it seemed that world weather patterns were becoming more unstable, more variable. In 2001, the IPCC firmed up its conclusion that human-induced climate change was already occurring, and raised its estimation of the likely range (1.4–5.8 °C) of temperature increase during the twenty-first century (IPCC, 2001).

The prospect that climate change and other environmental changes will affect population health poses radical challenges to scientists; fortunately, this has arisen at a time of growing interest among epidemiologists in studying and understanding the population-level influences on patterns of health and disease. These strivings to understand population disease risks and profiles within a larger contextual framework – be it social, economic, cultural or environmental – will, hopefully, be mutually reinforcing. After all, they share a recognition that there are complex underlying social, cultural and environmental systems which, when perturbed or changed, may alter the pattern of health outcomes. In this respect they recognize the *ecological* dimension of disease occurrence – that is, as changes occur in the systems that constitute the milieu of human population existence, so the prospects for health and disease are altered.

The exploration of these systems-based risks to human health seems far removed from the tidy examples that abound in textbooks of epidemiology and public health research. Yet there are real and urgent questions being posed to scientists here. The wider public and its decision-makers are seeking from scientists useful estimates of the likely population health consequences of these great and unfamiliar changes in the modern world. Illustrative of this expectation is that the World Health Organization's second estimation of the "global burden of disease", conducted during 2000–2001, included an estimation of the burden attributable to climate change scenarios over the coming decades. Similarly, the United Nations Development Program, in seeking to identify "global public health goods", has paid particular attention to large-scale environmental changes as manifestations of losses in fundamentally important "public health goods" – losses of common-property environmental assets that are likely to impact most on the world's poor and vulnerable populations, and are likely to compound over the coming generations.

Clearly, there is a major task for health scientists in this topic area. This book seeks to identify the nature and scope of the problem, and to explore the conceptual and methodological approaches to studying these relationships, modelling their future realization, providing estimates of health impacts and communicating the attendant uncertainties. The next section of this opening chapter overviews the strategies available for studying and estimating the health impacts of climate change.

1.5 Challenges to population health research

The great majority of researchers are *empiricists* by training and tradition, studying the past and the present by direct observation. By definition, empirical methods cannot be used to study the future. To the extent that the advent of global environmental change obliges scientists to estimate future impacts, should current or foreseeable trends continue, then empiricism must be supplemented by predictive modelling. Epidemiologists, whose primary task is to identify risks to health from recent or current behaviours, exposures or other circumstances, are not much oriented to asking questions about health impacts several decades hence. That is beyond the time horizon and methodological repertoire of the standard textbook.

Western science has long set great store by *reductionism* – the assumption that one can understand the working of the whole by studying the component parts. Further, western science classically conducts such studies, preferably by deliberate experiment, by holding constant the context (i.e. other background factors) so as to more clearly describe and quantify some specific relationship. However, we cannot meaningfully study a complex dynamic system, such as an ecosystem or the world's climate system, by reducing it to a set of parts, assuming that each part is amenable to separate study.

Yet, these contextual difficulties aside, population health scientists must find ways to estimate the potential health consequences of current social and environmental trajectories. Not only is this an interesting scientific task, but – crucially – it will assist society in seeking a sustainable future. Clearly, elucidating these risks to population health from environmental changes such as long-term changes in global climatic patterns, depletion of stratospheric ozone and biodiversity loss poses a special research challenge (see Chapters 2 and 3). For a start, these environmental changes entail unusually large spatial scales. They also entail temporal scales that extend decades, or further, into the future. Some entail irreversible changes. While some direct impacts on health would result – such as the health consequences of increased floods and heatwaves due to global climate change, or increases in skin cancer due to ozone depletion – many of the impacts would result from disruption of the ecological processes that are central to food-producing ecosystems or to the ecology of infectious-disease pathogens. That is, many of the causal relationships are neither simple nor immediate.

1.5.1 Concepts

A fundamental characteristic of this topic area is the pervasive combination of complexity and uncertainty that confronts scientists. Policy-makers, too, must therefore adjust to working with incomplete information and with making "uncertainty-based" policy decisions. They must jettison misplaced assumptions that scientists can provide final and precise truths. Relatedly, society at large will have to come to terms with the Precautionary Principle, in order to minimize the chance of low-probability but potentially devastating outcomes. When the science is uncertain or infeasible and the stakes are potentially high, better to be safe than sorry. While scientists dislike "false positives" (hence their reflex invocation of statistical significance tests), society's interest lies in not being caught out by science's "false negatives".

Several aspects of the complexity and uncertainty of this research domain are dealt with specifically in three of the subsequent sections. Those aspects are: (i) complexity and surprises, (ii) uncertainties, and (iii) determinants of population vulnerability, and adaptive capacity, to these environmental changes.

1.5.1.1 Complexity and surprises

Predicting the impact of a changing world on human health is a hard task and requires an interdisciplinary approach drawn from the fields of evolution, biogeography, ecology and social sciences, and it relies on various methodologies such as mathematical modelling as well as historical and political analysis (see later). When even a simple change occurs in the physical environment, its effects percolate through a complex network of physical, biological and social interactions, that feed back and feed forwards. Sometimes the immediate effect of a change is different from the long-term effect, sometimes the local changes may be different from the region-wide alterations. The same environmental change may have quite different effects in different places or times. Therefore, the study of the consequences of environmental change is a study of the short- and long-term dynamics of complex systems, a domain where our common sense intuitions are often unreliable and new intuitions have to be developed in order to make sense of often paradoxical observations (see Chapter 4).

1.5.1.2 Uncertainties

The prediction of environmental change and its health impacts encounters uncertainties at various levels. Some of the uncertainties are of a scientific kind, referring to deficient understanding of actual processes; for example, knowing whether increased cloud cover arising from global warming would have a positive or a negative feedback effect. Some of the uncertainties refer to the conceptualization and construction of mathematical models in which the specification of linked processes may be uncertain or whose key parameter values are uncertain (see also Chapter 8). For example, what is the linkage between changes in temperature, humidity and surface water in the determination of mosquito breeding, survival and biting behaviour? Some uncertainties are essentially epistemological, referring to what we can and cannot reasonably foresee about the structure and behaviour of future societies, including for example their future patterns of greenhouse gas emissions. And, finally, there is of course the familiar source of uncertainty that arises from sampling variation, and which leads to the need for confidence intervals around point estimates.

Human societies have, of course, some experience of uncertainty-based policy-making. We avoid locating housing developments around nuclear power plants because of the recognized finite but unquantifiable risk of serious accident. We have taken various actions to prevent the final extinction of many species of plants and animals, in part because of concerns about likely but uncertain knock-on consequences for the functioning of ecological systems. Yet it is also clear that many such decisions are delayed or otherwise hampered by a lack of information about quantifiable risks, and hence, also, a lack of information about the likely economic costs to society. There is a need to reduce the gap between these two domains, the risk-based and the uncertainty-based policy-making. At least that need will exist while we come to terms with the as-yet unfamiliar inevitability of a substantial amount of uncertainty, as a property of the systems and processes in which changes are occurring (see Chapter 12).

1.5.1.3 Vulnerability and adaptation

Human populations vary in their vulnerability to health hazards. A population's vulnerability is a function of the extent to which a health outcome is sensitive to climate change and of the population's capacity to adapt to the new climate conditions. The vulnerability of a population depends on factors such as population density, level of economic development, food availability, local environmental conditions, pre-existing health status, and the quality and availability of public health care.

Adaptation refers to actions taken to lessen the impact of the (anticipated) climate change. There is a hierarchy of control strategies that can help to protect population health. These strategies are categorized as: (i) administrative or legislative; (ii) engineering; or (iii) personal (behavioural). Legislative or regulatory action can be taken by government, requiring compliance by all, or by designated classes of, persons. Alternatively, an adaptive action may be encouraged on a voluntary basis, via advocacy, education or economic incentives. The former type of action would normally be taken at a supranational, national or community level; the latter would range from supranational to individual levels. Adaptation strategies will be either reactive, in response to observed climate impacts, or anticipatory, in order to reduce vulnerability to such impacts (see Chapter 11).

1.5.2 Research methods

Next to the conceptual challenges we have to face, the assessment of the risks to population health from global environmental change requires several complementary research strategies. Research into the health impacts of these environmental changes can be conducted within three domains, and there is a variety of methods that can be used within each domain (see Chapter 5). The three categories of research are:

(i) The use of historical and other analogue situations which, as (presumed) manifestations of existing natural environmental variability, are thought likely to foreshadow future aspects of environmental change. These empirical studies help to fill knowledge gaps, and strengthen our capacity to forecast future health impacts in response to changing environmental–climatic circumstances.

(ii) The seeking of early evidence of changes in health risk indicators or health status occurring in response to actual environmental change. Attention should be paid to sensitive, early-responding, systems and processes.

(iii) By using existing empirical knowledge and theory to model future health outcomes in relation to prescribed scenarios of environmental change. This is referred to as scenario-based health risk assessment.

1.5.2.1 Analogue studies

Empirically based knowledge about the relationship between climate and health outcomes is a prerequisite to any formal attempt to forecast how future climate change is likely to affect human health. In fact, we cannot know in advance the exact configurations of the future world. Indeed, we should assume that in some respects the future will be unlike the present, both in its overall format and in the component relationships between now-familiar variables which, in future, will occur at unfamiliar levels. (For example, will the rate of evolution of drug resistance in malarial parasites increase as temperatures rise and generation time shortens? Will new pests and pathogens emerge in agriculture, thereby reducing harvest yields, as climatic conditions change? And will the North Atlantic deep-water formation system – part of the heat-transferring oceanic "conveyor belt" – weaken as ocean temperatures rise several degrees centigrade?) Nevertheless, our best guide to foreseeing the future is to have studied and understood the past and present (see Chapter 6).

1.5.2.2 Empirical studies of early health effects

If recent global climate trends continue, and it becomes more certain that this process is the beginning of anthropogenic climate change, then epidemiologists must seek early evidence of impacts on health. Such things as patterns of heat-related deaths, the seasonality of allergic disorders, and the geographical range and seasonality of particularly climate-sensitive infectious diseases can be expected to begin to change.

There is evidence that the global climate change over the past quarter-century has begun to affect patterns of plant growth and distribution, particularly at mid-latitudes and in many mountain regions, e.g. the Alps (Grabherr *et al.*, 1994). There is also good evidence of climate-related changes in the distribution and behaviour of animal species both within Europe and elsewhere. For example, the northern limit of the distribution of tick vectors for tick-borne encephalitis moved north in Sweden between 1980 and 1994. Further analysis shows that changes in the distribution and density of that tick species over time have been correlated with changes in seasonal temperatures and human disease (Lindgren *et al.*, 2000; 2001).

There is little evidence yet of changes in human population health that can be attributed to the observed recent changes in climate (principally the warming that has occurred over the last 20 years). The debate has primarily focused on malaria in the highlands (Epstein *et al.*, 1998; Reiter, 1998; Hay et al., 2002). Although many highland regions, particularly in Africa, have experienced a resurgence of malaria, the existence of many co-varying factors (e.g. land-use change, population movement) and too few time-series datasets has impeded formal assessment of the climate–malaria relationships. So, too, has the variable and often poor quality of the available data (see Chapter 7).

1.5.2.3 Modelling

Modelling is often used by epidemiologists to analyse empirical data; for example, to gain insights into the underlying dynamics of observed infectious-disease epidemics such as HIV. The estimation of the future health impacts of projected scenarios of climate change poses some particular challenges, because of both the complexity of the task and the difficulties in validating the model against relevant historical datasets and, relatedly, in then calibrating it against external observations. Several modelling approaches are used, particularly empirical-statistical models, process-based models and integrated models. The choice of model depends on several factors, such as the purpose of the study and the type of data available (see Chapters 4 and 8).

1.5.2.4 Geographical Information Systems and remote sensing

Remote sensed (RS) data from weather satellites can be used to monitor changes in temperature and precipitation in order to predict continental and global patterns of disease outbreaks. Higher resolution satellite data can been used in a landscape epidemiological approach to model patterns of disease-transmission risk at local to regional scales. A comprehensive model of disease risk due to, for example, climate change should incorporate the temporal aspects of the climate models integrated with the spatial forecasting made possible by the use of Geographical Information Systems (GIS) technologies and spatial analyses. RS and GIS technologies provide unprecedented amounts of data and data-management capabilities (see Chapter 9).

1.5.2.5 Monitoring

A range of national, regional and international organizations routinely collect relevant data, most obviously those monitoring environmental conditions, and (usually separately) health status. While these systems constitute a potentially powerful resource, most were implemented for purposes other than studying environmental change effects on health. Monitoring is "the continuous or repeated observation, measurement and evaluation of health and/or environmental data for defined purposes, according to prearranged schedules in space and time, using comparable methods for sensing and data collection." Environmental change/health monitoring should be directed towards the following aims: (i) early detection of the health impacts of global environmental change; (ii) improved quantitative analysis of the relationships between environment and health; (iii) improved analysis of population vulnerability; (iv) prediction of future health impacts of environmental change, and validation of predictions; and (v) assessment of the effectiveness of adaptation strategies. From the above it becomes clear that monitoring will also be an important component in the other methods mentioned earlier (see Chapter 10).

1.6 Conclusions

The advent of global environmental change, with its complexities, uncertainties and displacement into the future, brings new challenges and tasks for science, the public and policy-makers. The advent of this research task also poses a political and moral dilemma. We already face many serious and continuing local environmental health hazards. Poor populations around the world are exposed to unsafe drinking water, which is microbiologically contaminated or, in the case of Bangladesh, contains toxic levels of arsenic. Environmental lead has been widely dispersed in the modern world, via industry, traffic exhausts and old house-paints; it continues to blight child intellectual development. Urban populations face continuing hazards from air pollution. All of these environmental health issues must continue to command our attention. Yet, now, we must also extend the agenda of research and policy advice to include the larger-framed environmental change issues as emerging hazards to the health of current and future populations. This, as has been made clear in this book, will entail not just an expansion of effort but a widening of the repertoire of science.

We are entering a century in which science must increasingly engage in issues relating to the processes and consequences of changes to ecological systems, be they the systems of the natural biosphere, the biophysical systems of global climate, or the increasingly large and complex social systems in which we live our lives. While we do our best as scientists and policy-makers to understand and ameliorate the present, we must, increasingly, look to the need to anticipate the future – and seek a socially and ecologically sustainable path to it.

References

Cassen, R. & Visaria, P. (1999). India: looking ahead to one and a half billion people. *British Medical Journal*, **319**, 995–7.

Cohen, J. (1995) *How Many People Can the Earth Support?* New York: Norton.

Epstein, P. R. (1999). Climate and Health. *Science*, **285**, 347–8.

Epstein, P. R., Diaz, H. F., Elias, S., Grabherr, G., Graham, N. E., Martens, P., Mosley-Thompson, E. & Susskind, J. (1998). Biological and physical signs of climate change: focus on mosquito-borne diseases. *Bulletin of the American Meteorological Society*, **79**, 409–17.

Gleick, P. (2000). *The World's Water: The Biennial Report on Freshwater Resources 2000–2001*. Washington D.C.: Island Press.

Grabherr, G., Gottfried, N. & Pauli, H. (1994). Climate effects on mountain plants. *Nature*, **369**, 447.

Hay, S. I. *et al.* (2002). Climate change and the resurgence of malaria in the East African highlands. *Nature*, **415**, 905–9.

Homer-Dixon, T. F. (1994). Environmental scarcities and violent conflict: evidence from cases. *International Security*, **19**, 5–40.

Intergovernmental Panel on Climate Change (WGI). (1996). *Climate Change, 1995 – The Science of Climate Change: Contribution of Working Group I to the Second Assessment Report of the Intergovernmental Panel on Climate Change*, ed. J. T. Houghton, L. G. Meira Filho, B.A. Callander BA *et al.* Cambridge: Cambridge University Press.

Intergovernmental Panel on Climate Change. (2001). *Climate Change 2001: The Scientific Basis*. Cambridge: Cambridge University Press.

Lindgren, E., Tälleklint, L. & Polfeld, T. (2000). Impact of climatic change on the northern latitude limit and population density of the disease-transmitting European tick *Ixodes ricans*. *Environmental Health Perspectives*, **108**, 119–23.

Lindgren, E. & Gustafson, R. (2001). Tick-borne encephalitis in Sweden and climate change. *Lancet*, **358**, 16–18.

Loh, J., Randers, J., MacGillivray, A., Kapos, V., Jenkins, M., Groombridgez, B. & Cox, N. (1998). *Living Planet Report, 1998*. Gland, Switzerland: WWF International, Switzerland; London: New Economics Foundation; Cambridge: World Conservation Monitoring Centre.

Martens, P. *et al.* (1996). The impact of ozone depletion on skin cancer incidence: an assessment of the Netherlands and Australia. *Environmental Modelling and Assessment*, **1**, 229–40.

Myers, N. (1997). Biodiversity's genetic library. In *Nature's Services: Societal Dependence on Natural Ecosystems*, ed. G. C. Daily. Washington DC: Island Press.

Raleigh, V. S. (1999). World population and health in transition. *British Medical Journal*, **319**, 981–4.

Reiter, P. (1998). Global warming and vector-borne diseases [letter]. *Lancet*, **351**, 188.

Slaper, H. *et al.* (1996). Estimates of ozone depletion and skin cancer incidence to examine the Vienna Convention achievements. *Nature*, **384**, 256–8.

Tenenbaum, D. J. (1998). Northern overexposure. *Environmental Health Perspectives*, **106**, A64–A69.

United Nations Environment Program. (1999). *Global Environment Outlook 2000*. London: Earthscan.

Vitousek, P. M., Mooney, H. A., Lubchenco, J. & Melillo, J. M. (1997). Human domination of Earth's ecosystems. *Science*, **277**, 494–9.

Vos, G. *et al.* (2000). Health effects of endocrine-disrupting chemicals on wildlife, with special reference to the European situation. *Critical Reviews in Toxicology*, **30**, 71–133.

Watson, R. T., Dixon, J. A., Hamburg, S. P. *et al.* (eds) (1998). *Protecting our Planet. Securing our Future. Linkages Among Global Environmental Issues and Human Needs*. UNEP/USNASA/World Bank.

2

Historical connections between climate, medical thought and human health

ANN G. CARMICHAEL & MILLICENT FLEMING MORAN

Traditional Western environmental medicine acquired renewed significance during the 1990s. Significant global climate change is likely to occur during the twenty-first century, and will alter the needs for population health maintenance as well as the resources available for the management of disease crises. In the past, environmental medicine held that human health and disease could not be assessed independently of climate and place. Interactions between changing climate and human health were thus assumed. Those who hope to recover a measure of this more ancient stance towards medicine question the utility of framing future epidemiology in narrow clinical paradigms. Advocates of a more global epidemiology turn away from the study of risk factors and therapy, in favour of larger environmental models of health and disease.

The study of the history of disease and biometeorology during the last century carried the expansive environmental perspective far more than did clinical and community epidemiology. History can have relevance now for those crafting a new global epidemiological vision. Medicine's former interest in weather and climate directed investigation and intervention towards population health maintenance. Withdrawing from grand and costly goals, western medicine increasingly focused on individuals and local environmental hazards, even in the arena of public health. The heroes and often-told stories of medical history relayed by medical and scientific practitioners accentuate this narrowed perspective. By questioning accepted and self-congratulatory historical constructions of the past, epidemiologists excavate new foundations for the future.

2.1 Environmental medicine before epidemiology

Direct human connection to all creation was the premise of most medical systems in antiquity (Glacken, 1967; Tuan, 1968; Smith, 1979; Zhang, 1993). Thus personal hygiene and public sanitation bore religious significance. Nevertheless, the corpus of writing attributed to Hippocrates and his school (fourth century, b.c.e.) clearly linked some responsibility for disease to observable natural phenomena, the healing power of nature and the necessity of medical interventions to mimic nature's processes in healing. "Dis-ease" was caused by an imbalance of the body's humours, which were the microcosmic reflection of a larger macrocosm. At the time, itinerant Hippocratic physicians gained social protection by invoking explanations for disease that were politically, culturally and theologically neutral. "Nature", whether ruled by capricious gods or not, was indifferent to kings and commoners. Hippocratic medicine therefore focused on predicting the course and outcome of an illness through detailed observations of clinical symptoms, understanding that winds, waters and seasons could make some diagnoses more likely. Environmental problems were local; disease occurred in the individual (Smith, 1979; Lloyd, 1983).

Public health intervention was not the primary objective of Hippocratic medicine. A rationale for public health required the explanation and prediction of atypical disease and death patterns localized in time and space. Aristotle (ca. 330 b.c.e.) provided such a foundation, linking cosmological, meteorological and terrestrial phenomena, because his physics denied the existence of vacuums. Anything that moved – as in the heavens – thus had a direct or indirect causal relationship to events on Earth. Galen (died 210 c.e.) synthesized Hippocratic and Aristotelian ideas, elevating empirically based medicine to the stature of a science (Hannaway, 1993; Grant, 1994). For the next thousand years Galenic medicine was in turn gradually welded to the great monotheistic religious traditions, which provided the necessary imperative for public health intervention. Monotheistic societies could not support a medical system that expressly and intentionally limited the right to health to a privileged elite. Muslim and Byzantine Christian societies invented hospitals and pharmacies to meet such goals. Only in western Christendom did public health thinking assume the tasks of monitoring and altering local environments in order to preserve or restore the health of all people (Ranger and Slack, 1992).

While the transformation in social health objectives in medieval and early modern Europe can be attributed to more inclusive values of such faith traditions, unquestionably the appearance of a novel health crisis of unprecedented dimension shaped other new responses. The bubonic plague pandemic of 1347–1350 was regarded as both a punishment for collective human sins and the result of unusual celestial events in 1345. Yet the progress of the first plague pandemic followed a clear geographical–temporal progression through the Middle East, North Africa and Europe. Recurring plagues with the same peculiar human pathology and the same geographical–temporal spread as in 1348 belied the Aristotelian model of the disease's origins. Similarly the divine plan in God's ongoing vengeance was difficult to fathom. Plague returned and disappeared inexplicably; novel experience challenged received wisdom. New theories and practices to predict local outbreaks thus eroded ancient physics, and secular acceptance of differential mortality defied religious ideals (Slack, 1985; Pullan, 1992; Jones, 1996). Observably local epidemics, with discernible patterns over space and time, shifted attention in the search for causes from the universal to the particular, from the remote to the proximate. As a consequence the Hippocratic writings of antiquity enjoyed a new vogue in early modern Europe (Smith, 1979).

Until the last plagues in Europe in the early 1700s, medical theories and practices tried to accommodate ancient science. Medical practitioners and state governments were realigned to very different religious and political objectives during these centuries (Jordanova, 1979; Riley, 1987; Jones, 1996). Medical men devised ever more complicated explanations for clinically distinctive epidemic diseases. City and state governors meanwhile created public health departments in order to minimize the political and financial risks of catastrophic plagues. Their efforts protected a privileged few. While their rhetoric continued to draw heavily upon inclusive and community-oriented religious tenets, the moral and explanatory dissonance of such a patchwork approach to the public's health grew louder.

2.1.1 The Scientific Revolution and the "terraqueous globe"

Just as recurrent, multi-regional epidemics altered how Europeans responded to health crises and led to the imperative of some form of overtly "public" health, so was the Scientific Revolution equally crucial for the creation of nonAristotelian environmental science. The larger cosmological premises of ancient physics were effectively dismantled by Copernicus, Galileo, Descartes, Newton and other natural philosophers of the sixteenth to the eighteenth centuries. Aided by the revelations of exploration and overseas navigation, men of learning came to see the environment as dependent upon physical forces within a self-contained "terraqueous globe". The Earth was no longer at the centre of the universe; therefore, large-scale changes in the physical environment could no longer be explained by the influence of heavenly bodies, or be subjected to supernatural intervention. Local environments that mediated human health could be actively changed by human intervention. Humans finally appropriated responsibility for managing the planet (R. Porter, 1981; Riley, 1987).

The "Enlightenment" of the eighteenth century built upon these premises, searching for physical laws governing health, disease, meteors, tides, epidemics and all sorts of weather disasters or regional peculiarities. Better prediction and understanding – especially by drawing upon mathematics in service to scientific knowledge – permitted many investigators to cling to the hope that a vast array of new observations made on a global scale would expose a rational, divine purpose and plan to creation (Glacken, 1967). Well into the twentieth century some respected scientific practitioners still held on to the notion that variations in human health, character and physical appearance could be correlated with cycles or patterns in climate (Fleming, 1998).

Medical environmentalism and expanding European cultural conquest of the globe stimulated both geography and meteorology as new earth sciences, which parted company with medicine and the life sciences rather quickly (Cassedy, 1969; Riley, 1987). In 1800, collecting both medical and meteorological information – and mapping the data – was an activity dominated by physicians bent on producing medical topographies. By the 1840s and 1850s, study of climates and places had rapidly diminishing connection to medicine or the maintenance of health. Only outside Europe were the links between the study of disease and climate robustly pursued.

By 1800 the new environmental medicine gravitated around new specific issues: (i) whether economically costly and commercially disruptive quarantines were ever really necessary; (ii) whether diseases unfamiliar to Europeans before exploration, colonization and the conquest of other continents and peoples were caused by variations in terrestrial and meteorological factors, or by variations in human physiology; and (iii) whether "laws" of epidemics and endemic disease could be devised, similar to the those that Newton and his followers had effected in physics. Did life differ fundamentally from nonlife? Was new chemical knowledge useful in understanding the cause and origin of disease, as, for example, in the identification of specific poisonous substances in the air, water or soil?

2.1.2 Climate and disease during the early modern centuries: perceptions and realities

One of the most interesting aspects of this great social and intellectual transformation is that it took place during a time period of significant, but unnoticed, global climate change. Frequently called the "Little Ice Age", the period from approximately 1550 to 1850 witnessed global cooling from 1 to 2 °C on average and glacial advances observable from one generation to the next (Grove, 1988). Contemporary observers naturally saw best what seemed dramatic and unfamiliar. Unable to see changes in the European climate with the retrospective advantage that we have, physicians and scientists of this activist, enlightened era focused instead on their discovery of global climatic diversity: they linked cultural, racial and class-based perceptions of peoples to regional diseases; they framed a comfortable model of human health fortunes as consequent to civilization or its absence. Even though global cooling in this period was not on so drastic a scale as current global warming projections, the conclusions and actions issuing from their investigations on place, race and disease specificity became relevant to the ways modern nation-states crafted and defended public health policy.

Some radical and irreversible changes were easily observed as they occurred. Early modern observers noticed on-going demographic growth: the planet-altering modern rise of the human population began in this period (Livi Bacci, 2000). Urban growth and human migration patterns drew the attention of "statists" (those who first produced "statistics") to both accumulating refuse and geographical variation in the prevalence of familiar diseases (Porter, 1986). From 1600 to 1850, rapid urbanization in Eurasia, Europeans' discovery of new diseases, climates and peoples, and novel sources of wealth created far heavier demands on maintaining an acceptable level of public health (Grove, 1995). Collecting such information indeed led to manifestos for broad social reform, such as the redesign of prisons, the construction of sewer and drainage systems, the early management of water resources, and national vital registration systems (Riley, 1987). Simultaneously the data were used to affirm the association of some locales and populations with an increased risk of disease and death.

However, because the techniques, terminologies and ideologies of emergent scientific medicine were changing at the very same time that we can demonstrate – independently of records produced directly by humans – global climate change, the evidence for human health and disease before the mid-nineteenth century is exceedingly difficult to evaluate (Bradley and Jones, 1992). Sustained interdisciplinary attention to all available data before that time will probably provide few unambiguous and determinate links between disease change and climate change 300–400 years ago. The observations that were recorded, however, permit a firm conclusion that the era was significantly cooler than in the later twentieth century; that bubonic plague receded permanently from Europe and the Mediterranean basin; that the control of smallpox, first with variolation, then with vaccination, had a measurable positive impact on early phases of the modern rise of population; and that infectious diseases commonplace in Europe had a devastating, ongoing impact on the survival of indigenous peoples outside Eurasia (Kunitz, 2000).

2.2 Thinking locally, acting globally

For much of the twentieth century, epidemiologists cobbled together a story of their particular disciplinary origins, beginning with a time-worn appeal to Hippocrates' *"Airs, Waters, Places"*. The Hippocratic text provides an important ecologic frame of reference, here excerpted by John Last:

Whosoever should study medicine aright must learn the following subjects . . . First consider the seasons of the year; . . . secondly the warm and cold winds, both common to every country and . . . to particular [localities] . . . [A physician] should consider [a district's] situation . . . aspect to the winds . . . the water supply . . . the soil . . . [whether thickly covered with] vegetation . . . or exposed . . . Lastly consider the life of the inhabitants themselves . . . .

(Last, 1993).

The ancient synthesis was global, in that it appealed to universal and ubiquitous parameters of health and disease. A more geographical global view emerged during the age of exploration. The early modern re-formulation of medical environmentalism prompted the quest for a unifying, supra-human deterministic theory of climatic influence on both disease and human nature. The revised version in part justified European colonial expansion, while employing paradigms of miasma and contagion to explain local epidemic outbreaks (Crosby, 1986; Grove, 1995). Observation of human behaviour and interactions with the environment persisted, now without reference to astrological events, human transgression or demonic device.

2.2.1 Miasm versus contagion

A debate between environmental miasmatists and those believing that important epidemic diseases were attributed to contagion began 200 years ago with mercantile objections to the costs of rigid maritime quarantines (Bynum, 1993; Pelling, 1993; Baldwin, 2000). Still looking for overarching patterns in disease experience, these proto-epidemiologists localized diseases to peoples (groups, races, classes) and places in quite different ways than the Ancients did. In the early nineteenth century fear of filth and pollution thus translated the labels of miasm and contagion to particular people and locales (Pelling, 1993; Tesh, 1995). Europeans deemed some regions unsuitable for their own habitation without aggressive environmental management. High morbidity and mortality among indigenous peoples in these same regions were attributed to their diets, their hereditary "defects", or their unenlightened customs (Anderson, 1996; Harrison, 1996; Worboys, 1996). Many of their afflictions were labelled "colonial diseases" or "diseases of the tropics".

The discussion of miasm and contagion theories marshalled debates about the proper objectives and methods of environmental medicine. Decisions about the ways investigation should proceed in turn led to the rapid development of sciences essential to the practice of modern epidemiology. In many cases the assumption that one construction was a better starting point for analysis or intervention than another led to change within both miasm and contagionist positions (Kunitz, 1987; Rosenberg, 1992; Tesh, 1995; Hamlin, 1998; Baldwin, 2000). The left-hand column of Table 2.1 illustrates the substance of debates that focused nineteenth-century discussion about the proper study of epidemic diseases. Out of these debates the disciplines in the right-hand column began to take shape. The pursuit of one particular method or objective fuelled the emergence of the new sciences that underpin epidemiology today.

The table should not be interpreted narrowly – i.e. that the investigatory stances listed in the "miasm" and "contagion" column correspond to proponents of these two positions. Likewise, the table is not comprehensive in its listing of scientific

Table 2.1. *Emerging scientific disciplines emerge from nineteenth-century debates on epidemic diseases and their causes*

Opposing approaches within the nineteenth-century debate about the causes and control of epidemics	New disciplines and areas of scientific investigation in the nineteenth century emerging from these debates
miasm – contagion	Epidemiology
stochastic – deterministic	Laboratory microbiology
prediction – explanation	Mathematical statistics
observation – intervention	Meteorology, cartography, geology
indirect – direct	Demography
forces – agents	Physical/mathematical modelling
dynamic – static	Organic chemistry
local – global	Biology and evolution

disciplines. Contagionists could certainly entertain dynamic models, advocate localist observation, understand that causes of disease operate indirectly or through (chemical, environmental, behavioural) forces indirectly, and so on. Nevertheless polarized assumptions in our models continue to fuel debate within the epidemiological community, as well as in the larger political and medical arenas. Thus the diad constructions remain unaltered in modern epidemiology. When considering many of the emerging infectious diseases today, the utility of "eradication" versus environmental containment quickly evokes a similarly messy combination of methods, objectives and explanations. Highly threatening diseases, such as the global spread of human immunodeficiency virus (HIV) infection in the last two decades, likewise sharpen scientific debate about the level of danger and the appropriate medical and public health responses. In the early nineteenth century, these conflicting scientific opinions centred on the sudden appearance of global cholera epidemics in the 1830s (Baldwin, 2000).

2.2.2 Enter cholera, enter Snow

Previously unknown in Europe and the Americas, cholera quickly challenged simple notions that outbreaks were attributable to local causes. Some immutable substance or organism had to be responsible for the re-creation of a specific disease pathology in different environments. Rigorous isolation of the ill, called back from long-unused plague policy, neither contained cholera at the population level nor protected individuals mindful of anti-contagion rules. Contagionists reasoned that a material cause could potentially be identified or isolated through chemical analysis or with the microscope. Yet, even if cholera were caused by a specific chemical ferment (the prevailing notion of scientific contagionists before the 1850s), further study of airs, waters and places predicted neither its arrival nor its virulence (Baldwin, 2000). Contagionists lacked evidence to prove that the new disease, coming from outside of Europe, existed independently of local conditions facilitating its spread.

Thus Western debate about cholera in the first pandemic wave, 1831–1832, reinforced local miasmatists (Hamlin, 1998). Focus on filth and local conditions did not explain why a disease, that was previously observed only in British India, was now spreading quickly through temperate zones. "Predisposing" local causes, whether the behavioural habits of particular individuals or the chemical properties of discrete soils and water sources, seemed better predictors of outbreaks, but the introduction of an alien substance seemed necessary as well. Most western scientists shared a conviction that "Asiatic cholera" was a different disease when not in Asia or in Asian bodies. To Europeans, cholera was an emergent disease, yet they cited local and behavioural factors to explain why one individual in a community would be stricken and another spared.

By the 1850s the ecological study of disease patterns was an enterprise shared by both miasmatists and contagionists. For example, proximity to putrid waters, by measuring linear distance or altitude from a polluted source, was no longer an approach limited to anti-contagionists. The Sanitarian movement's large-scale sewage, water lines, drainage and housing projects were based on a miasmatist approach, and account for Europe's steepest decline in infectious disease mortality (Lilienfeld & Stolley, 1994; Livi Bacci, 2000).

At the same time contagionists produced distinctive triumphs in the theoretical arena. The approach John Snow took in 1849 and again in 1854 illustrates well a new strategy for dealing with cholera as a disease transmitted by a specific agent. At the clinical, individual level, Snow believed that an ingested, cholera-causing toxin caused massive fluid loss from the intestines. At the population level Snow believed that district-by-district data of cholera mortality could be integrated with other, more environmentally oriented analyses. Without the collection and publication of death statistics by cause, age and place, begun by William Farr director of the General Register Office, Snow's investigations could not have focused so clearly on the differences in cholera mortality within sections of London. Farr's first cholera report of 1848–1849 provided 400 pages of data, analysis and suggestions for prevention. Simultaneously John Snow published his now celebrated hypothesis that the pollution of water was involved in the transmission of cholera, and Farr quickly adopted a legislative agenda for water purification (Eyler, 1979 pp. 111–22).

Between late August and September 7, 1854, 500 London fatalities had been reported. Using General Registrar's records, Snow moved from district counts of total mortality to individual case-level reports of cholera mortality, geographically re-presenting the outbreak: Snow identified and mapped the locations of decedents' homes. A noticeable cluster of cases appeared in Golden Square, in the Soho district

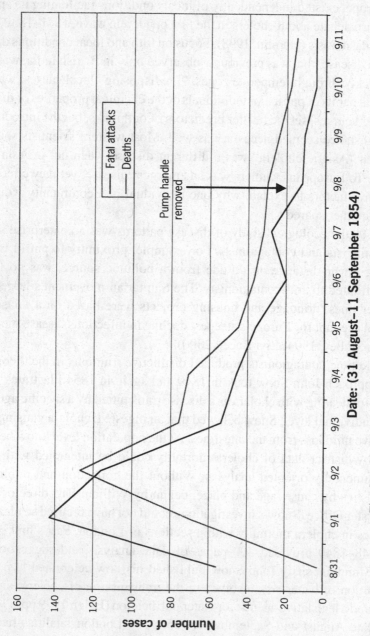

Date: (31 August–11 September 1854)

Fig. 2.1 Cholera in Golden Square, 1854, based on John Snow's data.

of London where Snow lived. Most of the victims lived within 250 yards of the Broad Street pump, where they habitually drank water from the only public source in the immediate area (Brody *et al.*, 1999; McLeod, 2000).

Today epidemiologists and historians hail Snow's proto-germ theory, noting his hypothesis of the pump's contamination by an "agent". Yet most heroic accounts stop with Snow's appeal to remove the pump handle, where, by most accounts, the outbreak ceased in mid-September. In his own writings, Snow worked from earlier cases to evaluate the cholera "contagion" among its victims, testing both common vehicle (water) and person-to-person modes of transmission. Believing that cholera began with the ingestion of "morbid material", Snow linked accidental contamination to water polluted with human excrement (Fig. 2.1).

For modern epidemiologists Snow's gumshoe investigation is not only a case-finding exercise, coupled to comparison of sick and healthy individuals in the same disease context; it is the first well-known use of a "case–control" method. Snow compares sick and healthy individuals residing within the same household or living environment. When inspection of the pump water for visible contamination produced no useful results, Snow persisted, reviewing the histories of those who remained healthy. Those who failed to acquire cholera despite being near to the fatal pump were equally interesting to Snow. He thus provided a controlled comparison of presumed exposure to the affected pump. Inmates in a workhouse and 70 brewery workers all remained free from cholera, even though all resided close to the pump. Snow showed that the workhouse had its own well; the brewery workers drank beer. One last enigmatic case outside the affected neighbourhood also fitted: the former resident imported her water from the pump to her home.

Not even then did Snow believe that his detective work proved the necessary connection to the pump's water. Farr had noted fewer deaths in districts served by the Lambert water company. Lambert had changed its intake source, so that it was now upstream of the city's sewage discharge into the Thames. When Snow visited each decedent's home, he hoped to determine whether Lambert or the Southwark and Vauxhall company provided water to the household. Thus Snow moved from district-level case confirmation, to analysis of household water-service provision, and finally to individual-level analyses to identify the source of exposure for 300 000 cases throughout the city. Not relying on verbal and company testimonies exclusively, he tested for chlorine content, showing that the Southwark and Vauxhall water contained more seawater than Lambert water.

Snow's "natural experiment" became the most important outbreak investigation in modern epidemiology, even though his work was not well known at the time (Brody *et al.*, 1999). Knowing each company's service population, he could link water companies to cholera casualties; thus Snow calculated the first cholera incidence rates. Snow published sensible guidelines to interrupt cholera transmission, advising city residents to boil drinking water, isolate cholera victims and protect foods from contaminated persons and water sources. To provide against future out-breaks he supported drainage, water purification and, with a nod to the contagionist tradition, attention to "persons and ships arriving from infected places". Thus joint contagionist and sanitarian methods were commonly adopted by port health author-ities in European and North American cities. Paccini, and later Koch, identified the cholera agent, *Vibrio cholerae*, after such anti-cholera practices made epidemics a lesser threat (Howard-Jones, 1972).

In the 1990s stories of Snow's work on cholera became something of a lightning rod, focusing the energy of debate on the priorities and foundations of modern epidemiology (Vandenbroucke, 1988, 1989; Vandenbroucke *et al.*, 1991; Brody *et al.*, 1999). Snow's utility as a "founding father" of the discipline relies upon assimilation of his work to the subsequent germ theory of disease. As environ-mentalism gathers greater momentum, Snow's fate as a model of epidemiological perspective is suddenly uncertain.

2.2.3 Global environmentalism, 1850–1914

The *localist* methods and objectives of epidemiological investigation, common to both miasmatists and contagionists by the 1850s, enjoyed and continues to enjoy widespread support within medicine, epidemiology and public health. Local con-texts in disease experience became particularly important with the bacteriological revolution of the 1880s. Nevertheless two overtly global agendas pertaining to later epidemiology emerged within medicine and science during the middle decades of the nineteenth century: global meteorology and international cooperation in the control of a few feared epidemic diseases. Global meteorological science moved rapidly away from its origins in medicine, forging a comprehensive physics of the "terraqueous globe" (Nebeker, 1995). Simultaneous weather reporting throughout North America and Europe was an objective of early meteorologists, but was of great use to agricultural, military and commercial interests, in that identifying pat-terns of the movement of storms might lead to better prediction of severe weather. Many sanitarians thus hoped to make theoretical contributions to the understanding of disease origins and geographical distribution, through the correlation of more reliable meteorological data with the incidence and prevalence of environmentally caused diseases (Fleming, 1998). But most of their efforts never rose above "a chaotic statistical fog" (Davis, 2001, p. 213).

Meteorological successes of the period before the First World War did not lead to a new global perspective in environmental medicine. But there are some impor-tant exceptions to this generalization. British medical officers in India maintained that cholera had unique properties in the subcontinent, and they were among the few to read and advance the study of Snow's work. To serve their colonial goals, investigators kept detailed records linking distinctive Indian diseases to distinctive Indian weather, especially the monsoon rains (Harrison, 1996). British investigators in India not only used Snow's cholera study, but they continued the pursuit of older environmentalist research that had large pay-offs in meteorological science a century later. From the Indian Meteorological Service's data on local environments and inter-annual change, in the 1920s Sir Gilbert Walker forged a synthetic meteorological description of a "Southern Oscillation" in global oceanic sea surface temperatures (Glantz, 1996). Most of the data were collected systematically over the late nineteenth century in response to widespread and devastating famines and epidemics, cataclysms discrete in time and linked to the failure of monsoon rains, but nearly global in geographical expanse. Monsoon-related famines spanned Africa, south Asia and South America (Davis, 2001). The connection of climate change to massive mortality was simply too great to be explained by references to local customs of cleanliness, or the lack of it.

Malaria was most prominent among the microbial killers during the famine-crisis years, but never became a significant threat to the rapidly industrializing, temperate-zone "first" world (Dyson, 1991). Instead, the ancient fear of bubonic plague and the newer scourges of yellow fever and cholera prompted global-control efforts: the second global agenda of the post-Snow decades. In 1851, the first International Sanitary Congress met in Paris, focusing on diseases believed to be introduced from nonEuropean settings and to be alien to European climates: cholera, yellow fever and bubonic plague. Thirteen meetings of these International Congresses met before the First World War, but very little discussion on any other disease problems worried the delegates (Goodman, 1952; Howard-Jones, 1981; Bynum, 1993). The "miasm versus contagion" debate, already anachronistic within scientific communities, was still an important issue in these largely political debates. Restrictions and governance of international trade made the issue vital to Britain and her economic opponents, for to control cholera was equated to the control of British trade with India. The policies of the International Sanitary Congresses preserved older, globalist concerns and even today restrict much international health law to the feared epidemic diseases of the nineteenth century: plague, yellow fever and cholera (Fidler, 1999).

Medical meteorology and international controls of a few feared diseases were global ventures: famines and epidemics with clear correlation to particular climate changes, and a political will to control with international diplomacy a limited number of dramatic epidemic diseases. The global perspective harmonized better with the original definition of epidemiology as the study of epidemics. The environmental perspective that had fuelled the sanitarians' reforms was localist, and deftly sidelined any broader agenda. As Snow "conquered" cholera by removing a pump handle, the most prominent proponent of environmentalism in Europe, Munich chemist Max von Pettenkofer, published his very different analysis of cholera transmission. Von Pettenkofer's contributions stimulated much of the science of sanitation that improved health in European cities during the mid-nineteenth century, but his name is almost forgotten today. Many who do know of his work heap scorn and ridicule upon him (Winslow, 1926; Sigerist, 1941). Yet his attention to environmental chemistry was in many ways equal to Snow's triumphs in early case-tracing, and both are worthy ancestors of today's environmental epidemiology.

2.2.4 The West takes the rest

By the end of the nineteenth century mortality and morbidity rates had fallen rapidly throughout the industrialized world, distinguishing the health experience of Europeans and Euro-Americans dramatically from that of the less privileged peoples of the planet. The contrast reinforced their notions of their own racial, national and cultural superiority. Nonwestern peoples and the subalterns in Europe and North America were increasingly represented as dangerous threats to healthy Western hegemony. Inequities in disease experience had once been viewed as predispositions governed by local miasma or the result of moral and behavioural choices. Predisposition facilitated disease in individuals who were subjected to identical climatological factors as their neighbours who remained disease free.

In the late twentieth century Mervyn Susser characterized this complex period surrounding the acceptance of bacteriology the first "revolution" in epidemiology. After 1882, younger epidemiologists accepted the fundamental determinism of the germ theory. Older practitioners of epidemiology, trained before Robert Koch's successes, took futile issue with the deterministic premises of one-germ/one-disease theory, and quite a few undertook massive histories in order to pursue the study of epidemics over time (e.g. Charles Creighton, Alfonso Corradi and August Hirsch). To others, such as Francis Crookshank, the limited success of microbiology during the Great War and the influenza pandemic begged a cautious approach to solving infectious disease problems with bacteriology alone. Moreover, neither sanitary science nor the evolving international epidemic controls explained the changing incidence of parasitic and arthropod-borne diseases. A broader ecological perspective was required. Crookshank chided the optimists who did not grasp the philosophical underpinnings of epidemiological investigation:

Statistics and field-observations are matters of method, not principle . . . The *cathedra* of the professed epidemiologists seems to have wobbled uncertainly on two legs: one the belief that specific diseases were real things to be found in the post-mortem room; the other, that epidemics were to be abolished by sanitation. Both derived by *a posteriori* methods out of Victorian realism. Within our own memory an attempt has been made to secure equilibrium by the insertion of a third support – that of positivist bacteriology

(*Crookshank, 1922 p. 24*)

Crookshank still believed that the objectives of epidemiology included larger issues: explaining why epidemics appeared and disappeared; why they appeared in one place or time rather than another; and why some diseases clustered together, often under particular seasonal influences. Younger contemporaries had given up this broad quest, forgetting "the infinitely great in the contemplation of the infinitely little" (Crookshank, 1922 pp. 23–24).

British and American epidemiologists trained just two decades later in the 1900–1920 period – such as Charles-Edward Avery Winslow, Wade Hampton Frost, Charles Chapin, Edgar Sidenstricker and Sir Arthur Newsholme – incorporated a firm understanding of the germ theory into their elaboration of new methods in epidemiology. Winslow, for example, found that the frontiers of twentieth-century epidemiology lay with the germ theorists' focus on specific diseases, rather than with further study of the environment:

During the half century that has passed the communicable and environmental diseases have been so substantially reduced that the problems of the future are heart disease, the acute respiratory diseases and cancer. We face a new situation and we must adopt new methods if we are to meet it with any measure of success . . .

(Winslow, 1926)

Between Crookshank and Winslow we glimpse the passing of an era, as causal models of disease became ever more localized. During this same time of the germ theory's ascension, the control of epidemic disease moved from being local to being global. By emphasizing the progressive path that epidemiology took after the Second World War, we ignore the extent to which leading epidemiologists and sanitarians a hundred years ago debated the utility of a global environmental view.

2.3 Conquering heroes, stubborn foes

Until the 1920s epidemiologists and sanitarians debated reductionist approaches to infectious disease. Laboratory-based bacteriology left no respectable hold for ancient environmental views of health and disease, though a few pungent voices continued to advance environmental determinism (Petersen, 1934–37; Fleming, 1998). At the same time the germ theory was not as expansive as the contagion theory; models of disease cause and disease transmission were now separated. Diseases caused by germs were not necessarily contagious; the means of control now had to be determined for each infectious disease.

Until the final decades of the nineteenth century, epidemiologists and sanitarians found miasmatic premises useful, and so returned to both local environmental sources of corrupt, decomposing matter, and to local climatic and meteorological variations. Such a tilt to miasmatic views of disease causation trained their attention on aspects of the environment that readily disgust the senses – filth, stenches, chemical odours, vermin and even unwashed human bodies.

Measurable improvements in health had accompanied advances in sanitary engineering, giving localist versions of the miasm theory a new currency in public health departments (Tomes, 1998). However, the globalist claims about diseases and climates that sanitarians once advanced resisted both empirical confirmation and concrete strategies for intervention. An expansive view of public health needs also carried costs that nations were not yet ready to pay. To the extent that miasms could be construed as the product of human poverty, curing disease would require social reform. Proponents and opponents of the germ theory alike knew that radical governmental and philanthropic "cures" for poverty could be avoided by narrowed focus on specific microbial or chemical causes of disease.

Thus only those environments that presented a danger to the general health of a region's environment were seen as the proper targets of public sanitation. Eliminating specific disease threats within a limited geographical area was especially attractive if the rich were to get richer and the poor no poorer. Over the twentieth century, the relationship between health and wealth would become a doctrine of enlightened development: employing a healthier labour force was good for business; protecting developers from the dangerous environments of the "third" world removed risks to the "first".

2.3.1 Tropical medicine and the germ theory misfits

During the first generation of the germ theory's ascendance, the world outside Europe and North America became a world teeming with menacing microbes and unwashed, uncivilized humans. Converting heathens to western sanitary ways was an ultimate goal of "constructive imperialism", but solving the problem of diseases in the tropics with European-style sanitation was seen as an exercise in futility (Worboys, 1993). Here too the germ theory offered a way to redefine the problem of disease and narrow the objectives and targets of intervention. Hardening and exasperated colonial attitudes towards disease and famine crises in Asia, South America and Africa won the day. A few who resisted full conversion to the great germ gospel objected on humanitarian grounds; others held fast to older sanitary projects.

The most lasting impact on the development of epidemiological practices for this other world came from those who re-framed diseases in the tropics as "tropical diseases" (Worboys, 1996). Diseases such as malaria, yellow fever, bubonic plague, sleeping sickness and filariasis were widely distributed over the globe, but were more common in tropical and subtropical regions. Moreover, patterns of epidemics of these diseases reaffirmed climate and weather connections, and reinforced environmental association rather than a relationship with socioeconomic status. The new germ consciousness could be linked to old environmental consciousness and the emerging third world seen as a special case for disease investigation and control.

Before the 1890s, few such theories attracted the interest of prominent scientific investigators, with the exception of the yellow fever–mosquito connection by Louis Beaupherthuy in Panama (1853), and Carlos Finlay in Cuba (1870s and 1880s). Yellow fever, like cholera, was new to Europeans of the nineteenth century. Appearing most dramatically during Atlantic slave trafficking, yellow fever became a feared component of exaggerated European mortality during colonization and imperialism in Africa and one of the quarantinable diseases addressed by International Sanitary Congresses (Coleman, 1987; Bynum, 1993). Quarantine was repeatedly shown to be futile in the control of yellow fever (Powell, 1970; Coleman, 1987). Similarly plague, sleeping sickness and filariasis served intermittently as epidemic and endemic diseases. By the time of the germ theory, dreaded epidemic diseases that failed to respond to contagion practices added depth to the debate about the role played by the physical environment in their production and transmission (Farley, 1991; Cunningham, 1992).

Malaria, in contrast, long defied contagionist solutions. Present in west Africa and Spanish colonics in the seventeenth century, by the late eighteenth century it affected ports throughout British, Dutch and Iberian trade routes. Even in long-settled regions of Europe and the British Isles, malaria was resurgent during the early modern period, largely because of population and agricultural expansion (Bruce-Chwatt & de Zulucta, 1980; Dobson, 1997). For the most part, Europeans met the old foe in military, colonial settings, where it caused considerable morbidity and mortality (Dyson, 1991). Because the risk of infection and death was also more common in tropical and subtropical colonial settings, diseases in hot climates were redefined as exotic "tropical diseases" rather than as fevers made worse for Europeans living in tropical settings (Worboys, 1993; Bynum & Overy, 1998; Haynes, 2000). Race had been a sufficient explanation of differential mortality in the past; thus, colonials transplanted to the tropics would be expected to have difficulty surviving tropical fevers.

Military physicians working in the "field" rather than the laboratory were the architects of tropical medicine, and were quick to point out that vector-borne diseases and diseases caused by parasites or viruses resisted easy assimilation to germ theory and localist epidemiology. The process of identifying causal agents, the interpretation of risk and resistance, and paths of transmission did not follow the common rules and patterns elaborated by Koch's followers (Worboys, 1996).

The saga of tropical medicine, explored in many recent studies, illustrates important and persistent elements of the history of vector-borne diseases and the growing importance of tropical medicine to the development of epidemiology. First, even though evidence was gathered in the "field", much of the work rewarded by fame and fortune was done in centres far removed from the tropics. Second, the notion of tropical diseases was advanced as natural and as a distinguishing characteristic of colonial settings. Distinctive settings would require a different sort of environmental sanitary management from that developed for urban slums. The identification of the mosquito vector promised reconciliation between environmentalists and germ theorists, who initially focused only on the plasmodium. At just this point Manson argued that parasites and bacteria differed in too many ways to make the assimilation of tropical medicine to the germ theory unproblematic. The confounding factor was climate: tropical climates made the environment an obligatory element of tropical medicine by definition alone, if not by the weight of evidence.

So defined, domestic occurrences of malaria or other "tropical diseases" seemed to require tactics that treated a region or subaltern population in the ways that colonial medicine was administered. Hierarchically conceived, militarily managed campaigns administered environmental cures. Malaria's resurgence in Italy during the second half of the nineteenth century clearly accompanied building and even land reclamation designed to eradicate swamps and marshes (Snowden, 1999). Pre-war malariologists Angelo Celli and Giovanni Grassi battled over the efficacy of quinine prophylaxis versus larvicides, battles that were overtaken by outsiders in the 1920s (the Rockefeller Foundation and the League of Nations Health Organization), whose different approach to malaria control was endorsed by Mussolini's Fascist state (Evans, 1989; Snowden, 1999). Similarly in the U.S., segments of the rural, agricultural south became targets of domestic health campaigns against tropical diseases (Ettling, 1981; Humphreys, 1996).

Similar, externally imposed, militarized campaigns allied to government and business interests supported the interwar development of tropical epidemiology, as in Fred L. Soper's Rockefeller-sponsored campaigns in Brazil (Löwy, 1997). The distance imposed between those making careers in tropical medicine and those funding humanitarian or development-oriented medical interventions in tropical regions on the one hand, and those collecting necessary data and implementing control regimens on the "other", strongly divided this branch of epidemiology's disciplinary development from that focused on the health problems of those in the first world.

2.3.2 Epidemiology at war: 1920–1950

Americans joined the European discussion about tropical medicine during their wars in Cuba and the Philippines (1898–1903) and in building the Panama Canal (completed in 1914). Both ventures hurled newly trained germ hunters and military physicians onto a global stage, battling for international professional stature on this new front. The U.S. Army deployed yellow fever commissions during and after the Spanish American War, taking a domestic war on yellow fever in the southern

U.S. abroad. Infectious disease among both troops and islanders became a veritable apocalypse in the conquest of the Philippine Islands (De Bevoise, 1995).

The army did not at first intend to protect U.S. troops by battling insect vectors. Surgeon General Sternberg, Major Walter Reed and their Army Yellow Fever Commission colleagues wanted to test Giuseppe Sanarelli's hypothesis that *Leptospira* bacilli caused yellow fever (Monath, 1998). Failing to confirm Sanarelli's results, they demonstrated that a microscopically invisible, blood-borne agent or toxin caused yellow fever. At the same time British medical officers, colleagues of Ronald Ross at Liverpool, were working in Havana and spurred Reed towards vector control. By 1900, Finlay's mosquito hypothesis was upheld by Henry Rose Carter's demonstration that the pathogen's extrinsic two-week incubation period in mosquitoes necessarily preceded the transmission of a living agent to humans. Within four years Carter and Juan Guiteras demonstrated that killing mosquito larvae was the quickest way to manage an epidemic. Knowing the entomology and ecology of *Aedes aegypti* vectors enabled Reed to craft (and claim the credit for) anti-mosquito protocols.

These "sanitary methods" were then successfully adapted to Panama by William Gorgas, Havana's Chief Sanitary Officer and later Surgeon General of the U.S. Army during the First World War. Colonial medical institutes allied to germ theory research were in no way restricted to the British and, after the Great War, the Americans. The Pasteur Institutes throughout the French colonies served much the same function and were especially successful in exporting their methods and ideas without trying to assume control – Oswaldo Cruz and his early 1900s institute in Brazil is the best known. German-speaking researchers were less numerous among permanent institutions, but figured prominently in the teams of investigators sent to study every eruption of an exotic epidemic disease.

Americans not only mastered military medical imperialism before 1920, they came to dominate much of the framework of tropical epidemiology in the twentieth century. More than the American military, American money opened the world to American epidemiologists. The Rockefeller Foundation, chartered in 1913, created an International Health Board (IHB), borrowing the idea and models from international sanitary commissions, including the Pan American Health Organization (Fee, 1987). From the outset the Rockefeller Foundation utilized the expertise of prominent military sanitarians. The Foundation's agenda was dominated by control of a feared disease, yellow fever, but structured on domestic campaigns against hookworm and pellagra in the rural southern states. Elsewhere campaigns against tropical diseases at this time were national, colonial efforts, carried out by tropical disease experts in locales where they had clear national jurisdiction – such as the U.S. assumed in Cuba and the Philippines at the end of the war with Spain. Thus militarization and internationalization of tropical medicine were innovations of the IHB.

Early participants in the IHB's campaigns against yellow fever and hookworm diseases in the Americas saw themselves in a model of internationalism and humanitarian medical intervention, and indeed the same kinds of efforts quickly characterized the new League of Nations Health Organization created out of the political and economic rubble of the Great War. In the 1920s, therefore, numerous international agencies embarked on campaigns to eradicate specific diseases in underdeveloped regions, domestic or overseas, justifying their missions on humanitarian grounds, but implementing their programmes with military models of organization (Evans, 1989). Fred Lowe Soper, hired by the IHD in early 1920, set off for Brazil, immersing himself in the task of hookworm eradication under a programme created by a private foundation in one country and the national health ministry of another, a programme "bridging the gap between science and application of science in promoting the welfare of mankind throughout the world" (Soper, 1977, p. 30). Soper describes the experience he accumulated, first with the hookworm campaign, then in aggressive yellow fever eradication efforts, and finally in a full-scale battle against a "foreign", "invading" mosquito in northeast Brazil, *Anopheles gambiae*. Soper felt that he discovered jungle yellow fever (Brazilian epidemiologists had advanced, but not named, the hypothesis decades before). The discovery, to whomever credit is given, led to the recognition that far broader environmental study and management would be necessary to eradicate yellow fever from the Americas.

Through the 1940s the weapons used against tropical infectious disease followed the growth of medical and pharmaceutical advances. Americans contributed with antibiotics to reduce the chronic disability and endemic nature of tuberculosis and syphilis. The development of dichlorodiphenyltrichloroethane (DDT) and other biocides in the Second World War advanced the control of malaria but, unlike with yellow fever, vaccines were not an option. Gordon regarded insecticide control of arthropod vectors as important as the impact of antibiotics on the world's infectious diseases. He pointed to the utility of insecticides for the war effort in the control of typhus and malaria, both on the home front and in military settings. Part of the legacy of U.S. malaria control was the creation of the Communicable Disease Center, now the Centers for Disease Control (CDC) in Atlanta (Humphreys, 1996).

Privately funded research created and tested successful vaccines; for example, Max Theiler's prototype yellow fever vaccine for the Rockefeller International Health Division, tested in Brazil during the 1930s and then on U.S. soldiers during the Second World War (Sawyer *et al.*, 1944; Monath, 1998). After the war, volunteers and voluntary contributions to the private National Foundation for Infantile Paralysis sponsored the first community trial of Salk's polio vaccine (Fee & Rosenkrantz, 1991). The effects of mass vaccination on herd immunity in turn required from epidemiologists new surveillance policies and new diagnostic antibody tests, all of which shaped views of the dynamics of "mass disease". Epidemiologists had to reconsider their own role in the control of mass disease too. The publicly recognized success of controlling polio may have encouraged the optimism that feared third world diseases might also be conquered.

2.3.3 When was the revolution in epidemiology?

By the mid-twentieth century, epidemiologist John Gordon looked back to the First World War era as a watershed period in epidemiology's maturation as a discipline. Not only had the quickness of success in germ discoveries slowed considerably, "the main objective was being lost". That objective, he maintained, was medical ecology, understanding "disease as an ecological process". Gordon believed that the leading edge of epidemiological theory was born in the all-too-apparent failures of the germ theory and sanitary science to deal with pandemic influenza, encephalitis and polio. One of the most forceful spokespersons of the interwar period was Theobald Smith, who pointed out the steady, dramatic decline in mortality over the previous 50 years, an interval also notable for advances in transportation technology, agriculture and human migration – processes long understood to be the source of new epidemics and high mortality rates. Thus for him improvement was general and due to "community" development, but epidemiologists poorly understood the factors that led to human betterment (Kunitz, 1987).

With hindsight we can also see that the epidemiology of the 1930s and 1940s acquired the habits and methods of controlling specific diseases through hierarchical, even dictatorial, decision-making. In the early 1950s, Gordon praised disease eradication efforts in the war and interwar years. He worried that Cold War era epidemiologists were abandoning the dynamic view of infectious disease aetiology – descriptions of the carrier effect, of secondary attack rates, of host-agent parasitism and symbiosis, of the population dynamics of tuberculosis and other chronic infectious diseases, and the biology of seasonal, racial and geographic patterns. Disease eradication had been enabled not by technology alone, but by the concerted and cooperative research observation and analysis between the Great War and the end of the Second World War. Epidemiology had, in his view, assumed a rightful position as the explanatory science of disease in populations, equally important to clinical medicine's focus on disease in individuals.

Mervyn Susser's later (1973) formulation of the stages of epidemiology's disciplinary development accepts very little of Gordon's view of the early twentieth century. Susser subsumed the entire period from the 1920s to the early 1950s into a mopping-up phase of the 1880s germ theory revolution. Susser marked the beginning of epidemiology's next era in his own generation's shift from infectious to chronic diseases, from "observational", empirical field epidemiology to the construction of prospective epidemiological investigations, rigorous statistical method, and the training of epidemiologists in their own doctoral programmes. Finally, the new breed of epidemiologists, such as Susser himself, needed no primary training in medicine or public health.

Thus while old hands like Gordon were calling for a broader view of ecological forces in disease, especially for new studies of human influences on the environment, epidemiologists moved towards a model of scientific research leading to clinical interventions. Investigations of chronic diseases were henceforth to focus on the individual and on "risk factor" analysis. Disciplinary leaders needed to craft methodological and statistical tools for case–control and cohort design. Demographer statisticians Bradford Hill and Richard Doll offered the methodological framework most attractive to the brave new post-war generation.

A brief review of the choices and objectives of epidemiologists from the 1920s to the 1950s is thus useful, given that the contrast between Gordon's and Susser's views of this period is so vivid. There are some common themes, however. Both Gordon and Susser shift from the germ hunters and colonial tropical disease investigators, to a perspective of epidemiology led by North American (especially U.S.) interests and, secondarily, British investigators. Gordon, one of the premiere epidemiologists in the Johns Hopkins programme, does so explicitly, for he undertook a history of American epidemiology in the twentieth century, and particularly emphasizes wartime developments. Susser constructed an Anglo-American alliance, praising Hill's crucial role in providing a firmer scientific scaffolding for epidemiology's coming-of-age. The entry of Americans into the story of epidemiology is not an entirely chauvinistic perspective for either chronicler.

2.4 1950s–1989: two worlds – two programmes of post-war epidemiology

Susser attributes the reason for the early 1950s change of orientation in epidemiology to the disappearance of the original germ theory targets: acute, epidemic, infectious diseases. But infectious disease mortality began to decline in the nineteenth century, long before the war-era technological advances in disease eradication, and long before the disciplinary focus shifted to chronic disease morbidity and mortality (Kunitz, 1987).

2.4.1 Horizontal epidemiology

Epidemiologists' new domestic agendas of the Cold War era turned away from hierarchical, interventionist, expansive overseas ventures, to the suffering patients outside their office and laboratory doors. At just this point Gordon reminded colleagues that the "host–agent–environment" ecological triad implied first a multi-factorial focus on causation; second a long-term equilibrium and survival of both human host and a disease agent (not the latter's eradication); and third a dynamic (versus static) state of host–disease interaction.

Gordon's triad became an icon for the discipline, as multi-causal models of all three factors are universally accepted by modern epidemiologists. Yet in 1950, the "environment" leg of Gordon's triad appeared to be losing importance to the younger academic epidemiologists. The new U.S. National Institutes of Health courted civilian epidemiologists for clinical studies of chronic diseases, especially heart disease, cancer and "environmental" exposures, including both tobacco smoking and chemical toxins. Medical schools grew as research institutions with federal monies for both training and research. As René Dubos was becoming the most prominent medical spokesperson for an ecological perspective of human health and disease, many American epidemiologists were migrating towards academia, or to industry, to expand successful technologies (Dubos, 1959; Terris, 1959, cited in Fee & Acheson, 1991).

Their "study populations" were patients seeking care. The "community" studies included volunteers from the catchment area of large urban hospitals, not exotic tropical peoples. Where once aetiological studies tracked cases during short-lived outbreaks, studies now extended over 20–30 years with thousands of pre-selected subjects, studies that required new survey, computing and statistical techniques under much slower, and less controlled, observational conditions. Given these constraints, researchers focused more on what was observable and controllable: the individual and his or her behaviours. Protocols to measure risk behaviours and physiological changes in blood pressure and weight could be standardized and then used to predict which combination of individual characteristics best explained the likelihood of, for example, a future heart attack.

Rather than a revolution in epidemiology, much of this programme could be described as an updated version of the nineteenth-century sanitarians' environmental engineering (Kunitz, 1987). The earlier programmes were essentially localist and focused on environmental miasms. In the new miasms, the environments were within human bodies – epidemiologists were far more interested in individual genetic predisposition to particular chronic illnesses than ever before – and the target of health intervention might be described as lying within a horizontal plane. Particular habits and behaviour, particular genetic heritage, particular occupational or situational risks might be shown to distinguish one individual from another in a common, generally healthy environment. Applications to improve health derived from horizontal epidemiology were focused on demographic and sociological surveys, educational campaigns, ongoing clinical assessment and determined clinical follow-up.

"Horizontal" epidemiology created new-fashioned social epidemiology, focused on measures of the "fundamental" causes of inequities in health status, such as social class, quality of life, social capital and indices of social disruption. Guided by the assumption that there was nothing new in the world that epidemiologists had to survey, the objective, beginning in one's own back yard, was to minimize individual differences in health and disease. Ecological epidemiology was based on finer-grained understanding of social ecologies.

2.4.2 *Vertical epidemiology*

The advent of antibiotics and insecticides (primarily DDT) during the Second World War allowed unprecedented control of infectious and vector-borne diseases. It is easy to document the resurgence of malaria and yellow fever throughout the World War (Bruce–Chwatt & de Zulueta, 1980). Nevertheless the sheer power of new technological weapons against vectors, and military experimentation with new vaccines encouraged American epidemiologists to read the data optimistically. Successful control during the war provided a global model for vertically organized, technologically focused, one-disease control campaigns. Control strategies for both malaria and yellow fever emerged from long-standing collaboration among epidemiologists, field biologists, clinicians, military surgeons and international health personnel (Gordon, 1952).

Military and civilian authorities previously addressed changing aspects of the host–environment–agent/vector disease triangle with methods adapted to each environmental setting. The locales where epidemiologists battled disease determined the character of hostilities. Climate concerns followed local and seasonal variations for insecticide applications, the use of defoliants, and/or the prescriptions for vaccines or pharmaceuticals. Recognition of the human influence on disease control was even more apparent in the war's aftermath, as refugees strained health infrastructures and increased host pools. New vectors entered deforested battle sites and moved relentlessly with refugees' resettlement. Epidemiological investigation of long-known infectious diseases followed U.S. military interests in scrub typhus in Malaysia (Gordon, 1952), the "new" disease of Hantan fever in Korea, and later drug-resistant malaria in Vietnam (Packard, 1997).

Meanwhile, powerful multi-national interests fostered industrialization and consumerism, marketing health expenditures in the nondomestic arena as essential for economic "development". "Investing in health" became – and to a considerable extent remains – the economic development strategy for investments in developing nations. Social scientists and technical experts of the 1960s and 1970s were involved in early health-related projects of the new Agency for International Development (1961), of the Food and Agriculture Organization (FAO), relief programmes, population control programmes, health and nutritional programmes – all deemed necessary to support these broad social changes (Lee, 1985).

Technical assistance to control endemic disease served the expansion of western political influence in the third world, by averting rural-urban migration, land tenure conflict and malnutrition, especially due to "explosive" population growth. In 1965, the U.S. Senate heard George McGovern urge additional monies for "technical assistance, [which] is not entirely a charitable undertaking ... It is one of the best economic, political and moral investments we can make ..." John D. Rockefeller asserted, "Population growth is like a lingering, wasting illness ... [endangering] the future of life on the planet, or more important perhaps, life as we want it to be." (Fee and Acheson, 1991 p. 263). Development and food production required a more central focus on climate, ecology and needed health care to ensure a stable labour force. Technology again provided scientific agriculture through the Green Revolution developments in crop production, fertilizers and insecticides.

In the 1970s the World Bank's approval of development project loans required the inclusion of "requisite preventative health care measures", and evaluation of a project's impact on the "human ecology" of affected local populations (Lee, 1985 p. 58). Some development projects were notorious for exacerbating local endemic diseases, or introducing new hazards, as in the introduction of schistosomiasis in irrigation (Mubarak, 1982), onchocerciasis in hydroelectric projects, or the emergence of malaria, or trypanosomiasis with new transportation corridors and agricultural projects (Lee, 1985, pp. 54–87). Human influence on disease control became more apparent as displaced populations strained local health infrastructure, increased susceptible host pools, and/or allowed new vectors to emerged (Prothero, 1961).

Whether multi-national corporations and investors actually believed their own presumption of a simple link between top-down health intervention and the economic success of development programmes has been the subject of considerable discussion and doubt among recent historians and anthropologists (Farid, 1980; Packard & Brown, 1997). But attention to the traditional meanings of the term "ecological" certainly characterized the epidemiologists who aided these efforts. Population control programmes, health and nutrition programmes and specific disease-eradication programmes were the medical-epidemiological face of health-through-development/development-through-health initiatives. The control of endemic disease served political and economic goals of bringing labour to under-populated territories, slowing urban migration, as well as other objectives. When programme officials developed local sanitation or primary care projects, the technical means were already available, and did not necessarily require epidemiologists to understand the prevailing local diseases. Thus internationally sponsored health efforts aided, rather than threatened, local elites.

To some extent "vertical" epidemiology of the Cold War period emphasized continuity over the preceding century – from the International Sanitary Congresses through the Rockefeller and League of Nations efforts, into the 1948 creation of the World Health Organization. Nevertheless the age of individual heroes was over. In many cases local people understood the seasonal connections between vectors, rainfall and outdoor work exposure. Warfare and development projects promoted population movement, breaking the link that tradition and culture gave to indigenous populations under "development". The team approach taken in disease control, especially in development campaigns of the 1950s and 1960s, broke down the heroic model of one-person, one-disease, one-strategy, one-voice disease control.

2.4.3 Problems in the epidemiology of "disease campaigns"

Susser described the post-1950s as a new era of "black box epidemiology", reflecting the discipline's shift to focus on chronic diseases (Susser & Susser, 1996). The actual transition to a chronic disease regime began in the late nineteenth century (Omran, 1971). But for Anglo-American epidemiologists the problems it posed for method and theory emerged just as meteorologists and oceanographers were beginning to discern global climatological patterns, which in turn predicted widespread changes in the incidence and prevalence of many infectious diseases (Nebeker, 1995; Glantz, 1996; Reiter, 2001). Without question "horizontal", methodological epidemiology achieved success, especially in the development of multivariable models to identify factors predicting an individual's likelihood of, for example, a heart attack. Thus, the chronic disease focus forced the consideration of multiple host and agent causes, rather than the more vertical, agent-elimination approach seen in infectious disease campaigns. Such models did not simply predict individual risk, but were rather applied "horizontal" efforts to lower population risk through reducing several risk factors concurrently, in the hope of preventing those diseases.

As in the earlier campaigns against infectious disease, many epidemiologists prior to the 1990s accepted the medical model's focus on individual risk assessment and technological medical intervention. Models for rarer diseases such as cancer were of necessity static estimates of individual risk. In chronic disease, the clinical focus and its medical school funding led to fewer physicians in public health research and training, and a halving of public health enrollments in the first decade after the Second World War (Fee & Rosenkrantz, 1991 p. 252). The discipline's absorption in methodology and statistical tools strained relations with public health professionals seeking other models for environmental influences, both social and physical, on human health. Social epidemiologists in particular faulted "risk factor" epidemiology for ignoring the "fundamental" causes of disease, such as inequities in social class, quality of life, social capital and indices of social disruption (Pearce, 1996).

From 1970 onwards, malaria and other infectious diseases have remained in a constant state of disequilibrium, due to the transport of vectors and infected/immune persons to new areas, agent drug resistance and human encroachment on agent habitats (Last, 1998 p. 195). The equilibrium observed between vectors and their environment was disrupted as vectors became insecticide resistant, or found new breeding grounds and habitat ranges. Warfare and development projects, however, both promoted population movement and manmade changes in the local environments. Deforestation, mono-crop agriculture, housing change, colonization, new domestic crops and animals and inadequate health infrastructure all complicated any single strategy to control vector-borne disease. The team approach taken in disease control in development projects began to break down the heroic model of one-person, one-viewpoint of disease control.

Epidemiologists' attempts to link noncommunicable disease to environmental exposures were limited by the lack of data an individual exposures. Mortality data were unavailable for the entire U.S. until 1933. Only in the mid-1960s did national morbidity data expand to include on-going population measures of health for prevention and service planning. The linkage of mortality to occupational exposures and health-care usage came only in the late 1970s, with the advent of environmental and occupational safety regulation (Last, 1998). Environmental exposure-outcome data for individuals (versus groups or localities) were notoriously difficult to find, especially for rare, slow developing diseases such as cancers. Within environmental epidemiology early attention to climatic variables mostly occurred in two areas: air quality and health (Broekers, 1975; GEMS, 1982), and solar radiation effects on cancer (Lee, 1982). The seeming emergence of "new" infectious diseases in the 1980s shook American confidence in the medical model of disease control, especially in the face of unknown threats such as acquired immunodeficiency syndrome (AIDS) (Fee & Fox, 1988 pp. 316–43; Thacker & Berkelman, 1988). Increasingly common reports of the re-emergence of "controlled" diseases such as dengue, viral encephalitis, malaria and cholera identified failures in surveillance and control, and disease expertise in many public health agencies. Re-emergence followed vector insecticide resistance, chemotherapy resistance by microbial agents, and a host of human behaviours that promoted both patterns. More broadly, environmental factors became increasingly important as "development" efforts to eliminate diseases such as malaria failed. In these cases the role of climate change has been explored by collaborating biologists and climatologists, especially with regard to changes in vector distribution.

2.5 Conclusion

As cholera deaths mounted in the Soho district of London in early September 1854, rumours circulated that work on city sewers had opened ancient burial pits, especially the poisonous ones of the great plague two centuries before (Brody *et al.*, 1999). To exonerate the Metropolitan Commission of Sewers, Edmund

Cooper drew maps of the streets and sewer lines, adding small dark rectangles where cholera deaths had occurred. In his celebrated presentation to the Epidemiological Society of London the following December, John Snow used precisely this kind of visual aid – that is, he borrowed Cooper's idea. More than that, Snow had not done his house-to-house survey about cholera deaths when he urged removal of the Broad Street pump handle on September 7. He had long believed that cholera was a contagion; he steadily elaborated this conviction (though not in writing) between 1849 and 1854. Snow went house-to-house in order to prove that he *had been right* to urge the removal of the handle (Paneth *et al.*, 1998).

For most of the twentieth century epidemiologists have either ignored these minutiae about Snow or found them pedagogically unhelpful. Snow was virtually unknown even in late nineteenth-century Britain. Not until Wade Hampton Frost republished his 1855 monograph on cholera did Snow become a pedagogical hero in the training of clinically oriented epidemiologists. Thus it is interesting that, rather suddenly in the late 1980s and 1990s, some epidemiologists began to question Snow's place and importance in the story of epidemiology. Several such sceptics undertook, alone or in collaboration with historians and historical geographers, original archival research as well as patient scholarly inquiry into published historical scholarship (Vandenbroucke *et al.*, 1991; Brody *et al.*, 1999; McLeod, 2000). John Snow's work remains of singular importance in the reassessment, but in ways that reflect some of epidemiologists' current debates about the future of the discipline.

How is Snow's image being remade? Vandenbrouke (1989) emphasized the process of Snow's "discovery" as a deductive one, rather than a case built on hypothesis-framing from inductive clinical and environmental observations. Epidemiologists, he argued, prefer to believe that Snow was a model for the practice of modern "black box" epidemiology, which they in turn laud as patient empirical data collection and methodological rigour. If Snow is exposed as just another energetic Victorian doctor peddling a pet idea that, later on, happened to fit well with the germ theory of disease, "black box" epidemiology also may later be seen as a comfortable academic niche for those who are blind to larger social and environmental purposes in epidemiology.

A fuller use of the ecological approach linking individual risk with larger "environmental systems" variables returned to eco-epidemiology, to use Susser's relatively recent term (Susser, 1994). Scientific evidence of global climatic change, population pressures, and re-emergent infectious diseases was strengthened prior to the 1990s. We argue that the present impetus for a global environmental epidemiology emerges from the practitioners' recognition of failures in the discipline's traditional mission of disease control. Strategies have been too limited to understand rapidly evolving disease patterns. We increasingly recognize that our global community has no borders, no effective barricades against disease. The most recent report of the World Health Organization steps away from earlier disease-campaign approach to an emphasis on the primary needs of whole communities. Even our past successes in health improvements are only maintained as long as health infrastructure can keep up with local population and economic changes. In the face of broad-scale health needs, disease-specific campaigns are too narrow and costly (WHO, 2000).

References

Anderson, W. (1996). Immunities of empire: race, disease, and the new tropical medicine, 1900–1920. *Bulletin of the History of Medicine*, **70**, 94–118.

Baldwin, P. (2000). *Contagion and the State in Europe, 1830–1930*. Cambridge: Cambridge University Press.

Bradley, R. S. & Jones, P. D. (1992). *Climate since A. D. 1500*. New York: Routledge.

Brody, H., Rip, M. R., Vinten-Johansen, P., Paaneth, N. & Rachman, S. (1999). Map-making and myth-making in Broad Street: the London cholera epidemic, 1854. *Lancet*, **356**, 64–8.

Broekers, S. (1975). Climate change: are we on the brink of a pronounced global warming? *Science*, **189**, 461.

Bruce-Chwatt, L. & de Zulucta, J. (1980). *The Rise and Fall of Malaria in Europe: a Historico-epidemiological Study*. Cambridge: Cambridge University Press.

Bynum, W. (1993). Policing hearts of darkness: aspects of the International Sanitary Conferences. *History and Philosophy of the Life Sciences*, **15**, 421–34.

Bynum, W. F. & Overy, C. ed. (1998). *The Beast in the Mosquito: The Correspondence of Ronald Ross and Patrick Manson*. The Wellcome Institute Series in the History of Medicine. Amsterdam: Rodopi.

Cassedy, J. H. (1969). Meteorology and medicine in colonial America: beginnings of the experimental approach. *Journal of the History of Medicine and Allied Sciences*, **24**, 193–204.

Coleman, W. (1987). *Yellow Fever in the North: the Methods of Early Epidemiology*. Madison, WI: University of Wisconsin Press.

Crookshank, F. G. (1922). *Influenza: Essays by Several Authors*. London: Heinemann, pp. 1–81.

Crosby, A. (1986). *Ecological Imperialism: the Biological Expansion of Europe*. Cambridge: Cambridge University Press.

Cunningham, A. (1992). Transforming plague: the laboratory and the identity of infectious disease. In *The Laboratory Revolution in Medicine*, ed. A. Cunningham & P. Williams, pp. 209–44. Cambridge: Cambridge University Press.

Davis, M. (2001). *Late Victorian Holocausts: El Niño Famines and the Making of the Third World*. London Verso.

De Bevoise, K. (1995). *Agents of Apocalypse: Epidemic Disease in the Colonial Philippines*. Princeton, NJ: Princeton University Press.

Dobson, M. J. (1997). *Contours of Death and Disease in Early Modern England*. Cambridge: Cambridge University Press.

Dubos, R. (1959). *Mirage of Health: Utopias, Progress, and Biological Change*. New York: Harper and Row.

Dyson, T. (1991). On the demography of south Asian famines. *Population Studies*, **45**, 5–25, 279–97.

Ettling, J. (1981). *The Germ of Laziness: Rockefeller Philanthropy and Public Health in the New South*. Cambridge, MA: Harvard University Press.

Evans, H. (1989). European malaria policy in the 1920s and 1930s: the epidemiology of minutiae. *Isis*, **80**, 40–59.

Eyler, J. (1979). *Victorian Social Medicine: The Ideas and Methods of William Farr*. Baltimore, MD: Johns Hopkins University Press.

Farid, M. A. (1980). The malaria programme – from euphoria to anarchy. *World Health Forum*, **1**, 8–33.

Farley, J. (1991). *Bilharzia: a History of Imperial Tropical Medicine*. Cambridge: Cambridge University Press.

Fee, E. (1987). *Disease and Discovery: a History of the Johns Hopkins School of Hygiene and Public Health, 1916–1939*. Baltimore, MD: Johns Hopkins University Press.

Fee, E. & Acheson, R. M. ed. (1991). *A History of Education in Public Health*. New York: Oxford University Press.

Fee, E. & Fox, D. M. (1988). *AIDS: The Burdens of History*. Los Angeles, CA: University of California Press.

Fee, E. Rosenkrantz, B. (1991). Professional education for public health in the United States. In *A History of Education in Public Health*, ed. E. Fee & R. M. Acheson. New York: Oxford University Press.

Fidler, D. P. (1999). *International Law and Infectious Diseases*. New York: Oxford University Press.

Fleming, J. R. ed. (1998). The climatic determinism of Ellsworth Huntington. *Historical Perspectives on Climate Change*. Oxford: Oxford University Press.

GEMS (Global Environmental Monitoring System) (1982) *Estimating Human Exposure to Air Pollutants*. WHO Offset Publication # 69.

Glacken, C. J. (1967). *Traces on the Rhodian Shore: Nature and Culture in Western Thought from Ancient Times to the End of the Eighteenth Century*. Berkeley, CA: University of California Press.

Glantz, M. H. (1996). *Currents of Change: El Niño's Impact on Climate and Society*. Cambridge: Cambridge University Press.

Goodman, N. (1952). *International Health Organizations and their Work*. Philadelphia PA, Blakiston.

Gordon, J. E. (1952). The twentieth century: yesterday, today and tomorrow. In *A History of American Epidemiology*, ed. C.-E. A. Winslow *et al.*, pp. 114–167. St. Louis, MD: Mosby.

Grant, E. (1994). *Planets, Stars and Orbs: the Medieval Cosmos, 1200–1687*, pp. 630–8. Cambridge: Cambridge University Press.

Grove, J. M. (1988). *The Little Ice Age*. London: Methuen.

Grove, R. (1995). *Green Imperialism*. Cambridge: Cambridge University Press.

Hamlin, C. (1998). *Public Health and Social Justice in the Age of Chadwick: Britain, 1800–1854*. New York: Cambridge University Press.

Hannaway, C. (1993). Environment and miasmata. In *Companion Encyclopedia of the History of Medicine*, ed. W. F. Bynum & R. Porter, pp. 292–308. London: Routledge.

Harrison, M. (1996). A question of locality: the identify of cholera in British India, 1860–1890. In *Warm Climates and Western Medicine: the Emergence of Tropical Medicine, 1500–1900*, ed. D. Arnold, pp. 133–59. Atlanta, GA: Rodopi.

Haynes, D. M. (2000). Framing tropical disease in London: Patrick Manson, *Filaria perstans*, and the Uganda Sleeping Sickness Epidemic, 1891–1902. *Social History of Medicine*, **13**, 167–93.

Howard-Jones, N. (1972). Choleranomalies: the unhistory of medicine as exemplified by cholera. *Perspectives in Biology and Medicine*, **15**, 422–33.

Howard-Jones, N. (1981). The World Health Organization in historical perspective. *Perspectives in Biology and Medicine*, **24**, 467–82.

Humphreys, M. (1996). Kicking a dying dog: DDT and the demise of malaria in the American South, 1942–1950. *Isis*, **87**, 1–17.

Jones, C. (1996). Plague and its metaphors in early modern France. *Representations*, **53**, 97–127.

Jordanova, L. J. (1979). Earth science and environmental medicine: the synthesis of the late enlightenment. In *Images of the Earth: Essays in the History of the Environmental Sciences*, ed. L. Jordanova & R. S. Porter, pp. 119–46. London: British Society for the History of Science.

Kunitz, S. J. (1987). Explanations and ideologies of mortality patterns. *Population and Development Review*, **13**, 379–407.

Kunitz, S. J. (2000). Globalization, states and the health of indigenous peoples. *American Journal of Public Health*, **90**, 1531–39.

Last, J. M. (1993). Global change: ozone depletion, greenhouse warming, and public health. *Annual Reviews in Public Health*, **14**, 115–36.

Last, J. M. (1998). *Public Health and Human Ecology*. Stamford: Appleton-Lange.

Lee, J. A. (1982). Melanoma and exposure to sunlight. *Epidemiological Reviews*, **4**, 110–36.

Lee, J. A. (1985). *The Environment, Public Health, and Human Ecology: Considerations for Economic Development*. Baltimore, MD: Johns Hopkins University Press, for the World Bank.

Lilienfeld, D. E. & Stolley, P. D. (1994). *Foundations of Epidemiology*. New York: Oxford University Press.

Livi Bacci, M. (2000). *The Population of Europe: a History* [translated by C. Ipsen & C. Ipsen]. London: Blackwell.

Lloyd, G. E. R. (ed.) (1983). *Hippocratic Writings*, section on Medicine, translated by J. Chadwick & W. N. Mann. New York City, NY: Penguin.

Löwy, I. (1997). What / who should be controlled? Opposition to yellow fever campaigns in Brazil, 1900–39. In *Western Medicine as Contested Knowledge*, ed. A. Cunningham & B. Andrews, pp. 124–46. Manchester: Manchester University Press.

McLeod, K. S. (2000). Our sense of Snow: the myth of John Snow in medical geography. *Social Science and Medicine*, **50**, 923–35.

Monath, T. P. (1998). Milestones in the conquest of yellow fever. In *Microbe Hunters: Then and Now*, ed. H. Koprowski & M. Oldstone, pp. 95–111. Bloomington, IL: Mid-Ed Press.

Mubarek, A. B. (1982). The shistosomiasis problem in Egypt. *American Journal of Tropical Medicine and Hygiene*, **31**, 87–91.

Nebeker, F. (1995). *Calculating the Weather: Meteorology in the Twentieth Century*. San Diego, CA: Academic Press.

Omran, A. (1971). The epidemiological transition: a theory of the epidemiology of population change. *Milbank Memorial Fund Quarterly*, **49**, 509–38.

Packard, R. M. (1997). Malaria dreams: postwar visions of health and development in the Third World. *Medical Anthropology*, **17**, 279–96.

Packard, R. M. & Brown, P. J. (1997). Rethinking health, development and malaria: historicizing a cultural model in international health. *Medical Anthropology*, **17**, 181–94.

Paneth, N., Johansen, P. V., Brody, H. & Rip, M. (1998). A rivalry of foulness: official and unofficial investigations of the London Cholera epidemic of 1854. *American Journal of Public Health*, **88**, 1545–53.

Pearce, N. (1996). Traditional epidemiology, modern epidemiology, and public health. *American Journal of Public Health*, **86**, 678–83.

Pelling, M. (1993). Contagion / germ theory / specificity. In *Companion Encyclopedia of the History of Medicine*, 2 vols, ed. W. F. Bynum & R. Porter, pp. 292–308. London: Routledge.

Petersen, W. F. (1934–37). *The Patient and the Weather*. Ann Arbor, MI: Edwards Brothers, Inc.

Porter, R. (1981). The terraqueous globe. In *The Ferment of Knowledge*, ed. R. Porter & G. S. Rousseau, pp. 285–324. Cambridge: Cambridge University Press.

Porter, T. M. (1986). *The Rise of Statistical Thinking: 1820–1900*. Princeton, NJ: Prinoeton University Press.

Powell, J. H. (1970). *Bring Out Your Dead: the Great Plague of Yellow Fever in Philadelphia in 1793*. New York: Arno Press.

Prothero, R. (1961). Population movement and problems of malaria eradication in Africa. *Bulletin of the WHO*, **24**, 405–25.

Pullan, B. (1992). Plague and perceptions of the poor in early modern Italy. In *Epidemics and Ideas: Essays on the Historical Perception of Pestilence*, ed. T. Ranger & P. Slack, pp. 101–23. Cambridge: Cambridge University Press.

Ranger, T. & Slack, P. ed. (1992). *Epidemics and Ideas: Essays on the Historical Perception of Pestilence*. Cambridge: Cambridge University Press.

Reiter, P. (2001). Climate change and mosquito-borne disease. *Environmental Health Perspectives*, **109**, 141–61.

Riley, J. C. (1987). *The Eighteenth-Century Campaign to Avoid Disease*. New York: St. Martin's Press.

Rosenberg, C. E. ed. (1992). Explaining epidemics. *Explaining Epidemics and Other Studies in the History of Medicine*, pp. 293–304. Cambridge: Cambridge University Press.

Sawyer, W. A., Meyer, K. F., Eaton, M. D., Bauer, J. H. Putnam, P. & Schwentker, F. F. (1944). Jaundice in Army personnel in the western region of the United States and its relation to vaccination against yellow fever. *American Journal of Hygiene*, **39**, 337–432.

Sigerist, H. E. (1941). The value of health to a city: two lectures, delivered in 1873, by Max von Pettenkofer, translated from the German with an introduction. *Bulletin of the History of Medicine*, **10**, 593–613.

Slack, P. (1985). *The Impact of Plague in Tudor and Stuart England*. London: Routledge and Kegan Paul.

Smith, W. D. (1979). *The Hippocratic Tradition*. Ithaca: Cornell University Press.

Snowden, F. M. (1999). Fields of death: malaria in Italy, 1861–1962. *Modern Italy*, **4**, 25–57.

Soper, F. L. (1977). *Ventures in World Health: the Memoirs of Fred Lowe Soper*, ed. J. Duffy. Scientific Publication No. 355. Washington, D.C.: Pan American Health Organization; WHO.

Susser, M. (1973). *Causal Thinking in the Health Sciences, Concepts and Strategies for Epidemiology*. New York: Oxford.

Susser, M. (1994). The logic in ecological: II. The logic of design. *American Journal of Public Health*, **84**, 830–5.

Susser, M. & Susser, E. (1996). Choosing a future for epidemiology, I: eras and paradigms. *American Journal of Public Health*, **86**, 668–73.

Terris, M. (1959). The changing face of public health. *American Journal of Public Health*, **49**, 1119, reprinted in Fee & Acheson, 1991.

Tesh, S. N. (1995). Miasma and "social factors" in disease causality: lessons from the nineteenth century. *Journal of Health Politics, Policy and Law*, **20**, 1001–24.

Thacker, S. B. & Berkelman, (1988). Public health surveillance in the United States. *Epidemiological Reviews*, **10**, 164–90.

Tomes, N. (1998). *The Gospel of Germs: Men, Women and the Microbe in American Life*. Cambridge, MA: Harvard University Press.

Tuan, Y.-F. (1968). Discrepancies between environmental attitude and behaviour: examples from Europe and China. *Canadian Geographer*, **12**, 176–91.

Vandenbroucke, J. P. (1988). Which John Snow should set the example for clinical epidemiology? *Journal of Clinical Epidemiology*, **41**, 1215–16.

Vandenbroucke, J. P. (1989). Those who were wrong. *American Journal of Epidemiology*, **130**, 3–5.

Vandenbroucke, J. P. Eelkman Rooda, H. M. & Benkars, H. (1991). What made John Snow a hero? *American Journal of Epidemiology*, **133**, 967–73.

WHO (2000). World Health Report.

Winslow, C.-E. A. (1926). Public health at the crossroads. *American Journal of Public Health*, **16**, 1075–85.

Worboys, M. (1993). Tropical diseases. In *Companion Encyclopedia of the History of Medicine*, ed. W. F. Bynum & R. Porter, 2 vol. pp. 511–36. London: Routledge.

Worboys, M. (1996). Germs, malaria, and the invention of Mansonian tropical medicine: from diseases in the tropics to tropical diseases. In *Warm Climates and Western Medicine: the Emergence of Tropical Medicine, 1500–1900*, ed. D. Arnold, pp. 181–207. Atlanta, GA: Rodopi.

Zhang, T. Y. (1993). Ancient Chinese Concepts of Infectious Epidemic Disease. In *History of Epidemiology: Proceedings of the 13th International Symposium on the Comparative History of Medicine East and West*, ed. Y. Kawakita, S. Sakai & Y. Otsuka. Tokyo: Ishiyaku EuroAmerica.

Bibliography

Acheson, R. & Poole, P. (1991). The London School of Hygiene and Tropical Medicine: a child of many parents. *Medical History*, **35**, 385–408.

Arnold, D. (1993). Medicine and colonialism. In *Companion Encyclopedia of the History of Medicine*, 2nd vol. ed. W. F. Bynum & R. Porter, pp. 1393–416. London: Routledge.

Beaglehole, R. & Bonita, R. (1997). *Public Health at the Crossroads: Achievements and Prospects*. New York City, NY: Cambridge University Press.

Bouma, J. J. S. H. J. W. (1987). A short history of human biometeorology. *Experientia*, **43**, 2–6.

Brown, P. J. (1997). Malaria, miseria, and underpopulation in Sardinia: the 'malaria blocks development' cultural model. *Medical Anthropology*, **17**, 239–54.

Cassedy, J. H. (1986). *Medicine and American Growth*. Madison, WI: University of Wisconsin Press.

Chapin, C. V. (1917). *How to Avoid Infection*. Cambridge, MA: Harvard University Press.

Cipolla, C. M. (1976). *Public Health and the Medical Profession in the Renaissance*. New York City, NY: Cambridge University Press.

Delaporte, F. (1991). *The History of Yellow Fever: an Essay on the Birth of Tropical Medicine*. Cambridge, MA: MIT Press.

Evans, A. S. (1978). Causation and disease: a chronological journey. *American Journal of Epidemiology*, **108**, 249–58.

Evans, R. J. (1987). *Death in Hamburg: Society and Politics in the Cholera Years, 1830–1910*. New York City, NY: Oxford University Press.

Eyler, J. (1997). *Sir Arthur Newsholme and State Medicine*. New York City, NY: Cambridge University Press.

Fox, D. (1989). Policy and epidemiology: financing services for the chronically ill and disabled, 1930–1990. *Milbank Memorial Fund Quarterly*, **67**, Suppl 2, 1–31.

Frost, W. H. (1941). Epidemiology. Reprinted in *The Papers of Wade Hampton Frost*, ed. K. F. Maxcy. New York City, NY: The Commonwealth Fund.

Godlewska, A. M. C. (1999). *Geography Unbound: French Geographic Science from Cassini to Humboldt*. Chicago, IL: University of Chicago Press.

Gordon, J. E. (1950). Epidemiology – old and new. *Journal of the Michigan State Medical Society*, **49**, 194–9.

Hamlin, C. (1995). Finding a function for public health: disease theory or political philosophy? [Commentary on Tesh] *Journal of Health Politics, Policy and Law*, **20**, 1025–31.

Harrison, G. (1978). *Mosquitoes, Malaria and Man: a History of Hostilities since 1880*. New York City, NY: Dutton.

Hill, A. B. (1953). Observation and experiment. *New England Journal of Medicine*, **248**, 995–1001.

Krieger, N. (1994). Epidemiology and the web of causation: has anyone seen the spider? *Social Science and Medicine*, **39**, 887–903.

Les Benedict, M. (1970). Contagion and the constitution: quarantine agitation from 1859 to 1866. *Journal of the History of Medicine and Allied Sciences*, **25**, 177–93.

Levine, M. M. & Thacker, S. B. (eds.) (1996). Emerging and re-emerging infections. *Epidemiological Reviews*, **18**, 1–97.

Lilienfeld, D. E. (1979).The greening of epidemiology: sanitary physicians and the London Epidemiological Society 1830–1870. *Bulletin of the History of Medicine*, **52**, 503–28.

Locher, W. (1993). Pettenkofer and epidemiology: erroneous concepts – beneficial results. In *History of Epidemiology: Proceedings of the 13th International Symposium on the Comparative History of Medicine East and West*, ed. Y. Kawakita, S. Sakai & Y. Otsuka, pp. 93–120. Tokyo: Ishiyaku EuroAmerica.

Loomis, D. & Wing, S. (1990). Is molecular epidemiology a germ theory for the end of the twentieth century? *International Journal of Epidemiology*, **19**, 1–3.

Löwy, I. (1990). Yellow fever in Rio de Janeiro and the Pasteur Institute Mission 1901–1905: the transfer of science to the periphery. *Medical History*, **34**, 144–63.

MacPherson, K. L. (1998). Cholera in China, 1820–1930. In *Sediments of Time: Environment and Society in Chinese History*, pp. 487–519. New York City, NY: Cambridge University Press.

Martens, P. & Hall, L. (2000). Malaria on the move: human population movement and malaria transmission. *Emerging Infectious Diseases*, **6**, 103–9.

McMichael, A. J. (1994). Molecular epidemiology: new pathway or new travelling companion? *American Journal of Epidemiology*, **140**, 1–11.

McMichael, A. J. (1999). Prisoners of the proximate: loosening the constraints on epidemiology in an age of change. *American Journal of Epidemiology*, **149**, 887–97.

Packard, R. M. & Gadehla, P. (1997). A land filled with mosquitoes: Fred L. Soper, the Rockefeller Foundation, and the *Anopheles gambiae* invasion of Brazil. *Medical Anthropology*, **17**, 215–38.

Roemer, M. (1993). Internationalism in medicine and public health. In *Companion Encyclopedia of the History of Medicine*, vol. 2, pp. 1417–35, ed. W. F. Bynum & R. Porter. London: Routledge.

Shy, C. M. (1997). The failure of academic epidemiology: witness for the prosecution. *American Journal of Epidemiology*, **145**, 479–84.

Skrabanek, P. (1992). The poverty of epidemiology. *Perspectives in Biology and Medicine*, **35**, 182–8.

Snow, J. (1936). *Snow on Cholera*, a reprint of two papers by John Snow, M.D., ed. W. Hampton Frost. New York: The Commonwealth Fund.

Susser, M. (1985). Epidemiology in the United States after World War II: the evolution of technique. *Epidemiologic Reviews*, **7**, 47–77.

Susser, M. (1986). The logic of Sir Karl Popper and the practice of epidemiology, *American Journal of Epidemiology*, **124**, 711–18.

Susser, M. (1998). Does risk factor epidemiology put epidemiology at risk? Peering into the future, with commentaries by K. McPherson, C. Pool & K. Rothman, J. P. Mackenbach, M. Kogevinas, P. Vineis, & A. Morabia. *Journal of Epidemiology and Community Health*, **52**, 608–18.

White, K. L. (1991). *Healing the Schism. Epidemiology, Medicine, and the Public's Health*. New York: Springer-Verlag.

Winkelstein, W. (1995). A new perspective on John Snow's communicable disease theory. *American Journal of Epidemiology*, **142**, S3–S9.

Winslow, C. E. A. (1980) [1943]. *The Conquest of Epidemic Diseases*. Madison, WI: University of Wisconsin Press.

3

The contribution of global environmental factors to ill-health

KIRK R. SMITH & MANISH A. DESAI

3.1 Introduction

In order to better understand the potential impact of global environmental change on human health it is necessary to accomplish two contrasting tasks. First, it is important to widen the conceptual structure by which health risks are evaluated to include a broader array of more distal risk factors than has been common in public health in recent years. In addition, to focus the analysis in terms that facilitate meaningful comparisons with other important risks to health, there is need to structure the analysis in absolute measures of ill-health and in terms of standard and emerging decision-making tools. Progress in both these arenas will be needed to effectively guide intervention policies.

To approach these tasks, we follow a temporal progression. First, we briefly examine historical views of human health and the environment to show that the challenges now created by global environmental change actually extend contemporary public health's scope into realms previously embraced by the field. We next offer an analytical structure for addressing the linkages and pathways between multiple social and ecological processes acting at various scales, ultimately influencing health. Issues of causality and capability lead us to examine how a disease-based, resource-effectiveness paradigm might be expanded to understand environment–health connections today. We quantify the current contribution of environmental risk factors to ill-health, and from this basis peer into the future. We discuss how attributable and avoidable risk calculations relate to public health planning, and the implications of incorporating considerations of net present value and sustainability. Lastly, we explore how these instruments might serve as a guide for orienting interventions to reduce the impact of global environmental change on health.

3.2 Environment and health: prologue

Ancient and indigenous knowledge systems have long recognized the crucial role played by the environment in causing and curing illness (Shahi *et al.*, 1997).

Growing public health concerns about global environmental change similarly emphasize the relationships, often complex, between environment and health (Colwell, 1996; Daily & Ehrlich, 1996; McMichael, 1997; McMichael *et al.*, 1998; Pimentel *et al.*, 1998). The first half of this section surveys Western[1] conceptualizations of this interplay. The past several centuries have witnessed the evolution of generally narrower views of the relationship between the environment and human health, a trend that is now tending to reverse. The successes and failures of this trajectory provide insight into the relative merits of various public health paradigms and accompanying intervention strategies. In order to assist in construction of broader public health agendas, we present a multilevel causal web framework for stimulating new theories and new methods in public health in the second half of this section (Corvalán *et al.*, 1999). Moreover, public health will have to move beyond extending and refining models of causality and towards harnessing resources to implement promising yet challenging interventions. Reflecting on the past and present can help operationalize "political will" (Reich, 1994) by provoking a critical examination of the structures and processes that drive the allocation of resources for public health purposes.

3.2.1 Historical survey

Paradigms constitute the milieu within which people define their relationships to one another and their shared environment. Ideas concerning the relationship between environment and health have changed dramatically with the radical transformation of human enterprise. These ideas are interwoven with developments in social, political and economic thought, as well as cultural beliefs and practices. The ensuing discussion focuses on Western views of environment and health, which now tend to dominate the globe (Smith, 2000). Yet, there are also many locally specific interpretations of this worldview. The hybridization of the Western worldview with other systems of thought, furthermore, has been convoluted and controversial. Clearly, a full appreciation of this multitextured coevolution, a cultural and intellectual history, requires far greater breadth and depth than can be covered here. The topic is more substantially treated in studies of the history, sociology and anthropology of public health (Trostle, 1986; Cohen, 1989; Inhorn & Brown, 1997; Hahn, 1999; Albrecht *et al.*, 2000, Melosi, 2000). Nonetheless, an abridged examination of the progression of dominant beliefs about environment and health sheds light on paradigm shifts and accompanying movements in policy and practice within public health. Understanding this history can help to encourage theories and methods appropriate to the challenges ahead.

Medieval Europe perceived the environment's impact on health, alternatively benevolent and malevolent, to reflect God's will. Christian beliefs simultaneously

---

[1] West and Western in this chapter refer broadly to Europe and North America.

condemned and championed the sick, a pattern also evident in the more intervention-based Greco-Arabic, Indian and Chinese traditions (Pedersen, 1996). By the fifteenth and sixteenth centuries, it was thought that humoral imbalances resulting from unfavourable environmental conditions, "miasmas", precipitated illness (Cipolla, 1992). During the seventeenth and eighteenth centuries, postulated though unconfirmed "contagions" were also thought to spread disease. The two competing theories of miasmatists and contagionists were sometimes merged into a complementary view that miasmas controlled epidemic propagation from one region to another, whereas contagions promoted the spread within a region (Shahi *et al.*, 1997). The progressive waves of diseases that marched across the European continent appeared to support beliefs in miasmas and/or contagions. Exchanges between the environment and the body, both viewed as sources of filth (Douglas, 1966), largely explained the origins of ill-health. The environment was viewed as uncontrollable and beyond human influence, whereas the body could be controlled (Trostle, 1986). As a result, interventions focused on prevention and treatment at the bodily level (Foucalt, 1975; Armstrong, 1993). The individual-based perspective accounting for the origins of health and disease is a recurrent theme in public health and underlies modern biomedicine.

The Enlightenment triggered a major philosophical shift in the dominant view towards the environment, moving away from passive fatalism and towards engaged activism with mixed consequences for health. The power of scientific and technological advances enabled nature to be manipulated to an unparalleled degree, stimulating rapid anthropogenic socioecological changes. Indeed, concern over environmental health risks with overtones familiar today played a prominent role in the greatest literature of the time, e.g. Goethe's *Faust* (Binswanger & Smith, 2000).

During the early nineteenth century, amidst the growing pestilence of urbanization and industrialization, a very different approach to the environment and health took shape. As a humane interest in the well-being of populations took root, Villermé in France, Virchow in Germany, and Farr in Great Britain (Farr, 1885; Coleman, 1983; Boyd, 1991) noted the correlations between mortality and poverty, occupation and environmental conditions. The progenitors of modern public health emphasized the importance of promoting hygiene and sanitation in order to improve well-being. Programmes championed by Chadwick and contemporaries centred on mitigating the dirt, crowding and social ills thought to generate miasmas or contagions (Chadwick, 1842). The emerging social sciences were drawn upon to explain disease patterns. Massive infrastructure projects and cleanliness campaigns transformed major cities in Europe and North America (Cohen, 1989; Melosi, 2000). Europe favoured government-led interventions, partly to legitimize newly consolidated nation-states, whereas the U.S. emphasized social activism based upon Puritan mores. Although the justification for these efforts differed somewhat across settings, the simultaneous marshalling of multiple institutions of society was a notable feature of early public health initiatives throughout the Western world (Porter, 1999). Significant reductions in morbidity and mortality resulted from strategies that focused on factors ranging from the proximal and individual, such as encouraging hand washing, to the distal and societal, such as constructing sewerage networks. Validating this strategy, crude death rates in Europe dropped roughly fourfold during the eighteenth and nineteenth centuries (Szreter, 1988).

Contrary to popular belief, most of these efforts predated the advent of germ theory. The heralded leaps in health associated with the introduction of modern public health did not happen because of rapid advances in microbiology. Rather, the early sanitarians and hygienists perceived the linkages between environmental conditions and poor health; for example, the pioneering epidemiological work of John Snow (Snow, 1855). Public health practitioners often attributed disease causation to levels farther up the causal chain than direct biological mechanisms, and they sought to alter those variables: drainage systems, water supply, housing and nutrition. Persuasive arguments reasoned that investments in social welfare would lead to improvements in the health, and thus productivity, of workers. Disease was posited as both a cause and an effect of poor socioeconomic development (Bloom & Canning, 2000). In the nineteenth century there were major investigations into the conditions of the poor, which continue to influence social science analyses of health and illness today (Engels, 1987). Although the distribution of health benefits was not uniform across classes, coalitions emerged and activism characterized this liberal era (Hamlin, 1998). In marked contrast, the implementation of sanitation and hygiene measures in colonial territories followed widely varying paths, with efforts generally dictated by imperial priorities. The squalor once pervasive throughout densely populated regions of the West became associated with the rest of the world.

The tension between paradigms of ill-health, namely either individualistic responses to disease entities or societal and environmental forces, persisted and came to a head during the mid-to-late nineteenth century. During this era, Koch and Pasteur confirmed the existence of microbes and they and their colleagues established infection by these agents to be the cause of many diseases. Other scientists linked cancer to occupational exposures, and nutritional disorders to vitamin deficiencies. These were clearly important advances, yet the focus shifted from relatively holistic considerations of the environment and health towards approaches that sought in particular to identify and combat disease-causing organisms. The domain of public health, which had grown somewhat unknowingly to span these pathogens and, more deliberately, the political action necessary to control them, collapsed into the doctrine of specific aetiology. Reductionism, well rooted in the physical sciences, now took hold of the biological sciences as well. Vaccinations and antibiotics yielded another round of major health improvements, this time based on proximal interventions alone. The period produced magic bullets, some effective but others illusory. Moreover, it generated persistent yet partial models of disease causation. Public health strategies based on specific aetiology could not replace improvements in the quality of living conditions. The notion that the exposure of individuals to certain micro-organisms leads to specific diseases continues to permeate, if not dominate, thinking about cause–effect relationships between environmental change and pathological conditions (Farmer, 1999). Like all paradigms, specific aetiology highlighted certain processes and obscured others.

In the early twentieth century, Frost (1941), Greenwood (1935) and Sydenstricker (1933), seeking to explain the distribution of disease in populations, looked again at the role of social and environmental factors in disease causation. Yet these efforts did not wrest the environment from the periphery of public health concerns. Instead, the twentieth century generated new health priorities in the developed world, as changes in these societies had an impact on behaviour as well as the environment. As noncommunicable diseases (some of which are now known to have communicable components) grew in relative importance, the lifestyle model of disease causation grew to encompass cancer, heart disease and other chronic illnesses (Lalonde, 1974). Health was detached from the environment, and disease resulted, it came to be believed, from poor individual choices. The lifestyle model implied that atomistic individuals make incentive-based choices from among various options. Diet, sedentarism, stress, and the abuse of alcohol and tobacco became targets for improving public health in the West. Risk factor analysis broadened the causative models of specific aetiology, allowing for multifactorial pathways to address the complex aetiology of chronic diseases. Increasingly sophisticated study designs and statistical models became central to this mission. The cultural context that fuels these allegedly misinformed behaviours, however, received relatively scant attention (Lupton, 1995).

By the latter half of the twentieth century, the environment resurfaced in public health in the form of concerns about pollution. What was once thought to be a necessary evil of development came instead to be seen as a threat to the well-being of human and nonhuman life. Nuclear accidents, toxic dumps and oil spills helped environmentalists to galvanize public opinion around perceived health threats. Growing awareness about the ramifications of environmental degradation stimulated more interdisciplinary and multidisciplinary work, but the reductionist paradigm of science continued to dominate. Toxicology, cataloguing chemicals and their effects, assumed a position akin to microbiology earlier in the century. With respect to communicable and noncommunicable diseases alike, biomedicine's spectacular promise offered seductive technological elixirs to cure the world's ills. In fact, biomedicine's success at packaging, commodifying and exporting health has become self-evident, and is now the major formal explanatory framework for illness in all countries of the world. This paradigm continues to be extended. Techno-utopians look forward to a world of genetically enhanced humans, virtual therapy and fabricated ecologies (Joy, 2000). Meanwhile, public health professionals have "been busy reacting to our consumer society's procession of new, potentially hazardous exposures" (McMichael, 1999). Ironically, health concerns have had more influence on environmental policy than environmental concerns have had on health policy. As a result, environmental controls, even those ostensibly aimed at protecting health, have often not focused on the largest environmental health risks within developed countries with well-established environmental health infrastructures, let alone the most significant health risks globally (Smith, 2001).

Specific aetiology and risk factor analysis are incomplete models of disease causation (Krieger & Zierler, 1996). The focus on the proximate determinants of health has generated important insights into biological processes and stimulated equally valuable programmes to counter diseases. From promoting mass immunizations to cautioning against lead exposure, the list of achievements is long. This emphasis on proximate causes, however, has also largely ignored the higher-level processes that create individual-level risk factors and has limited public health's influence on the many sectors that influence well-being. In general, programmes have targeted individuals while ignoring the more powerful agents and processes promoting unhealthy practices. Its successes notwithstanding, overzealous adherence to specific aetiology contributed to the haphazard use of pesticides, the evolution of drug resistance and, most disturbingly, the ignorance of foundational processes driving illness and health. The population explosion of the twentieth century can be attributed in part to well-intentioned but insufficient interventions based on this narrow philosophy. Falling death rates in the developing world were not accompanied, as they had been in the developed world, by sufficient investment aimed at improving the quality of life, promoting human development, and dealing effectively with class and gender inequities – important determinants of fertility (Pedersen, 1996; Shahi *et al.*, 1997). Recent work on the impact of social cohesion (Daniels *et al.*, 1999) and income distribution (Wilkinson, 1996) on health, as well as other examples from social epidemiology and medical anthropology, offer an instructive alternative to this trend.

From a historical perspective, the leaps in longevity witnessed during the past two centuries have been due primarily to changes effected in the structural determinants of health: smaller families, improved nutrition, enhanced infrastructure and preventative measures (McKeown, 1976; Szreter, 1988). Although the miasma/contagion theory was also an incomplete model of disease causation, it suggested that macrolevel interventions could alleviate the disease burden of populations at risk. Early public health efforts sought cooperation from multiple societal actors, government, industry, professionals and citizens, to improve human environments.

Structural change within societies emerged from the convergence of forces involving but not limited to, the health sector. Health often improved dramatically through these intersectoral alliances. Such a philosophy motivated sustainable initiatives that not only reduced mortality and morbidity in the short-term, but helped provide for healthy and fulfilling lives for those saved in the long-term. Clearly it remains important to foster such cooperation today.

Throughout much of the contemporary world the intransigent persistence of infectious diseases and malnutrition, the expansion of toxic industries and periurban slums, the growth of inequities and injustices, and the rise of behavioural and chronic disorders point to the lack of multisectoral co-ordination to mitigate unhealthy environments. Addressing these ills has been difficult to achieve in developed and developing countries alike. There are no magic bullets to achieve multisectoral collaboration. Yet many in the public health establishment have been reluctant to engage broader social or environmental issues. As a result, public health issues have too often been manipulated to legitimize political claims and pave the path for development – with dubious consequences for health (Reich, 1994; Pedersen, 1996; Packard & Brown, 1997; Mantini-Briggs, 2000). Moreover, the uncritical application of public health theory has reinforced dominant modes of discourse, which have obscured the "pathogenic roles of social inequities" (Farmer, 1996) and environmental degradation (Das, 1995; Packard & Brown, 1997). As a number of prominent epidemiologists comment, reflecting on their discipline, the "central" science of public health, epidemiology must face the daunting challenge of extending its theories and methods to address these higher-level factors or else risk losing its relevance and utility (Rose, 1985; Krieger & Zierler, 1996; Pearce, 1996; Beaglehole & Bonita, 1997; Susser, 1998; McMichael, 1999).

Although global environmental change has helped motivate an extension of public health's perspective (Haines & McMichael, 1997; McMichael *et al.*, 1999), such a transition remains inchoate (Susser, 1998; Yach & Bettcher, 1998; McMichael, 1999). As the history of ideas concerning the relationship between health and environment reveals, public health is not contending with unfamiliar terrain. Much as industrialization and urbanization generated health dilemmas several centuries ago, today global environmental change and globalization are altering the systems upon which civilization depends for health. Many of the changes projected or underway threaten to adversely impact housing, water supply and nutrition, and exacerbate problems with disease vectors and pollution. The task ahead presents challenges akin to those confronted by the founders of public health. Thus it will be essential to both broaden and refocus the scope of public health to include indirect as well as direct causal mechanisms, and traditional as well as modern hazards.

The focus on physical, chemical and biological agents of the environment is now expanding to include political, economic and cultural conditions influencing the pathways from environment to health (Mustard, 1996; Hahn, 1999; Kawachi & Kennedy, 1999). The linkages between development and both the state of the environment and the quality of health has been the focus of much recent scholarship (Litsios, 1994; Warford, 1995; Shahi *et al.*, 1997; Corvalán *et al.*, 1999; Woodward *et al.*, 2000). The re-expansion of public health, furthermore, extends beyond questions of causality to issues of capability. As noted in the introduction, the imperative to expand the causal framework does not relieve any resulting system from the obligation to provide means to compare cost-effectiveness and other efficiency measures with those of alternative possible routes to achieve better health. Below, therefore, we explore a multidisciplinary approach to determining effectiveness by developing a multilevel causal web framework, utilizing a burden of disease perspective, and considering sustainability explicitly in public health decision-making.

3.2.2 Causal webs of environmental risk factors[2]

Clearly, the relationship between human health and the environment is complex. Shifts in environmentally mediated disease patterns over history, geography and the degree of development reflect the changing relationships between society, environment and health. This complexity calls for a more comprehensive elucidation of risk factors acting at different levels of the causal chain when analysing data and interpreting findings. Classical ways of analysing epidemiological data only at the immediate level of the actual associations measured do not encourage a broader analysis of the consequences for policy and prevention of the associations found. An examination of the pathways linking environment and health can help to identify and assess leverage points for intervention at various scales. Additionally, comparing and contrasting different pathways can help to discern important differences and similarities in the various ways in which the environment affects human health. Of course, there is no single best way of organizing and viewing relationships between the environment and health to reveal all the important interactions and possible entry points for public health actions. Several descriptions of the environment–health causal pathway have been proposed. Extended from these, and recognizing the links between environment, health and development as well as the need for specific actions at each step, a comprehensive framework can be devised. We hope to illustrate the utility of this framework and encourage the development of multilevel approaches to public health.

Exposure to environmental hazards is clearly the result of a complex set of events. The need to elucidate the pathways that give rise to these events motivates the framework presented in Fig. 3.1(*b*) – the driving force–pressure–state–exposure–effect

---

[2] The discussion in this section builds upon Corvalán *et al.*, 1999.

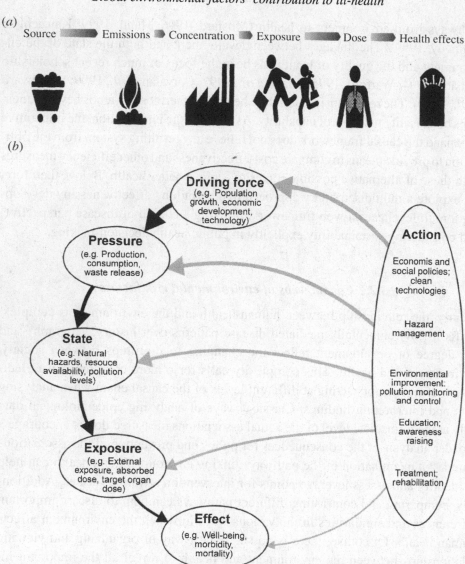

Fig. 3.1 Environmental health pathways. (*a*) Classic environmental pathway model. More distal causal steps could be identified further to the left, including social, economic and global environmental factors that lead to choices to take up particular activities. Such distal risk factors also operate to modify the links in the pathway shown here, changing the relationship between steps on the right. (*b*) DPSEE causal web framework – see text (optional).

hierarchy (DPSEE) taken from (Corvalán *et al.*, 1999). This framework explicitly recognizes that although exposure to an environmentally mediated health hazard may be the immediate cause of ill-health, the "driving forces" and "pressures" leading to environmental degradation suggest alternative and potentially more effective points for control of the hazard than the traditional interventions targeting "exposure" and "effect". The network of connections within the framework can be used to identify cause–effect pathways in either direction, depending on whether the framework is used to analyse the multiple health effects of a single driving force (e.g. transport policy relying on automobiles leading to increased motor-vehicle-related injuries and deleterious effects on the respiratory system, etc.) or the multiple causes of a single health effect (e.g. acute respiratory infections in children resulting from driving forces and pressures such as poverty, energy policies, crowding and land-use practices).

In particular, pathways raise questions relevant to public health about the underlying structures that mediate the connections between levels. For example, who experiences high rates of motor-vehicle-related injuries or what determines land-use practices? Different types of relationships between levels may also exist; for example, feedback loops between effects and driving forces. In general, higher levels of the framework are societal phenomena, intermediate levels are environmental factors and lower levels are health variables. As a result, higher levels of the framework typically fall within the scope of the social sciences, whereas lower levels of the framework ordinarily fall within the domain of the natural sciences. Of course, these distinctions are diminishing through collaborative and interdisciplinary work, a tendency that the framework seeks to encourage.

Within the DPSEE framework, driving forces create the conditions in which environmental health hazards can develop or be averted. Driving forces are generated by a society's aggregated pursuit of the basic necessities of life and the resulting appropriation and use of material and symbolic capital.[3] Driving forces include social phenomena, such as geopolitical relationships, public policy and incentive structures that determine trends in economic development, technological innovations, consumption patterns and population growth. The driving forces in turn generate different kinds of pressures on society, such as impoverishment, marginalization and rural flight, or on the environment, in forms such as waste from human settlements, depletion of natural resources and emission of pollutants from industrial and agricultural activities. These pressures can change the state of society, as seen when circumstances conducive to violence arise. Pressures may also alter the state of the environment when emissions or effluent increase the concentrations of hazards in air, soil, water or food. The pressures may be associated with all stages in the life cycle of industrial products, from initial resource extraction, to processing and distribution, to final consumption and disposal.

Whether the resultant states create a hazard to human health depends on many factors, including the degree to which humans are actually exposed. Exposure requires that people are present both when and where the environment becomes hazardous.

---

[3] Symbolic capital – the beliefs, attitudes and practices that give life to various worldviews and political economic structures.

Exposure thus refers to the intersection between people and environmental hazards in time and space. Traditionally, given known exposures and the knowledge of dose–response relationships, the health risks of specific hazards can be estimated within the confines of current knowledge (Fig. 3.1(*a*)). Although "hazard" describes the potential for causing harm to human health, it says nothing about the statistical probability that such harm will occur. In contrast, "risk" is a quantitative estimate of the probability of damage associated with an exposure.

Environmental hazards can have a wide range of effects varying in type, intensity and magnitude, depending on the type of hazard to which people are exposed, the level of exposure and the number of people affected. Most important diseases are associated with more than one type of exposure, and environmental hazards interact with genetic factors, nutrition and lifestyle in causing disease. Environmental hazards include novel perils, as well as the waxing and waning of more familiar dangers. These different environmental health threats can be divided into traditional hazards, associated with lack of development, and modern hazards, associated with unsustainable development (Smith, 1997). The distinction, though not strict, elucidates some important features shared by hazards, strategies used within each category, and the consequences of major transitions across the two categories. Global environmental change is believed to affect a number of pathways involving health, and the distribution and intensity of environmental risk factors from both categories, traditional and modern, may change (McMichael *et al.*, 1996). Efforts to predict or mitigate the deleterious consequences of such changes can benefit from an appreciation of our current experience with curbing environmental hazards.

Traditional hazards are related to poverty and insufficient development. They include: lack of access to safe drinking water; inadequate basic sanitation in the household and community; food contamination with pathogens; indoor air pollution from cooking and heating using biomass fuel or coal; inadequate solid waste disposal; occupational injury hazards in agriculture and cottage industries; natural disasters, including floods, droughts and earthquakes; and disease vectors, mainly insects and rodents. Modern hazards are related to rapid development that lacks health and environment safeguards and to the unsustainable consumption of natural resources. These hazards include water pollution from populated areas, industry and agribusiness; urban air pollution from automobiles, coal power stations and industry; solid and hazardous waste accumulation; chemical and radiation hazards from the introduction of industrial and agricultural technologies; emerging and re-emerging infectious disease hazards; deforestation, land degradation and other major ecological changes at local and regional levels; climate change; stratospheric ozone depletion; and transboundary pollution.[4]

---

[4] Sometimes modern hazards are further categorized into "community" and "regional/global" scales (Holdren & Smith, 2000).

The temporal and spatial differences between the two categories are most striking. Traditional hazards are often rather quickly expressed as disease. This enabled early public health practitioners to identify associations between social conditions and environment–health relationships without sophisticated study designs or statistical techniques. For many modern environmental health hazards, however, a long period may pass before the health effect is apparent. This comparatively lengthy latency increases the importance of understanding the hazard's environmental pathways. As a result of these differences, the time available to modulate health effects also differs between the two categories. In general, interventions can more rapidly arrest the consequences of traditional health hazards, whereas dealing with modern health hazards typically requires considerably more time and more reliance on risk assessment (see next section). In both cases, however, structural changes at the societal level are necessary to ultimately eliminate the creation of the hazard and not merely treat its symptoms. Each of the traditional and modern hazards is associated with a variety of aspects of economic and social development that the DPSEE perspective aims to clarify.

The framework is intended to highlight the important links between society, environment, health and development, and to help identify effective policies and actions to control and prevent ill-health. The particulars of environment–health relationships are specific to historical and geographical settings. Exploring possible pathways can help public health practitioners to understand whether public health priorities have been achieved and devise interventions that influence linkages between levels at various points on the causal chain. In today's context of global environmental change, the influence on the environment exerted by international relations, national governments, market forces and nongovernmental and nonmarket organizations is receiving particular attention. DPSEE allows us to add health to these investigations.

3.3 Environment in the current health context[5]

Next we seek to establish the nature and strength of current environment–health relationships, quantifying linkages in the lower half of the DPSEE diagram (Fig. 3.1(b)), the more traditional domain of environmental health. To find out how much of current or future ill-health might be attributable to environmental factors, we first need to carefully define what we mean by "environment", "ill-health" and "attributable". We then extend the discussion to the more policy-relevant "attributable risks". In the course of these exercises, we try to reflect current thinking about how best to deal with some of the broader issues inherent in these questions.

[5] Much of the discussion in this section derives from Smith *et al.*, 1999.

3.3.1 What is "environmental"?

The strict medical definition of environmental causes of diseases includes all those that are not genetic. This is the classic dichotomy between "nature" and "nurture", in which environmental factors include all those that affect an organism after conception regardless of whether they are mediated by social conditions or through environmental media. It could be argued even further, however, that genetic factors are actually also environmental, but merely on a different time scale. Thus, mutation, natural selection and other mechanisms of evolution have changed the genetic composition of humanity according to previously prevalent environmental conditions. In this context, i.e. in which current genes are seen as the outcome of previous environments, all diseases are entirely environmental.

Neither of these definitions ("diseases are entirely environmental" or "all non-genetic causes are environmental") is sufficiently useful for most purposes and both somewhat contradict common everyday understanding of what constitutes environmental factors. In terms of health, the most important difference from the common perception of the term "environmental" is whether to include "behavioural" or "lifestyle" factors. In particular, diet, including alcohol, and smoking are extremely significant risk factors for a range of important diseases and thus for total health status in many parts of the world. Yet, their inclusion as environmental risk factors would tend to overwhelm the other, more conventionally understood, environmental factors such as pollution (McKeown, 1976). Furthermore, the most effective interventions for mitigation tend to be somewhat different from those for more conventional environmental factors.

There are at least two other major risk categories that introduce ambiguity. Some of the important social risk factors for disease – including crime, stress, inequity and war, being external to the body – are sometimes included as environmental. In addition, natural hazards, such as earthquakes and inclement weather, although clearly "environmental" in some contexts, are sometimes not included because their frequency and scale have not generally been affected by human actions, although the extent of human vulnerability is affected by human actions. Thus, there is sometimes an unstated presumption that environmental health deals only with those aspects of the environment that are demonstrably affected by human activities and not those due to nature in the raw. Indeed, the term "natural" has come to imply clean and safe to many people. This view could only develop recently in rich countries, however, because most of humanity has spent most of history protecting itself from a range of hostile natural environmental conditions. In addition, of course, it is now becoming clear that human actions can affect "natural" processes even at large scales.

Fig. 3.2, from (Smith *et al.*, 1999), shows one way to represent the relationships among these major categories of risk factors. It may not be worthwhile trying to be very precise, because the demarcations between categories are not sharp.

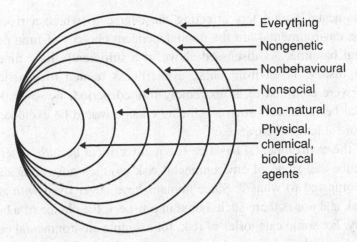

Fig. 3.2 What might be included in "environment". The figure offers a nested definitional hierarchy of what might be termed "environmental". Although useful, this simple representation is incomplete, e.g. it does not exhibit the pathways operating through the global environment that might modify social, behavioural or natural activities.

For example, diet, alcohol consumption and smoking, even though they are certainly functions of individual behaviour, are also functions of the social environment.

Another way to define the scope of environmental health is in terms of external stressors that affect bodily health. The usual routes mentioned are ingestion, inhalation and dermal penetration (Moeller, 1997). Clearly the eyes and ears ought to be added to this conventional list, however. UV light, laser light and noise, for example, are conventionally measured stressors that can cause ill-health without involving the lungs, gastrointestinal tract or skin.

An argument for the existence of significant environmental risk factors is the large difference in rates of the same diseases in different parts of the world. Although these differences can be deeply confounded by diet, smoking and other semi-environmental factors, as well as genetic variation in some cases, it offers the best route for distinguishing the true environmental component for many diseases.

Environmental risk factors are usually addressed only as to their effect in causing disease. Taking a burden of disease approach, however, adds the importance of variation in disease prognosis to the discussion. Clearly environment plays an important role here as well, and probably for nearly every kind of disease. Even diseases whose generation is independent of the current environmental conditions, e.g. genetic and sexually transmitted diseases, will take different courses depending on the environmental conditions to which their victims are subjected. Thus, like nutrition, the total burden, if not the incidence, of nearly every disease and injury is affected to some degree by environmental factors.

Two fundamental parameters affecting the degree to which a risk factor is defined to be environmental are the nonindependent choices of time period and environmental baseline. As discussed above, if a sufficiently long time horizon is taken, all disease is environmental, even illness related to genetic factors. If causation were confined to an extremely limited period, however, long-term environmental health threats such as climate change would be excluded. This is discussed more at length below.

Choice of the environmental baseline is equally critical because determination of the existence and scale of environmental risk requires answering directly or implicitly "compared to what?". Since humans have never lived with zero environmental risk and nor is there such a time in prospect, the choice of a baseline is often not easy. For some categories of risk, for example environmental circulation of synthetic chemicals, a baseline of zero may be suitable. However, for others it may not be; for example, airborne particulates, ionizing radiation, UV radiation and inclement weather. (See discussion later in Section 3.4.1.)

Box 3.1 summarizes our approaches in dealing with the question of what will be included as environmental risk factors.

BOX 3.1
What is meant here by "environmental"?

- We do not include genetic risk factors as we consider only current and future environments, not past ones, and do not consider forward changes in the environment over long enough periods that might affect genetic make-up.
- We do not include the major risk factors of diet and active smoking, but do include non-nutritional elements of diet, including food additives, infectious agents, pesticides, etc., and passive smoking (environmental tobacco smoke).
- We do, however, include behavioural factors related to personal and household hygiene, just as we do for behavioural factors leading to community and larger-scale pollution.
- We include a modest component of environmental risk in the direct and indirect risks of malnutrition to account for degraded soils, floods and other human-generated impacts on the quantity, quality and distribution of food.
- We assign a component of injuries to environmental factors based on the substantial variations in rates across the world, and also the recognition that even social nonphysical/chemical stressors are to some extent environmental.
- We include a small component of environmental risk for every disease category because of the environment's influence on disease outcome.
- We include health impacts of the natural environment, such as dust exposure and natural disasters, as well as changes to these caused by human activities such as desertification and global warming.

3.3.2 What is "ill-health"?

Although we sympathize with the ideals behind WHO's definition of health as not just the absence of disease, as proposed in its constitution in 1948, here we focus only on the disease portion, i.e. the environmental component of the total burden of disease, not of the total burden of ill-health. Neither do we try to address the conceptual and measurement difficulties inherent in defining disease. Rather, we use the Global Burden of Disease (GBD) database with relatively minor adjustments (Murray *et al.*, 1996). This database reflects the assistance of many experts to address a range of such conceptual and measurement issues. Among the most difficult of these issues were establishing a unit of measurement by which death, disease and injury can be combined; choosing appropriate groupings by age and sex; defining appropriate geographical boundaries; combining effects in different time periods; deciding whether effects at different ages should be weighted differently; reconciling the cause of death and disease when different diseases are involved (e.g. acquired immunodeficiency syndrome (AIDS) victims who contract tuberculosis); and establishing categories and weights for the wide range of disabilities associated with different diseases.

To make the dataset more manageable, we have confined it to those disease groupings that each cause at least 1 % of the GBD (see the first few columns in Table 3.1). The disability-adjusted life year (DALY) is used as the basic unit of ill-health. The total in DALYs is determined by summing the years of lost life due to premature deaths (YLL – years of lost life) and the weighted years of disability (YLD – years lost to disability) due to a particular disease or risk factor. Shown also are the separate DALY sums and populations for more-developed (MDC) and less-developed (LDC) countries, revealing the much larger burden of disease per capita in the latter. (The difference is even larger if age distributions are taken into account.)

The most important disease categories are ranked in Table 3.1 according to their percentage of the total global burden of disease. Shown are all those that each account for at least 1 % of the global total burden. As shown, together they account for about three-quarters (73 %) of the total, somewhat more in developing countries (LDCs – 74 %) and somewhat less in developed countries (MDCs – 63 %).

The only choices made in the GBD that we have altered slightly from the published GBD database relate to the grouping of diseases (see Box 3.2). There is no unique or absolute way of performing this grouping.[6] The way diseases are grouped, although often reflecting historical developments of understanding and convenience, can have significant implications for policy.

---

[6] Diseases in the GBD, for example, have been grouped, *inter alia*, according to biological criteria (e.g. helminths), presumed common mechanism (e.g. malignant neoplasms), organ system (e.g. cardiovascular), age (e.g. perinatal), sex (e.g. maternal conditions), type of intervention (e.g. child cluster), location (e.g. tropical cluster), cause (e.g. road accidents) and motive (e.g. intentional vs unintentional injuries).

Table 3.1. Major disease categories and major environmental risks

	% GBD	% MDC-BD	% LDC-BD	ERF attributable % Low	Middle	High	% GBD attributable to ERF	% MDC-BD attributable to ERF	% LDC-BD attributable to ERF	Outdoor air pollution	Housing conditions	Indoor air pollution	Sanitation & hygiene	Pollution/chemicals	Occupational factors	Carcinogenic infectious agents	Stress/degraded environment	Land management	Total number of ERF
Acute respiratory infections*	8.5	1.6	9.4	40	50	60	4.3	0.8	4.7	x	x	x							3
Diarrhoeal diseases*	7.2	0.3	8.1	80	85	90	6.1	0.3	6.9				x						1
Perinatal conditions*	6.7	1.9	7.3	10	15	20	1.0	0.3	1.1	x	x	x	x						4
Child cluster[a]	5.2	0.01	5.8	5	7.5	10	0.4	0.0	0.4										0
Cancer (malignant neoplasms)*	5.1	13.7	4	20	22.5	25	1.1	3.1	0.9			x		x	x	x			4
Depression	4.7	7.7	4.3	5	7.5	10	0.4	0.6	0.3								x		1
Malnutrition/anaemia (direct effects)	3.7	0.9	4.1	8	9	10	0.3	0.1	0.4				x						1
Heart (ischaemic disease)	3.4	9.9	2.5	8	9	10	0.3	0.9	0.2	x		x	x		x				4
Stroke (cerebrovascular disease)	2.8	5.9	2.4	8	9	10	0.3	0.5	0.2	x		x			x				3
Tuberculosis*	2.8	0.3	3.1	20	22.5	25	0.6	0.1	0.7		x	x							2
Road accidents*	2.5	4.4	2.2	25	27.5	30	0.7	1.2	0.6						x			x	2
Congenital anomalies	2.4	2.2	2.4	5	7.5	10	0.2	0.2	0.2					x					1
Malaria*	2.3	0	2.6	70	80	90	1.8	0.0	2.1		x							x	2
Maternal conditions	2.2	0.6	2.4	5	7.5	10	0.2	0.0	0.2		x				x				2
STD/HIV	2.2	1.3	2.3	5	7.5	10	0.1	0.1	0.1										0
COPD*	2.1	2.1	2.1	33	41.5	50	0.9	0.9	0.9	x		x			x				3
Falls*	1.9	1.5	2	25	27.5	30	0.5	0.4	0.6						x				1
War*	1.5	0.7	1.5	50	60	70	0.9	0.4	0.9								x	x	2
Suicide	1.4	2.3	1.2	5	5	5	0.1	0.1	0.1								x		1
Violence*	1.3	1.1	1.3	15	17.5	20	0.2	0.2	0.2						x		x		2

Table 3.1. (cont.)

	% GBD	% MDC-BD	% LDC-BD	ERF attributable %			% GBD attributable to ERF	% MDC-BD attributable to ERF	% LDC-BD attributable to ERF	Outdoor air pollution	Housing conditions	Indoor air pollution	Sanitation & hygiene	Pollution/chemicals	Occupational factors	Carcinogenic infectious agents	Stress/degraded environment	Land management	Total number of ERF
				Low	Middle	High													
Nondiarrhoeal food/water cluster*	1.2	0	1.3	80	85	90	1.0	0.0	1.1		x								1
Alcohol	1.2	4	0.8	5	7.5	10	0.1	0.3	0.1						x				1
Drowning*	1.1	0.5	1.2	25	27.5	30	0.3	0.1	0.3										1
Total	73.4	62.9	74.3				21.8	10.5	23.1										
*>10% attributable to ERF																			
Approximate contribution to GBD										1.9	3.1	2.7	6.4	0.7	2.6	0.3	1.0	1.6	20
GBD rank										5	2	3	1	8	4	9	7	6	
Approximate contribution to MDC-BD										1.0	0.7	1.8	0.5	1.0	2.9	0.8	1.0	0.7	10
MDC-BD rank										5	7	2	9	3	1	6	4	8	
Approximate contribution to LDC-BD										2.0	3.4	2.8	7.2	0.7	2.5	0.23	0.9	1.7	21
LDC-BD rank										5	2	3	1	8	4	9	7	6	

The first column lists the 23 disease categories accounting for more than 1 % of the total Global Burden of Disease (GBD) in 1990 (Murray et al., 1996). Columns 2–4 provide the fraction of the total burden of disease for the world more-developed countries (MDC-BD) and less-developed countries (LDC-BD). Columns 5–7 give the low-, medium-, and high-range estimates for the attributable risk percentage of disease for each disease category, as taken from Smith et al. (1999). Utilizing the mid-range estimate, we estimated the percentage of the GBD, MDC-BD and LDC-BD that results from environmental risk factors producing measurable impacts on disease burden. The number of environmental risk factors for each category in columns 8–10. The remaining column headings specify the principal proximate environmental risk factor, whereas a blank cell indicates no relationship. In the resulting matrix, a "x" indicates a significant relationship between the corresponding disease category and environmental risk factor. The percentage of GDB for each disease category (column 2) matches is totalled at the right side of the matrix, for each disease category, and at the bottom of the table. The percentage of GDB for each disease category that is attributable to environmental risk factors (column 8) is then divided equally among the associated risk factors. The resulting fractions of GDB are summed at the bottom of the table. This calculation is repeated in a similar fashion to derive an estimate of the approximate contribution to MDC-BD and LDC-BD for each major environmental risk factor.

a Measles, pertussis, polio, tetanus, diphtheria.

a Nondiarrhoeal food/water cluster; polio, tetanus, diphtheria. COPD, chronic obstructive pulmonary disease; ERF, environmental risk factors; GBD, Global Burden of Disease; LDC-BD less-developed countries, burden of disease; MDC-BD, more-developed countries, burden of disease.

BOX 3.2
Adjustments made in disease categories

We have made adjustments to the Global Burden of Disease in four groupings.

- Depression here includes only unipolar and bipolar depression.
- Heart disease here only includes ischaemic heart disease and is separated from the next category, cerebrovascular disease, because these conditions have quite different outcomes and are often investigated separately in environmental epidemiological studies.
- We have combined the separate GBD categories of "other sexually transmitted diseases" and "HIV" into one grouping called STD/HIV.
- We have created a category of disease called "nondiarrhoeal food/water cluster" which includes intestinal nematode infections and the tropical-cluster diseases (filariasis, leishmaniasis, schistosomiasis, trypanosomiasis, etc.) which account, respectively, for an additional 0.4 % and 0.8 % of the GBD. This would raise the visibility of food/water quality and hygiene as a potential intervention.

Inclusion of asthma would add another 0.8 % to the GBD for COPD in a category that could be called "chronic respiratory cluster". Given some indication of common risk factors, this might be appropriate. However, the still uncertain and mysterious apparent rise in asthma rates in some parts of the world probably argues for keeping it separate so that increases can be more easily highlighted.

This is illustrated by the most important grouping in Table 3.1, acute respiratory infections (ARI). Up until the 1970s, the many separate bacterial and viral diseases that constitute ARI were often listed in separate groupings, often inconsistently from one tabulation to the next. It was realized, however, that, although different micro-biologically, the group had common risk factors, common gross symptoms and common outcomes, particularly sharing a high risk of producing life-threatening pneumonia in developing country children (Jamison & World Bank, 1993). Grouping them together gave them much more visibility in health policy, particularly because together they exceed diarrhoeal diseases (another, but older, grouping) to become the single largest category.

The child cluster grouping (measles, tetanus, pertussis, polio and diphtheria) is another example, in which the importance of child vaccination programmes becomes strikingly emphasized by placing together the five diseases for which relatively inexpensive and reliable vaccines are available. In this case, of course, the diseases do not even have the same risk factors, symptoms or outcomes.

In both these groupings, the biological or medical classifications would be much less useful for policy than one focusing on common interventions.

3.3.3 What is "attributable risk"?

By "attributable environmental risk", we mean the percentage of a particular disease category that would be eliminated if environmental risk factors had previously been reduced to their lowest feasible values (Murray & Lopez, 1999).

Most important diseases with significant environmental components have multiple risk factors, a characteristic that on the one hand makes assessment more difficult, but on the other offers many routes for finding useful interventions, as shown in the previous DPSEE discussion (Fig. 3.1(*b*)). Somewhat counter-intuitively, the disease burden accountable to each risk factor depends on the order in which the risk factors are listed: the closer to first, the closer to the maximum possibly accountable to that factor. In other words, those that are considered first out-compete the others and reduce the remaining available risk to be allocated.

Risks only make sense when compared to alternatives. Determining attributable risk requires choosing a counter-factual level, an exposure scenario that might have been possible. Several sorts of counter-factual levels have been proposed (Murray & Lopez, 1999):

- Plausible minimum risk is the distribution of exposure among the set of plausible distributions that would minimize overall population risk. Plausible implies that the shape of the distribution is imaginable or possible in a society.[7]
- Feasible minimum risk is the distribution of exposure that exists or has existed in a population.
- Cost-effective minimum risk is the distribution of exposure that would result if all cost-effective interventions (against some reference value) could be implemented.

The only practical way to determine the contribution of a risk factor (attributable risk in health terminology), therefore, is in terms of how much the disease burden would be changed by reducing or increasing a particular risk factor that can be conceptualized and manipulated in relation to other existing risk factors. There may be hundreds of potential interventions, but those that have no practical meaning are not described, e.g. if only people could be taught to breathe less, they might not be so affected by air pollution (or assign a police officer to every car to reduce traffic accidents). A corollary is that there are value judgements involved in choosing which risk factors to examine; for example, whether income or social inequities are manipulatable risk factors in a public health context.

Therefore, it worth remembering that the total attributable risk for all the important risk factors viewed independently often adds to more than 100 % (Walter, 1976; 1978).[8] Each risk factor must be considered in light of the others. Except in

---

[7] They also define a theoretical minimum risk distribution, which would be the optimum distribution from a health standpoint even though never observed.

[8] With enough ingenuity, it might be argued that attributable risks are in fact always infinite.

relatively rare cases, as when a specific chemical is associated with a particular type of unusual cancer, the risk factors will interact rather than simply sum. Progress in reducing one risk factor will affect the remaining potential of the others (Rothman, 1976). Indeed, the more potentially useful interventions known, the greater the attributable risk. In contrast, the diseases for which known attributable risks add up to less than 100 % are the most troublesome. Breast cancer would seem to fit this description at present.

Furthermore, the statement that large fractions of ARI are attributable to air pollution, poor housing, crowding and chilling is not incompatible with the also incontrovertible truth that large fractions are also due to lack of breastfeeding, vitamin A deficiency and malnutrition in general. In addition, as the major categories such as malnutrition and environment become more differentiated, the total aggregated attributable risk for the general category becomes apparently larger. When all the sub-categories are combined together into each of the main categories of either nutrition or environment, they cannot be simply added because they interact. Thus, all possible improvements by nutrition alone will have a value less than if the attributable risks of breastfeeding, vitamin A, protein supplements, etc. are calculated separately and added together.

The most basic of risk factors, of course, is poverty. But what is poverty? Just lack of money at the household level is an insufficient criterion, because in most cases provision of money on its own does not lead to permanent and substantial improvements. If one assumes, however, that the alleviation of poverty would bring along with it the advantages of better education, nutrition, environment, health-care, etc., which have accompanied income improvements in most strata of developed countries, the attribution of disease in developing countries to poverty is almost a tautology. Yet surprisingly, the specific pathways by which poverty undermines health are either seldom studied or are altogether ignored. Clearly the relationship between poverty and health, often mediated by social and environmental conditions, are complex (Adeola, 2000).

Much of the history of public health can be viewed as success in pinpointing the specific sub-categories of attributable risk in the form of better nutrition, environment and health-care that could be effectively modified by education, technology, management and empowerment to achieve better health *before people become rich*. To propose poverty alleviation as the primary means to improve health is to ignore the huge potential improvements that can be achieved well before that far off day when poverty is eliminated. It also fails to recognize that improved health is itself a prerequisite for achieving and maintaining viable sustainable development. The causality goes both ways.

Box 3.3 summarizes some of the major points to remember about the term "attributable". Keeping these definitions of disease and environment in mind, and choosing "feasible minimum risk" as the counter-factual level, we present the percentage of each attributed to environmental risk factors in the fifth to tenth columns of Table 3.1.[9]

BOX 3.3
Characteristics of "attributable" risks

The following are some basic principles related to determining the fraction of a disease category that is attributable to environmental risk factors.

- All known attributable risks for a disease often add to more than 100 % and, consequently, a large attributable fraction claimed for one risk factor does not necessarily imply that another risk factor cannot also have a large attributable fraction.
- Size of individual attributable risks depends on the order in which different risk factors are examined, because if one risk factor is reduced, the remaining disease available to be reduced by other risk factors will decrease.
- Existence and size of an attributable risk factor presumes the existence of feasible intervention because there is an unlimited number of hypothetical risk factors that cannot be manipulated with current knowledge or feasible application of resources.
- Defining counter-factual levels can be important, i.e. it is often necessary to specify a minimum possible or reasonably attainable risk, particularly for environmental risk factors that cannot be feasibly reduced to zero levels because, for example, of natural background.

3.3.4 Global environment–health connections today

To identify how global environmental change influences the burden of disease associated with environment, we have extended the analysis further in Table 3.1 (see caption). We thus arrive at an estimate of the approximate contribution to the GBD and for developed (MDC-BD) and developing countries (LDC-BD) separately for each major environmental risk factor. This exercise is for illustrative purposes only.[10] Below is the resulting rankings of environmental risk factors according to how much of the GBD they cause:

(i) Sanitation/hygiene.
(ii) Housing.
(iii) Indoor air pollution.

---

[9] These values are taken from Smith *et al.*, 1999, except for the addition of the nondiarrhoeal food/water cluster, third from the bottom, which we have assigned an 80–90 % environmental portion.
[10] Ideally, each risk factor would be weighted by its relative importance for each disease category at each of the three levels of analysis (GBD, MDC-BD and LDC-BD).

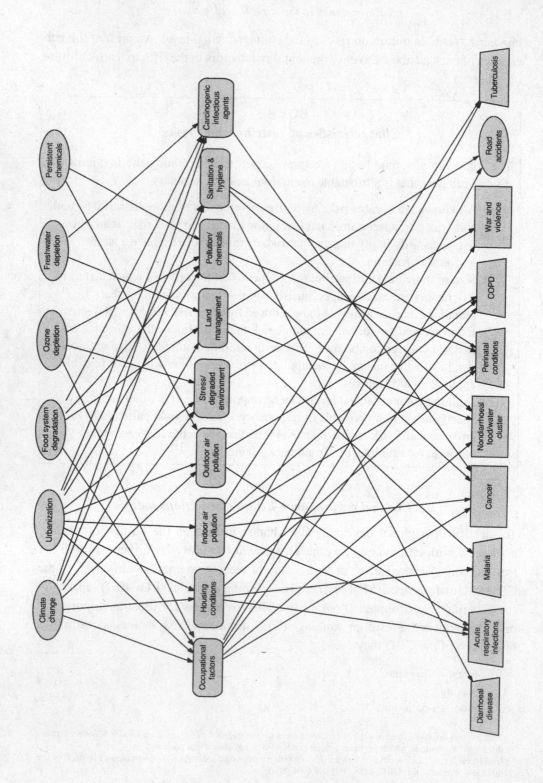

(iv) Occupational factors.
 (v) Outdoor air pollution.
(vi) Land management.
(vii) Stress/degraded environment.
(viii) Pollution/chemicals.
 (ix) Carcinogenic infectious agents.

The results reinforce the general importance of environmental hazards (Smith *et al.*, 1999) and confirm the conventional if often neglected wisdom that poor sanitation, hygiene and indoor air quality are the major environmental threats to well-being today (and, at least, in the near future).

This analysis is further elaborated in Fig. 3.3 (see caption) to link these nine major proximate environmental factors to global environmental change, the major distal environmental factors. In the figure, disease categories have been limited to those with both a significant contribution to the GBD and a significant degree of environmental association. Thus, for example, STD/HIV is left off because the environment is not a major risk factor, and drowning has been left out because it makes a relatively small contribution to the GBD. The arrows linking proximate risk factors to diseases match the matrix on the right side of Table 3.1. Finally, to make a link with global environmental change processes, we list the major categories of global environmental change taken from McMichael *et al.* (1998) above the proximate risk factors in the figure. The arrows in the upper portion indicate their connection with the important proximate risk factors.

This preliminary figure does not attempt to do so, but a full picture of the relationships in this causal web would involve identifying and quantifying the most important links (arrows). In this way, the relative importance of each of the distal risk factors (types of global environmental change) could be ascertained. Even in this simple version, it is instructive to follow the arrows in each direction to see which diseases are most closely linked to which proximate and distal factors and vice versa. It is also interesting to note that to control some disease conditions (e.g. perinatal) would seem to require paying attention to several environmental factors, while others (diarrhoea) have more limited pathways leading to them.

Fig. 3.3 Causal web diagram of global risk factors, proximate environmental risks and major disease outcomes. In the top row, major global environmental changes are ranked from left to right by the number of linkages to the row below. The middle row, major environmental risk factors defined by Smith *et al.* (1999), are ranked from left to right by the total number of linkages to disease categories (see rightmost column of Table 3.1). The bottom row, disease categories, is ranked from left to right by the percentage of attributable risk resulting from environmental factors (see column 6 of Table 3.1). Trapezoids indicate a disease category with a top ten difference in DALYs between the pessimistic and optimistic scenarios of the 2020 GBD projections (see Table 3.3) (Murray & Lopez, 1996). Rectangles indicate a disease category that is not appreciably different in DALYs between the two scenarios.

3.4 Environment and health: the future

Given the need to incorporate both wider views of the interactions between environment and health and more consistent and coherent indicators of ill-health itself, what might be done to illuminate some of the potential impacts of global environmental change processes on health and guide decision-making? To do so, here we examine approaches for dealing with three specific questions within this realm:

- What factors affect the difference between attributable risk and avoidable risk, i.e. the difference between the risk caused by a risk factor and the risk that can be reduced by manipulating that risk factor? They can be quite different.
- Even though global environmental change processes potentially affect nearly every disease in some way, the connections are weak in a number of cases and a large fraction of the GBD is actually due to a fairly small number of diseases and risk factors. What does this imply about how health may be affected by global environmental change processes?
- Recognizing that some of the major global environmental change processes are not yet seriously affecting health but threaten to do so before the mid-twenty-first century, what can be done now to blunt their potential impact in several decades?

3.4.1 Avoidable risk

From a policy standpoint, it is usually more critical to know *avoidable* rather than *attributable* risk. Avoidable risk is the burden of disease that would actually be eliminated by an intervention. In a steady-state situation of population and competing risks with reasonably acute effects from a particular environment risk factor, for example diarrhoea in young children from poor water/hygiene/sanitation (WHS), a reduction of risk factors today would lead to a reduction in disease tomorrow and in the future. In this case the size of attributable risk is similar to that of avoidable risk. In reality, however, these conditions are rarely met. In many parts of the world where childhood diarrhoea is a problem, for example, there are important changes occurring in all three principal parameters that contribution to the calculation of risk: exposure, population distribution and competing risks.

Consider, for illustration, calculating the avoidable diarrhoea risk for an intervention involving better WHS in a country like Thailand that has good prospects for continuing economic growth and has been moving rapidly through the demographic transition. (For simplicity, we ignore for the moment the potentially important differences in experiences among different population groups that is often hidden by using national averages.) In this case, the fraction of children under five is falling, along with the existence of competing risks such as malnutrition. In addition, even without any special intervention, there will be national, regional and household improvements in WHS. Thus, the true avoidable risk, namely the risk for which an

Table 3.2. *Examples of different national baseline conditions for environmental risk factors*

Environmental risk factor	Nation/region	Population distribution	Competing risks	Exposure status
ETS	Japan	Old – stable	Stable	Stable
ETS	USA	Old – ageing	Stable	Declining
ETS	Thailand	Middle – ageing	Declining	Rising
Poor WHS	West Africa	Young – stable	Rising	Stable
Severe storms	Bangladesh	Young – ageing	Declining	Rising

ETS, environmental tobacco smoke; WHS, water/hygiene/sanitation.

incremental intervention can take credit, will be substantially less over time than the current attributable risk. In a West African country, however, where economic growth may be negative, the demographic transition is proceeding slowly and public infrastructure related to WHS may be decaying, the avoidable risk may actually be higher than attributable risk, i.e. things are becoming worse in the baseline case. This latter condition applies to most human-engendered climate-change risks. The attributable risks today are probably close to zero, but the avoidable risks may be significant.

Shown in Table 3.2 are some examples of environmental risk factors and how they may be affected by changes in population, competing risks and exposures in the baseline case (without interventions). In each case, risk factors with similar attributable risks may have quite different avoidable risks.

Fig. 3.4(*a*) shows that with an acute disease and constant baseline conditions avoidable equals attributable risk. For a changing distribution, for example reducing WHS exposure through economic growth, Fig. 3.4(*b*) shows how avoidable is less than attributable risk. For chronic diseases, however, the situation is a bit different. Consider, for example, chronic obstructive pulmonary disease (COPD) from air pollution. Elimination of the exposure would not completely eliminate attributable COPD from past exposures for decades. This consideration would then be further complicated when population, competing risk and exposure distributions are not constant, as shown in Fig. 3.4(*c*). The resulting overall pattern and total of avoidable risk, as illustrated in the shaded area, could then be quite different from attributable risk.

3.4.1.1 Temporal factors modifying avoidable risk

There are several additional considerations that may change the relative attractiveness of interventions in the usual cost-effectiveness framework by affecting the numerator, i.e. the health benefit in DALYs or similar metric. For decisions involving

(a)

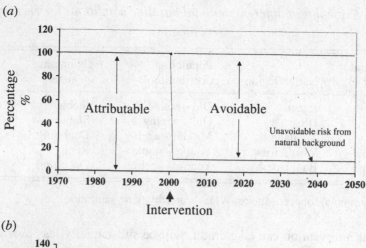

(b)

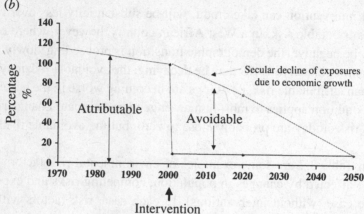

(c)

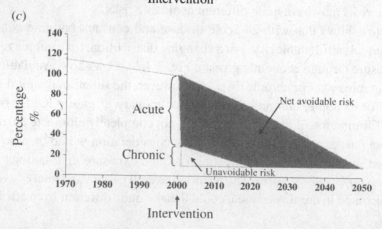

Fig. 3.4 Attributable versus avoidable risks: three versions. (a) For acute diseases with all relevant conditions remaining constant, avoidable essentially equals attributable risk, i.e. an effective intervention will immediately achieve a reduction in the attributable risk. (b) With endogenously lowering exposures, however, avoidable is less than attributable risk. (c) With part of the burden from chronic diseases, avoidable risk is even less because there is a component of unavoidable attributable risk.

choices among competing investments in health, there are good arguments for requiring the use of discounting. (See Box 3.4 for a brief review of the concept and usage of discounting and Net Present Value.) Without discounting (a zero discount rate), large resources would be devoted to clearly undesirable activities such as the eradication of minor diseases. The right size of the discount rate, however, is difficult to determine in a purely objective manner. For long-term phenomena, like global environmental change, it may seem appropriate to argue for low rates so those events decades hence are not completely discounted in today's decisions. About the lowest risk that is compatible with economic decision-making, however, is 3 % which is the rate used internally in the GBD calculations for DALYs.[11]

BOX 3.4
Discounting and Net Present Value

Consider two alternative programmes to invest the same amount of health funds:

1. To entirely eradicate a disease currently killing 12 000 children each year (something like a polio vaccination programme).
2. To reduce by 15 % a disease killing 500 000 children a year for 10 years (something like mosquito control for malaria).

Assuming that, like polio, there are no natural reservoirs, once eradicated from the human population the disease of Programme 1 is gone forever. Thus, the total number of deaths and DALYs averted could be considered infinite, or at least 12 000 per year as far into the future as we believe humanity might persist. Programme 2, however, can at most be counted as 750 000 deaths averted. Indeed, by comparison with any finite programme, we would prefer a programme with even a very small annual benefit, say one averted death per year, as long as it was permanent.

In reality, of course, we would not feel comfortable choosing to spend money to save hypothetical lives hundreds or thousands of years into the future while ignoring the problems of people today. But what about 50 years from now, or 20 years?

One way to handle this issue is to use time horizons, i.e. to say that we will count lives out to some fixed period, e.g. 20 or 100 years, but no further. In this case, therefore, at 100 years we would still favour programme 1 (1.2 million compared to 750 000 lives), but at a 20-year horizon, we would prefer programme 2 (240 000 compared to 750 000). Application of time horizons (20 100 500 years), for example, is how the United Nations' Intergovernmental Panel on Climate Change (IPCC) compares the future global warming commitments of different gases.

---

[11] Excellent summaries of the best current thinking by economists of the issues surrounding discounting in making decisions with long-term consequences are found in Lind (1982), Bruce *et al.* (1996) and Portney & Weyant (1999).

There are problems with time horizons, however. First, of course, is the arbitrariness. Why 20 years? Why not 25? Or 37? Second, a time horizon is illogical in that it implies we will care just as much as we do today about something 20 years from now, for example, but nothing about it in 21 years time. It represents a step function view of the future when there is no reason for one. As a result, some very strange decisions could be made; for example, to promote something that saved lives for up to 20 years and then caused many deaths afterwards from the toxic waste produced.

Since we are addressing decisions about alternative investments, there is a good argument for applying the discounting methods commonly used in economic and financial accounting. With discounting, the degree of value we assign to a future benefit or cost is smoothly reduced depending on the number of years it lies in the future. Thus, at a 10 % discount rate, $100 one year from now is worth only $90.91 today in our calculations. A $100 two years from now would be worth only $82.64 today.[12]

Some people are uncomfortable applying a discount rate to DALYs or deaths in the future, because they are so closely allied to human well-being. There are several arguments for doing so, however:

- Not doing so could produce clearly unacceptable decisions, as noted above.
- If no discounting is applied, it would often look best just to put any health project funds into financial investments at a reasonable interest rate so that we would have more money in ten years and could save even more lives then. Thus, very little would be done now.
- The future is uncertain, e.g. future generations may have needs, desires and capabilities unpredictably different from ours. We thus should focus on the near term when we understand the situation.
- If we do not do something to make the present generation healthy, we jeopardize the well-being of future generations born to it.
- It is not possible to avoid a discount rate; no discounting is just equivalent to a zero percentage rate. It is better to choose a rate appropriate to the situation.

The size of the discount rate, however, remains to be chosen. An extensive literature over many decades argues the issue, but for our purposes it can be summarized:

- For private investments, such as those by an industry or commercial bank, discount rates greater than 10 % are appropriate.
- For public investments by governments or other social institutions, rates of 6 % or less are more suitable. It is difficult, however, to justify rates less than 3 %. To give as much weight as reasonable to the future without asking the present to sacrifice unduly; therefore, standard calculation of the DALY involves application of a discount rate of 3 %, i.e., YLLs in the future are discounted back to the present at this rate.

---

[12] (Net Present Value)$_t = Q_t \times (1/(1+r)^t)$ where t is the number of years into the future, Q_t is the future value in year t, and r is the discount rate per year (e.g. for a 10 % discount rate, $r = 0.1$). For a stream of benefits or costs into the future, Net Present Value $= \Sigma_t \, Q_t \times (1/(1+r)^t)$.

How would discounting change the comparison of Programmes 1 and 2, above? Here are the results of applying different discount rates:

Net Present Value (NPV) of lives saved

Discount rate %	NPV of Programme 1	NPV of Programme 2
10	120 000	461 000
3	400 000	640 000
1.75	682 000	682 000
1	1 137 000	710 000
0	Infinite	750 000

At a discount rate of 10 % or even 3 %, Programme 2 is clearly preferable, i.e. it saves more lives. To choose Programme 1 would imply a discount rate less than 1.75 %, which is probably too low to be appropriate. It would entail imposing on the present generation an unacceptably high certain burden in trade for addressing the uncertain needs of future generations.

Some observers believe that decisions involving such pervasive phenomena over such long time periods as climate change need to be addressed in a special framework entirely different from the usual Net Present Value (NPV) procedures (Arrow *et al.*, 1996).[13] Although this has some attraction, it is difficult to imagine exactly how this would occur in practice since there is no sharp distinction between "normal" and "special" phenomena, but rather a continuum.[14] In any case, here we first confine ourselves to exploring some of the implications of the usual NPV methods for determining avoidable risk and then examine how the standard decision-making tools behind NPV calculations might be stretched to accommodate avoidable risk calculations focused on pervasive long-term phenomena such as climate change.

Three time-related factors lead to a reduction of avoidable risk in a traditional NPV calculation:

- Age: future life years of someone saved at an early age are discounted, thereby reducing the DALYs for younger deaths relative to older deaths. In the GBD, for example, although a girl dying at age five would lose 78 years of life expectancy, because of discounting[15] only about 36.6 years would be accounted to this death.

---

[13] NPV is determined by discounting the future stream of benefits, such as saved DALYs, and future stream of costs, such as expenses associated with an intervention, over the lifetime of the project back to the present, i.e. the decision point.

[14] There are alternatives to discount rates for dealing with long-term impacts. Truncations are a prominent example, as in the "rule against perpetuities" in English Common Law (Nottingham, 1685). This limits impact to roughly the potential length of one lifetime, i.e. 100 years. This would also fit with the choices made by scientific committees of the Intergovernmental Panel on Climate Change, i.e. a choice of truncations of 20, 100 or 500 years, rather than discount rates. Economists tend to find such step functions counterproductive, however (Arrow *et al.*, 1996).

[15] Partly due to age-weighting also (Murray & Lopez, 1996).

- Delay: diseases prevented in the future, through for example reducing the spread of vectors due to climate change 30 years from now, would be counted less than those prevented earlier in a present-value comparison of alternatives.
- Chronicity: as shown above, interventions addressing risk factors leading to chronic diseases create a different relationship between attributable and avoidable risks compared to those for acute diseases. The immediately avoidable risk from an intervention addressing a chronic disease is less than the attributable risk because the intervention will not immediately stop the expression of disease that is the result of many years of past exposure. Thus, a NPV calculation of the avoidable risk for a chronic disease intervention will give a lower value than one for an acute disease, even if they have the same attributable risk, because its full realization is delayed. For example, at a 3 % discount rate the total avoidable risk is 8.5 times the annual attributable risk for a ten-year intervention addressing acute disease. For chronic disease taking 20 years to develop, however, the NPV of total avoidable risk would be only 2.7 times the annual attributable risk.

Because these three factors (age, delay and chronicity) all affect present value in a similar fashion, they can be put into trade-offs with one another to illustrate some interesting hypothetical equivalencies. For example, saving the lives of 100 five-year-olds from an acute disease in 30 years is roughly equivalent, on a DALY basis,[16] to saving 100 55-year-olds today from an acute disease or saving 150 30-year-olds from a chronic disease. One hundred five-year-olds saved today, in contrast, would be equivalent to saving almost 370 65-year-olds in 30 years from acute diseases or 1200 from chronic diseases.[17]

We have briefly discussed six different factors that affect the calculation of avoidable risk from an intervention beyond the usual parameters of effectiveness and efficacy. Three of these (population distribution, competing risks and exposure trends) operate to affect the baseline. Three others interact with the discount rate (age, delay and chronicity). These have been discussed as they operate within the standard NPV model for determining avoidable risk. We now explore briefly how this standard model might be extended to better consider the special characteristics of pervasive long-term threats.

3.4.1.2 The SALY (sustainability-adjusted life year)

One of the more frustrating dilemmas of public health is that short-term interventions affecting acute conditions often seem to have great advantage over longer-term preventive measures in terms of NPV calculations. The NPV of oral rehydration therapy (ORT), for example, is far less than improvements in WHS in terms of dollars per DALY (World Bank, 1993; World Health Organization, 1996). And yet,

---

[16] About 15 YLL/death in present value.
[17] 3 % discount rate, 20-year latency for chronic diseases.

although ORT is effective against diarrhoea mortality, it does not stop diarrhoea morbidity and may indeed increase it. Thus, it hardly seems to be a sustainable intervention. In the context of allocating resources, however, sustainability itself seems often to retain little currency after the application of even modest discount rates.

This raises the question of whether a sort of SALY might be created, i.e. a sustainability-adjusted life year. Keeping with diarrhoea, for example, should a change in infrastructure, such as represented by improvements in water supply or sanitation, be granted a sustainability factor advantage in the present-value calculations over a therapy like ORT? The advantage might be given in the form of lower discount rates or other favourable weighting in the NPV calculations for interventions according to how well they met sustainability criteria.

One way of thinking about how to assign weights for the sustainability portion of a SALY could be in terms of the classic environmental pathway model, Fig. 3.1(*a*). At the right end of the pathway, therapies such as ORT that intervene after the disease is manifest might receive no sustainability factor advantage, i.e. be calculated under standard DALY and NPV terms in relation to the usual efficacy and effectiveness criteria. As interventions are considered further to the left in the diagram, a sustainability factor advantage might be introduced. Thus, sanitation and clean water infrastructure, which intervene further to the left, would be granted a sustainability credit. Alternatively, interventions with multiple linkages advantageous to health, as indicated by application of the DPSEE causal web framework of Fig. 3.1(*b*), might receive a sustainability credit. Of course, determining let alone quantifying sustainability is a challenging task, but one being attempted by a number of disciplines.

3.4.1.2.1 Sustainability as resilience. One way of linking a sustainability factor to the economic and decision-making literature would be to consider it as a function of, although not equivalent to, a resilience ranking. The lower the resilience, the lower the sustainability factor. Resilience in this sense is in terms of time buffer and multiple causality. Antibiotics work well to stop children from dying of ARI, but they depend on being available immediately at exactly the right time in order to do so. If anything happens to delay application even for a day or two, the very sick child dies, i.e. the intervention is not very resilient. Interventions, such as hygiene education, that work at the more distal end of the pathway can fail too, but the immediate risks of doing so are less irreversible and still allow for more emergency measures, such as antibiotics in this case, to be applied. In addition, chosen correctly, a more distal intervention will have multiple benefits that balance a possible miscalculation of the benefits to the endpoint of primary concern. Thus, local clean water supplies, if properly done, can have benefits in the form of reducing other diseases (e.g. hepatitis) and increasing other types of welfare (e.g. time savings). ORT does not.

If it is accepted for the moment that the degree to which an intervention works at the distal end of the environmental pathway is likely to be one important aspect of a SALY, how do interventions today to reduce the negative health effects of global environmental change in the coming decades fare? Arguably, such interventions should be assigned rather large sustainability factors for they operate decades in advance of potential disease.[18] There is plenty of time to try other avenues further to the right on the pathway, should distal interventions fail to work, and to accumulate parallel benefits.

3.4.1.2.2 Sustainability as portfolio management. One of the important aspects of projects with uncertain outcomes is that expected values usually accumulate, but risks can spread. This is part of the theory used in portfolio management by investment managers, for example. An investment is judged on its expected return and the variability in that return (uncertainty). An investment with high variability, all other things ignored, will be less preferred than one with the same expected value, but less variability. In practice, however, the individual variability of the investment is less important than how it affects the variability of the entire investment portfolio. In some cases, therefore, an investment with a high variability will be preferred to another with less, if it acts to reduce the variability of the overall portfolio of investments. Thus, gold was once recommended for many portfolios even though it had relatively low return and high variability, because it tended to react in the opposite fashion to the stock market as a whole, thus lowering the overall variability of a portfolio heavily invested in stocks.

This raises another possible way in which conventional decision-making might be applied to enhance the attractiveness of spending resources today to avoid the health burden of pervasive long-term phenomena such as climate change. Current interventions to reduce the size of climate change in 30 years, which often appear to have low return in terms of NPV calculations, might have an important place because they reduce the overall variability of the entire portfolio of health invest-ments. Since climate change and other global long-term phenomena can potentially have quite pervasive impacts on a range of societal and natural processes affect-ing health, there could well be substantial effects on the overall variability of the health-improvement investment portfolio. This possibility raises some interesting issues to explore and relates directly to the call by some epidemiologists to expand the scope of the discipline to be more population and globally focused (McMichael 2001).

---

[18] Standard economic thinking prefers adjustments to the benefits and/or costs each year rather than to the discount rate for such factors (Arrow *et al.*, 1996).

3.4.2 *Where do future dangers lie?*

As part of the GBD database (Murray *et al.*, 1996), projections of the GBD to 2020 were done under three sets of assumptions similar in population projections but differing in projections of economic and human capital (education) growth. These distal factors of human development influence more proximate ones affecting health, including investment in such public health infrastructure as sanitation and air pollution management, and, through education as well as personal income, household hygiene and nutrition. As a result, there is a significant difference in the GBD between the pessimistic and optimistic scenarios by 2020, reflecting differences in public health infrastructure, nutrition and other factors.

Table 3.3 shows the burdens of all major disease conditions in the pessimistic and optimistic scenarios divided into the three classic groupings of disease:

 (i) Group I: Infectious, perinatal, nutritional, maternal – sometimes called traditional.
 (ii) Group II: Chronic, noncommunicable, congenital – sometimes called modern.
(iii) Group III: Injuries – sometimes called nontransitional.

Note that not only is there an absolute difference in the totals and for some individual diseases, the relative rankings among diseases shift as well. Indeed, the choices that might lead the world to follow one or the other path can make a substantial difference in the character of ill-health by 2020.

Although some of the global environmental change factors shown at the top of Fig. 3.3 are probably affecting health today (ozone depletion, for example), most are not likely to be doing so strongly for several decades. Since the proximate risk factors they affect (middle row in Fig. 3.3) act principally to enhance the rates of diseases, the existing burden of a particular disease acts as a co-risk factor for the absolute burden produced. Thus, the more cases of diarrhoea that exist in 2020, the more diarrhoea will be caused by additional poor sanitation. This is likely to be true in absolute if not relative terms and, of course, in a burden of disease approach it is the absolute increase that counts.

Thus, it is possible to affect the distribution and quantity of disease in 2020 by decisions made today to allocate resources in the interim. By 2020 some diseases may then be experiencing upwards pressure from global environmental change factors. Thus, societies can affect their vulnerability to global environmental change in 2020 by investing during the interim in ways that reduce those particular diseases that are especially likely to be enhanced by the global environmental changes projected to be substantial in the coming decades.

A detailed analysis of this question is warranted, although difficult. As a start, we use the existing GBD database as summarized in Table 3.3 (see caption) to gain some insights. The last columns show three ways of comparing the pessimistic

Table 3.3. Differences between the optimistic and pessimistic scenarios for 2020 from the GBD

		Pessimistic			Optimistic			Rank differences	Pessimistic versus optimistic differences			
		DALYs	% of total	Rank	DALYs	% of total	Rank	>5	DALYs	Absolute difference rank	% of total for that scenario	Proportionate difference rank
												(Top 10/ Bottom 10)
Total		1 687 661	100	(Top 10)	1 295 628	100	(Top 10)		392 033	(Top 10)		
Group I	Tuberculosis	60 032	3.56	13	22 343	1.72	22	−9	37 689	2	1.83	1
	STDs	13 747	0.81	33	7 053	0.54	33	0	6 694	16	0.27	11
	HIV	51 187	3.03	16	22 797	1.76	21	−5	28 390	5	1.27	5
	Diarrhoeal diseases	69 872	4.14	7	33 127	2.56	13	−6	36 745	3	1.58	3
	Childhood cluster	52 403	3.11	15	24 889	1.92	20	−5	27 514	6	1.18	6
	Bacterial meningitis	3 800	0.23	38	1 562	0.12	38	0	2 238	27	0.10	15
	Hepatitis B & C	1 335	0.08	40	473	0.04	41	−1	862	33	0.04	19
	Malaria	29 598	1.75	23	14 154	1.09	27	−4	15 444	10	0.66	9
	Tropical cluster	6 989	0.41	36	2 046	0.16	37	−1	4 943	22	0.26	12
	Leprosy	145	0.01	44	126	0.01	42	2	19	38	0.00	26
	Dengue	288	0.02	42	85	0.01	44	−2	203	36	0.01	24
	Japanese encephalitis	272	0.02	43	91	0.01	43	0	181	37	0.01	25
	Trachoma	1 036	0.06	41	484	0.04	40	1	552	34	0.02	20
	Intestinal nematode infections	2 722	0.16	39	1 027	0.08	39	0	1 695	29	0.08	16
	Acute respiratory infections	81 152	4.81	5	38 867	3.00	12	−7	42 285	1	1.81	2
	Maternal conditions	15 539	0.92	30	3 243	0.25	36	−6	12 296	13	0.67	8
	Perinatal conditions	62 208	3.69	12	31 097	2.40	16	−4	31 111	4	1.29	4
	Nutritional deficiencies	32 883	1.95	20	14 740	1.14	26	−6	18 143	8	0.81	7
Group II	Malignant neoplasm	142 026	8.42	1	135 450	10.45	1	0	6 576	18	−2.04	43
	Diabetes mellitus	14 992	0.89	31	9 463	0.73	32	−1	5 529	21	0.16	14
	Endocrine disorders	6 899	0.41	37	5 014	0.39	35	2	1 885	28	0.02	22
	Depression	98 391	5.83	3	100 328	7.74	3	0	−1 937	43	−1.91	42
	Other neuropsychiatric disorders	107 850	6.39	2	102 992	7.95	2	0	4 858	23	−1.56	41

Table 3.3. (cont.)

	Pessimistic			Optimistic			Pessimistic versus optimistic differences				
									Absolute difference		
	DALYs	% of total	Rank	DALYs	% of total	Rank	Rank differences	DALYs	difference rank	% of total for that scenario	Proportionate difference rank
Sense organ diseases	23 087	1.37	26	21 760	1.68	23	3	1 327	30	−0.31	33
Ischaemic heart disease	89 958	5.33	**4**	81 574	6.30	**4**	0	8 384	15	−0.97	40
Cerebrovascular heart disease	69 290	4.11	**8**	59 739	4.61	**6**	2	9 551	14	−0.51	36
Other heart diseases	71 553	4.24	**6**	57 745	4.46	**7**	−1	13 808	12	−0.22	31
COPD	64 638	3.83	**10**	46 924	3.62	**9**	1	17 714	**9**	0.21	13
Asthma	15 691	0.93	28	11 088	0.86	31	−3	4 603	24	0.07	17
Other respiratory diseases	33 291	1.97	18	26 683	2.06	18	0	6 608	17	−0.09	29
Digestive diseases	62 564	3.71	11	41 093	3.17	11	0	21 471	7	0.54	**10**
Genito-urinary diseases	20 584	1.22	27	14 850	1.15	25	2	5 734	20	0.07	18
Musculo-skeletal diseases	31 246	1.85	21	30 930	2.39	17	4	316	35	−0.54	37
Congenital anomalies	33 270	1.97	19	32 022	2.47	14	**5**	1 248	32	−0.50	35
Oral conditions	12 321	0.73	34	12 687	0.98	28	6	−366	39	−0.25	32
Road traffic injuries	58 450	3.46	14	75 592	5.83	**5**	**9**	−17 142	44	−2.37	44
Poisonings	7 586	0.45	35	6 309	0.49	34	1	1 277	31	−0.04	28
Falls	26 207	1.55	24	19 896	1.54	24	0	6 311	19	0.02	23
Fires	14 335	0.85	32	11 147	0.86	30	2	3 188	26	−0.01	27
Drownings	15 614	0.93	29	11 676	0.90	29	0	3 938	25	0.02	21
Other unintentional	68 840	4.08	**9**	54 154	4.18	**8**	1	14 686	11	−0.10	30
Self-inflicted injuries	25 536	1.51	25	26 067	2.01	19	6	−531	40	−0.50	34
Violence	30 461	1.80	22	31 408	2.42	15	7	−947	41	−0.62	38
War	40 022	2.37	17	41 498	3.20	10	7	−1 476	42	−0.83	39

Group III applies to the lower block (Road traffic injuries through War).

The table summarizes projections from the optimistic and pessimistic scenario for 2020 from the GBD (Murray et al., 1996), highlighting differences in results between the two scenarios. Major disease categories are listed on the left side of the table. The centre of the table recapitulates the absolute disability-adjusted life years (DALYs) and the proportionate (to the entire GBD for each respective scenario) DALYs, along with rankings. These results in the top ten are in bold. The right side of the table gives differences between the two scenarios' projections (pessimistic minus optimistic) in absolute DALYs, proportionate DALYs and rankings. Rankings that shift by more than five spots and top ten changes in absolute DALYs rank are in bold. The penultimate column shows the differences in proportionate DALYs between each scenario. The final column ranks these results, with top ten differences (largest increases) in bold and bottom ten differences (largest decreases) in italics.

and optimistic scenarios disease by disease. The first examines the difference in rankings. The third examines the difference in the percentage of the total under each scenario. The middle columns address the absolute difference in DALYs between the two scenarios, which is arguably the best indicator of the differential impact of each disease for society as a whole.

Using this indicator, it can be seen that there is little difference between the DALY burden of many important diseases. For example, depression, although accounting for many DALYs, remains basically the same in both scenarios. It is a significant function of age distribution, which is essentially the same in both scenarios and does not depend on public investments in sanitation or education, for example. The same is true for heart disease and cancer, i.e. even though they account for large DALY totals, they are not significantly affected by choice of scenario.

The degree of difference resulting from choice of scenario is indicated by the ranking column adjacent to the DALY column. The top ten are:

   (i) ARI
  (ii) TB
 (iii) Diarrhoea
 (iv) Perinatal conditions
  (v) HIV
 (vi) Childhood cluster
(vii) Digestive diseases
(viii) Nutritional deficiencies
 (ix) COPD
  (x) Malaria

Leaving out the childhood cluster and HIV,[19] which are only weakly related to environmental factors, and digestive diseases (ulcer, cirrhosis, appendicitis), which are only moderately related, all the other disease categories have major environmental determinants, as indicated in Table 3.1 and reiterated in Fig. 3.3. To highlight that society's choices over the next few decades can make big differences in these diseases, these seven factors are represented by the trapeziodal-shaped box in the Fig. 3.3.

In Table 3.3, only one disease category is significantly larger in the optimistic scenario compared to the pessimistic scenario – road accidents. This presumably has to do with the slower development of both the size of vehicle fleets and the pace of safety improvements. To indicate its special status, it is represented by an oval-shaped box in Fig. 3.3. The three remaining diseases with important environmental determinants in Fig. 3.3, cancer, nondiarrhoeal cluster, and war/violence,[20] are not

---

[19] All the disease projections for 2020 in the GBD must be considered uncertain, of course, but those for HIV are particularly so.
[20] Another projection with a high degree of uncertainty.

significantly affected by the choice of scenario and are indicated by the rectangular-shaped box.

The conclusion of this exercise is that choices in the next few years can indeed significantly affect human vulnerability to global environmental change factors in 20 years. Recalling the discussion about the SALY above, a more rigorous analysis of the type just discussed could well be a way to assign sustainability weights. For example, based on Fig. 3.3, certain diseases could be chosen for special consideration because the larger they are later on, the larger our vulnerability to global environmental change factors. Thus, programmes aimed at these diseases, such as those subsumed in the optimistic scenario of the GBD, could be judged using NPV measures granting extra weighting to the DALYs saved, i.e. using SALYs.

3.5 Conclusion: beyond redefining and reconceptualizing

A wider array of environment–health relationships are beginning to be addressed by twenty-first century public health professionals than ever before. This broadening paradigm includes risk factors operating at unprecedented scales linked to global environmental change. In order to encompass the still-important environmental hazards of the past along with those of the future, methods are needed to evaluate the opportunities and constraints for affecting environmentally mediated diseases at multiple scales. Making the task of achieving public health even more important, in recent years the structural determinants of health, such as socioeconomic development and healthcare access, have been joined by the fundamental concerns of equity, justice, democracy and sustainability. Moreover, discourses surrounding well-being and ill-health are now beginning to incorporate phenomena from the molecular to the planetary, and often explicitly or implicitly invoke the environment. These concerns have served to highlight the complicated processes by which environmentally mediated hazards have an impact on human health, as well as other aspects of human welfare.

The spectre of global environmental change motivates the need for integrated action in order to combat resurging and emerging threats to human health. Ramifications will manifest through a complex web of interactions cascading down through global, regional, national and community levels, eventually affecting individuals. Comprehending the linkages between global and local processes shall prove to be a central challenge to achieving public health. The uncertainties involved are daunting and challenge conventional decision-making in public health. Awaiting the arrival of proximate-level hazards from global environmental change hardly seems prudent. Intervening at higher-level points in the causal web, however, may be impractical to rapidly affect health outcomes. Methods for weighing the different possibilities in a world of limited resources are critically needed (Martens,

1999). Addressing exposures and effects at the individual level will remain vital to maintaining public health. At the same time, neglecting the pressures and driving forces at the population level will do little to create sustainable public health. In order to compare policy levers acting at various levels, public health professionals can employ not only the standard efficacy and effectiveness criteria, but also political benefit-cost analyses, economic project appraisal tools and cultural appropriateness approaches. Despite lack of formal rigour, the judicious application of these calculations has frequently been critical to public health successes and often absent in public health failures. Such decision-making instruments will prove invaluable for addressing global environmental change.

An example may be drawn from climate change. Altered temperature and precipitation may not have a dramatic effect but may still perceptibly impact on sanitation and hygiene. The additional burden of disease from a small increase in diarrhoeal disease may outweigh the impacts from more concentrated but isolated malaria outbreaks. This suggestion is speculative but serves to highlight the fact that, on a global scale, small changes in the major causes of mortality and morbidity, most of which have significant environmental attributable risks, can far exceed large changes in intermediate or minor causes of mortality and morbidity. There are a number of linkages between global environmental change phenomena and disease categories such as ARI or perinatal conditions. Establishing the direction and strength of these relationships is an important but under-appreciated aspect of the potential health impact of global environmental change. Moving back up the causal chain entails not only an expansion of perspective but, if affecting these more complex relationships is to become part of public health practice, an expansion of the repertoire of tools as well.

Referring to climate change again, an emphasis on states or exposures (in DPSEE parlance) may also stress the altered temperature and precipitation patterns that expand vector habitat or the increased frequency of extreme climatic events that cause natural disasters. These are often cited and clearly important aspects of the possible effects of global environmental change on human health. Yet while useful, such an emphasis potentially neglects, or at least diverts attention away from, driving forces and pressures resulting from climate change, such as migration or exacerbated socioeconomic inequality, which may significantly undermine housing quality or access to health-care. If these driving forces or pressures are addressed, the health impacts incurred by dengue or floods could be significantly lessened. Mitigating the deleterious health impacts of migration and inequality remains at the periphery of public health at present, yet as the previous example suggests, these factors can have considerable influence on the environment–health nexus.

Sustainable development policies that are to ensure a healthy environment will need to focus on more long-term, broad-spectrum interventions that touch on the driving forces operating in society. In many developing countries, this means tackling traditionally "nonhealth" issues such as poverty, inequity, demographic changes, inadequate legal systems and poor information flows, which now contribute to biodiversity loss, land degradation, food insecurity and poor water and air quality. These "nonhealth" issues also play an important role in determining the sustainability of various intervention options. In developed countries, inequities are also important, as sizeable population groups live in squalor and relative poverty without access to the power and information needed for improving their health. In addition, it is difficult to envisage dealing effectively with global environmental change processes without a greater emphasis on reducing unsustainable consumption and curbing the use of nonrenewable fuels. All of these actions would have long-term and sustained beneficial effects on human health, often with major near-term health benefits as well (Wang & Smith, 1999; Holdren & Smith, 2000).

In a sort of pendulum motion, a new public health movement has begun recently to turn attention away from individual health and, as in the eighteenth century, focus once again on population health. As seen above, public health has slowly drifted away from environmental concerns, progressively narrowing its focus on the individual, deploying disease-centred intervention strategies based on selective case-management or specific disease-prevention technologies. The emergence of the new public health may be seen as a reaction mostly against the positivistic approach of the conventional biomedical model of disease causation prevalent during the last few decades. On a global level, the increasing exchange of ideas, materials, finances and organisms, and the accelerating integration of institutions and systems (social, economic, political, cultural, environmental, technological, etc.) i.e. post-modern interconnectedness, can alter the incidence and pattern of many illnesses in an unparalleled fashion. Confronted with a rapidly changing world and shifting paradigms in health–environment–development, the conventional model of environment and health is at a crossroads.

Although recent paradigms do not seem well suited in their current forms to address the global environmental change issues now on the horizon, it is also clear that they have brought some notable successes in expanding understanding of the determinants of health and finding ways to improve it. If the art and science of public health can be defined as finding ways to make the poor healthy before they become wealthy, much progress has been made, although much is still needed, especially in extending the benefits to all groups. It is important not to throw out the baby with the bathwater. The task, therefore, is not to abandon the disease and cost-effectiveness paradigm, but to expand and enhance it to encompass the new set of global environmental change risks that are now becoming apparent partly because such success has been achieved in dealing with the older and larger risks.

References

Adeola, F. C. (2000). Cross-national environmental injustice and human rights issues. *American Behavioral Scientist*, **43**, 686–706.

Albrecht, G. L., Fitzpatrick, R. & Scrimshaw, S. C. (eds.) (2000). *Handbook of Social Studies in Health and Medicine*. Thousand Oaks, CA: Sage Publications.

Armstrong, D. (1993). Public health spaces and the fabrication of identity. *Sociology*, **27**, 393–410.

Arrow, K. J., Parikh, J. & Pillet, G. (1996). Decision-making frameworks for addressing climate change. In *Climate Change – 1995: Economic and Social Dimensions*, ed. J. P. Bruce, H. Lee & E. F. Haites. Cambridge, UK: Cambridge University Press.

Beaglehole, R. & Bonita, R. (1997). *Public Health at the Crossroads: Achievements and Prospects*. New York City, NY: Cambridge University Press.

Binswanger, H. C. & Smith, K. R. (2000). Paracelsus and Goethe: founding fathers of environmental health. *Bulletin of the World Health Organization*, **78**, 1162–4.

Bloom, D. E. & Canning, D. (2000). The health and wealth of nations. *Science*, **287**, 1207–9.

Boyd, B. A. (1991). *Rudolf Virchow: the Scientist as Citizen*. New York City, NY: Garland.

Bruce, J. P., Lee, H. & Haites, E. F. (eds.) (1996). *Climate Change – 1995: Economic and Social Dimensions*. Cambridge, UK: Cambridge University Press.

Chadwick, E. (1842). *Report on the Sanitary Condition of the Labouring Population of Great Britain*. Edinburgh, UK: Edinburgh University Press.

Cipolla, C. M. (1992). *Miasmas and Disease: Public Health and the Environment in the Pre-Industrial Age*. New Haven, CN: Yale University Press.

Cohen, M. (1989). *Health and the Rise of Civilization*. New Haven, CN: Yale University Press.

Coleman, W. (1983). *Death is a Social Disease: Public Health and Political Economy in Early Industrial France*. Madison, WI: University of Wisconsin Press.

Colwell, R. R. (1996). Global climate change and infectious diseases. *Science*, **274**, 2025–31.

Corvalán, C. F., Kjellström, T. & Smith, K. R. (1999). Health, environment and sustainable development: identifying links and indicators to promote action. *Epidemiology*, **10**, 656–60.

Daily, G. C. & Ehrlich, P. R. (1996). Global change and human susceptibility to disease. *Annual Review of Energy and the Environment*, **21**, 125–44.

Daniels, N., Kennedy, B. P. & Kawachi, I. (1999). Why justice is good for our health: the social determinants of health inequalities. *Daedalus*, **128**, 215–51.

Das, V. (1995). *Critical Events: an Anthropological Perspective on Contemporary India*. New Delhi, India: Oxford University Press.

Douglas, M. (1966). *Purity and Danger: an Analysis of Concepts of Pollution and Taboo*. London, UK: Routledge Press.

Engels, F. (1987). *The Condition of the Working Class in England*. Harmondsworth, UK: Penguin Books.

Farmer, P. (1996). Social inequalities and emerging infectious diseases. *Emerging Infectious Diseases*, **2**, 259–69.

Farmer, P. (1999). *Infections and Inequities: the Modern Plagues*. Berkeley, CA: University of California Press.

Farr, W. (1885). *Vital Statistics: a Memorial Volume of Selections from the Reports and Writings of William Farr*. London, UK: The Sanitary Institute.

Foucalt, M. (1975). *The Birth of the Clinic: an Archeology of Medical Perception*. New York City, NY: Vintage Press.

Frost, W. H. (1941). *Papers of Wade Hampton Frost, M.D.: a Contribution to Epidemiologic Method.* New York City, NY: Commonwealth Fund.

Greenwood, M. (1935). *Epidemics and Crowd-Diseases: an Introduction to the Study of Epidemiology.* London, UK: Williams & Norgate.

Hahn, R. A. (ed.) (1999). *Anthropology in Public Health: Bridging Differences in Culture and Society.* New York City, NY: Oxford University Press.

Haines, A. & McMichael, A. J. (1997). Climate change and health: implications for research, monitoring, and policy. *British Medical Journal*, **315**, 870–4.

Hamlin, C. (1998). *Public Health and Social Justice in the Age of Chadwick: Britain, 1800–1854.* Cambridge, UK: Cambridge University Press.

Holdren, J. P. & Smith, K. R. (2000). Energy, Environment, and Health. In *World Energy Assessment*, ed. J. M. Goldemberg. New York City, NY: United Nations Development Programme.

Inhorn, M. C. & Brown, P. J. ed. (1997). *The Anthropology of Infectious Disease: International Health Perspectives.* Amsterdam, The Netherlands: Gordon and Breach Publishers.

Jamison, D. T. & World Bank (1993). *Disease Control Priorities in Developing Countries.* New York City, NY: Published for the World Bank by Oxford University Press.

Joy, B. (2000). Why the future doesn't need us. *Wired*, **2000**, 1–15.

Kawachi, I. & Kennedy, B. (1999) Income inequality and health: pathways and mechanisms. *Health Services Research*, **34**, 215–27.

Krieger, N. & Zierler, S. (1996). What explains the public's health? A call for epidemiologic theory. *Epidemiology*, **7**, 107–9.

Lalonde, M. (1974). *A New Perspective on the Health of Canadians.* Ottowa: Canadian Department of National Health and Welfare.

Lind, R. C. (1982). *Discounting for Time and Risk in Energy Policy.* Baltimore, MD: Distributed by the Johns Hopkins University Press for Resources for the Future.

Litsios, S. (1994). Sustainable development is healthy development. *World Health Forum*, **15**, 193–5.

Lupton, D. (1995). *The Imperative of Health: Public Health and the Regulated Body.* London, UK: Sage Press.

Mantini-Briggs, C. (2000). *Critical Perspectives on Health and Environment in Delta Amacuro, Venezuela.* Baltimore, MD: School of Hygiene and Public Health, Johns Hopkins University.

Martens, P. (1999). How will climate change affect human health? *American Scientist*, **87**, 534–41.

McKeown, T. (1976). *The Role of Medicine.* London, UK: Nuffield Provincial Hospitals Trust.

McMichael, A. J. (1997). Integrated assessment of potential health impact of global environmental change: prospects and limitations. *Environmental Modeling Assessment*, **2**, 129–37.

McMichael, A. J. (1999). Prisoners of the proximate: loosening the constraints on epidemiology in an age of change. *American Journal of Epidemiology*, **149**, 1–11.

McMichael, A. J. (2001). *Human Frontiers, Environments, and Disease: Past Patterns, Uncertain Futures.* New York City, NY: Cambridge University Press.

McMichael, A. J., Haines, A., Slooff R., & Kovats, S. (1996). *Climate Change and Human Health: an Assessment Prepared by a Task Group on Behalf of the World Health Organization, the World Major Metereological Organization and United Nations Environment Program.* Geneva: World Health Organization.

McMichael, A. J., Patz, J. & Kovats, R. S. (1998). Impacts of global environmental change on future health and health care in tropical countries. *British Medical Bulletin*, **54**, 475–88.

McMichael, A. J., Bolin, B., Costanza, R., Daily, G. C., Folke, C., Lindahl-Kiessling, K., *et al.* (1999). Globalization and the sustainability of human health. *Bioscience*, **49**, 205–10.

Melosi, M. V. (2000). *The Sanitary City: Urban Infrastructure in America from Colonial Times to the Present.* Baltimore, MD: Johns Hopkins University Press.

Moeller, D. W. (1997). *Environmental Health.* Cambridge, MA: Harvard University Press.

Murray, C. J. L. & Lopez A. D. (1996). *The Global Burden of Disease: a Comprehensive Assessment of Mortality and Disability from Diseases, Injuries, and Risk Factors in 1990 and Projected to 2020.* Cambridge, MA: Published by the Harvard School of Public Health on behalf of the World Health Organization and the World Bank, distributed by Harvard University Press.

Murray, C. J. & Lopez, A. D. (1999). On the comparable quantification of health risks: lessons from the Global Burden of Disease Study. *Epidemiology*, **10**, 594–605.

Mustard, J. F. (1996). Health and social capital. In *Health and Social Organization*, ed. D. Blane, E. Brunner & R. Wilkinson. London, UK: Routledge.

Nottingham, H. F. (1685). *Arguments...(on) Methods of Limiting the Trust of a Term for Years.* London: George Tatarshall.

Packard, R. M. & Brown, P. J. (1997). Rethinking health, development, and malaria: historicizing a cultural model in international health. *Medical Anthropology*, **17**, 181–94.

Pearce, N. (1996). Traditional epidemiology, modern epidemiology, and public health. *American Journal of Public Health*, **86**, 678–83.

Pedersen, D. (1996). Disease ecology at a crossroads: man-made environments, human rights and perpetual development utopias. *Social Science and Medicine*, **43**, 745–58.

Pimentel, D., Tort, M., D'Anna, L., Krawic, A., Berger, J., Rossman, J. *et al.* (1998). Ecology of increasing disease. *Bioscience*, **48**, 817–26.

Porter, D. (1999). *Health, Civilization, and the State: a History of Public Health from Ancient to Modern Times.* London, UK: Routledge.

Portney, P. R. & Weyant, J. P. (1999). *Discounting and Intergenerational Equity.* Washington, DC: Resources for the Future.

Reich, M. R. (1994). The political economy of health transitions in the Third World. In *Health and Social Change in International Perspective*, ed. L. C. Chen, A. Kleinman & N. C. Ware, pp. 413–52. Boston, MA: Harvard School of Public Health.

Rose, G. (1985). Sick individuals and sick populations. *International Journal of Epidemiology*, **14**, 32–38.

Rothman, K. J. (1976). Causes. *American Journal of Epidemiology*, **104**, 587–92.

Shahi, G. S., Levy, B. S., Binger, A., Kjellström T. & Lawrence, R. (eds.) (1997). *International Perspectives on Environment, Development, and Health: Toward a Sustainable World.* New York City, NY: Springer-Verlag.

Smith, K. R. (1997). Development, health, and the environmental risk transition. In *International Perspectives on Environment, Development, and Health: Toward a Sustainable World*, ed. G. S. Shahi, B. S. Levy, A. Binger, T. Kjellström & R. Lawrence, pp. 51–62. New York City, NY: Springer-Verlag.

Smith, K. R. (2000). Environmental health – for the rich or for all? *Bulletin of the World Health Organization*, **78**, 1156–61.

Smith, K. R. (2001). What the new administration should know about environment and health. *Environment*, **43**, 34–42.

Smith, K. R., Corvalán, C. F. & Kjellström, T. (1999). How much global ill health is attributable to environmental factors? *Epidemiology*, **10**, 573–84.

Snow, J. (1855). *On the Mode of Communication of Cholera*. London, John Churchill.

Susser, M. (1998). Does risk factor epidemiology put epidemiology at risk? Peering into the future. *Journal of Epidemiology and Community Health*, **52**, 608–11.

Sydenstricker, E. (1933). *Health in Environment*. New York City, NY: McGraw-Hill.

Szreter, S. (1988). The importance of social intervention in Britain's mortality decline c.1985–1914: a reinterpretation of the role of public health. *Social History of Medicine*, **1**, 1–37.

Trostle, J. (1986). Early work in anthropology and epidemiology: from social medicine to germ theory, 1840 to 1920. In *Anthropology and Epidemiology*, C. R. Janes, R. Stall & S. M. Gifford. Dordrecht, Germany: Reidel.

Walter, S. D. (1976). The estimation and interpretation of attributable risk in health research. *Biometrics*, **32**, 829–49.

Walter, S. D. (1978). Calculation of attributable risks from epidemiological data. *International Journal of Epidemiology*, **7**, 175–82.

Wang, X. & Smith, K. R. (1999). Near-term health benefits of greenhouse-gas reductions: health impacts in China. *Environmental Science and Technology*, **33**, 3056–61.

Warford, J. J. (1995). Environment, health and sustainable development: the role of economic instruments and policies. *Bulletin of the World Health Organization*, **73**, 387–95.

Wilkinson, R. G. (1996). *Unhealthy Societies: the Afflictions of Inequality*. New York City, NY: Routledge Press.

Woodward, A., Hales, S., Litidamu, N., Phillips, D. & Martin, J. (2000). Protecting human health in a changing world: the role of social and economic development. *Bulletin of the World Health Organization*, **78**, 1148–55.

World Bank (1993). *World Development Report 1993: Investing in Health*. University Press Book.

World Health Organization (1996). *Investing in Health Research and Development*. Geneva, Switzerland: World Health Organization.

Yach, D. & Bettcher, D. (1998). The globalization of public health I: threats and opportunities. *American Journal of Public Health*, **88**, 735–8.

4

Surprise, nonlinearity and complex behaviour

TAMARA AWERBUCH, ANTHONY E. KISZEWSKI & RICHARD LEVINS

4.1 Introduction

The world is stranger than we can imagine and surprises are inevitable in science; thus we found, for example, that pesticides increase pests, antibiotics can create pathogens, agricultural development creates hunger and flood control leads to flooding (Levins, 1995a,b). But some of these surprises could have been avoided if the problems had been posed so as to accommodate solutions in the context of The Whole, taking complexities into account. Predicting the impact of a changing world on human health is a hard task and requires an interdisciplinary approach drawn from the fields of evolution, biogeography, ecology and social sciences, and relies on various methodologies such as mathematical modelling and historical analysis (Awerbuch, 1994; Levins, 1995a,b; Awerbuch *et al.*, 1996; McMichael, 1997). Indeed, integrated assessment modelling of human health has been recommended as a global methodology to develop prevention strategies, educate policy makers and assess the impact of interventions (Martens, 1998; see also Chapter 8).

When even a simple change occurs in the physical environment, its effects percolate through a complex network of physical, biological and social interactions that feed back and feed forwards. Along some pathways the effects are attenuated and may even disappear; along others they are amplified and can show up at points far removed from their original entry into the system; along still other pathways the effects may be reversed so that, for example, heating may lower the temperature or adding nitrogen to a lake may reduce the nitrogen level (Levins & Lane, 1977). Sometimes the immediate effect of a change is different from its long-term effect, sometimes the local changes may be different from the region-wide alterations. The same environmental change may have quite different effects in different places or times.

Therefore, the study of the consequences of environmental change is a study of the short- and long-term dynamics of complex systems, a domain where our common-sense intuitions are often unreliable and new intuitions have to be developed in order to make sense of often paradoxical observations. Complexity itself has to be studied as an object of interest in its own right (Levins, 1973; Pattee, 1973; Puccia & Levins, 1985).

In this chapter we attempt to introduce some of the concepts that help grasp this complexity.

4.2 Complexities

4.2.1 Complex interactions

Organisms select, transform and define their own environments so that a physical factor such as "temperature" may have quite different biological meanings. It can determine the rate of development and indirectly the size of insects. This temperature relationship is especially interesting because of threshold effects. Each species of insect requires a certain amount of heat to complete each developmental stage and also develop as a whole (Wagner *et al.*, 1984). Let this quantity be R, measured in degree-days temperature above some threshold T^*, multiplied by the days at that temperature. Thus development time in days is:

$$D = R/(T - T^*), \qquad (4.1)$$

where T is the actual temperature of the insect's immediate environment. It is clear that the closer the actual temperature, T, is to threshold, the more sensitive is D to temperature changes. This is consistent with Schmalhausen's Law which states: "that a system at the boundary of its tolerance along any dimension of its existence is more vulnerable to small differences in circumstance along any dimension" (Schmalhausen, 1949). Suppose that $T^*=14°C$ and $T=15°C$. Then each day one degree is accumulated. If $R=100$ then development takes 100 days. Different species have different threshold temperatures T^*s, which are genetically determined. Suppose now that a second species has a threshold of $13°C$ and $R=200$. It requires two degrees each day per day; although the degree-day requirements are higher for the second species, both species emerge at the same time, after 100 days. Now consider the change of temperature from $15°C$ to $16°C$. The first species develops in 50 days while the second accumulates $3°C$ each day and needs 67 days to fully develop. Thus a small change of environment, such as $1°C$, can break the synchrony between two species (predator and prey, host and parasite, tree and herbivore), or the seasonal coordination of developmental processes among various species and pathogen transmission, as in the case of Lyme disease (Sandberg *et al.*, 1992; Awerbuch & Sandberg, 1995).

A temperature change can change the available growing season for plants and crop production, and therefore the composition of communities, including those

Fig. 4.1 Simulation of habitat competition across two orthogonal resource gradients. Conditions for parameter 1 vary stochastically from left (lowest concentration) to right (highest concentration). Conditions for parameter 2 vary from north to south. Organisms (red and green) populate sites that fall within their range of tolerance. Competitive outcomes are determined by least divergence from optimum conditions. For a colour version of this figure, see www.cambridge.org/9780521114028

of pests and pathogens that attack the plants (Coakley & Scherm, 1996). It can affect the suitability of nesting sites and of foraging sites of ants, the amount of time that people spend outdoors, the solubility of CO_2 and O_2 in water and therefore the abundance of dragonfly larvae that prey on developing mosquitoes, some of which are vectors of disease (i.e. *Anopheles* is the vector of malaria and *Aedes* species of Eastern Encephalitis and Dengue). It can lengthen or shorten the extrinsic incubation period of malaria parasites in mosquito vectors, powerfully affecting the force of transmission. It can determine the energy expenditure of organisms. It can alter the humidity and this can affect the patchiness of the habitat and the altitudinal gradient of day and night temperatures. Therefore, the first step in the analysis of environmental change is to translate physical climate factors into bioclimatological elements.

Uniform changes in temperature or other environmental factors across a landscape may thus result in nonuniform changes in the distributions of organisms. The intersection of diverse bioclimatological elements in part defines the boundaries of ecological niches. Organisms adapted to particular sets of conditions may become less competitive within a given habitat when conditions change. Consider a hypothetical distribution of two species whose relative competitive fitness varies with temperature and moisture. If those differences are large enough and environmental conditions vary across one or more gradients, then a mutually exclusive distribution is likely to develop (Fig. 4.1), perhaps with a region of overlap. When one element of the environment, such as temperature, is altered uniformly, then the local niche-space and competitive relationships may change in a complex manner, opening habitats that were previously closed or creating competitive disadvantages where fitness was once optimal (Fig. 4.2). Thus, uniform changes in a single bioclimatological element may translate into a nonuniform response that is expressed differently in different locales.

Complex interactions may also occur solely within the physical realm, leading to counterintuitive outcomes in response to change. For example, an increase in the

Fig. 4.2 Simulation of two-species competition for habitat across two orthogonal gradients. Parameter 1 remains unchanged from Fig. 4.1. Parameter 2 is increased uniformly across the entire matrix. For a colour version of this figure, see www.cambridge.org/9780521114028

mean daytime or night-time temperature of a region generally might be construed as increasing the threat of vector-borne disease throughout the entire area, especially where average temperatures were previously too low to support the development of parasites. In highland regions, however, the effect of increased global temperature must be considered in combination with changes in relative humidity. The rate of temperature decline with increasing altitude, or the adabiatic lapse rate, is highly responsive to atmospheric moisture (Doswell *et al.*, 1991). If drier conditions result from the increase in temperature, then the resulting increase in the slope of the lapse rate may reduce temperatures at higher altitudes, compensating for the overall increase in temperature across the broader region. Of course, reduced rainfall in such a case would also reduce the breeding of insect vectors that depend on precipitation. Thus, a single change in a bioclimatological element may lead to the opposite change in certain locales.

4.2.2 Modelling living systems

In living systems, especially where biological and social phenomena interact, our equations are not exact faithful photographic representations of the reality of interest. Our variables may be heterogeneous ensembles of quite distinct variables, or variables that we label as different may behave as a single variable; our constants may be changing; the connections among variables may shift and realign; processes we treat as instantaneous may occur only after delays. Deterministic models are perturbed by outside influences. Therefore, we have to look at even our best efforts at modelling with scepticism. This directs our attention to the development of robust models that are relatively insensitive to the details of the assumptions, some of which we describe below.

Models are intellectual constructs, objects that we create in order to indirectly study objects in nature and society, rather than studying them directly. In order to be useful, they have to be similar enough to what they represent so that what we learn from the model is applicable to the real world objects, but different in ways that make them more manageable, intelligible, testable and useful. Thus, model building involves two quite distinct processes: those that take us from the reality to the model and those that return us from the model to the reality. It is especially important here to separate those conclusions of the model that tell us about reality from those that tell us about the model itself. An example is the logistic equation often used for population growth in a limited environment:

$$dx/dt = r\,x(1 - x/k). \tag{4.2}$$

In this equation, x is the population size, r is the rate of growth in the absence of crowding and k is the carrying capacity of the environment for this population. This model grasps the important reality that populations do not grow forever and allows us to separate the initial growth rate from the eventual equilibrium: fast-growing populations may reach low equilibria, slow growers may reach high levels. It tells us that populations above carrying capacity will decline. It also gives us a criterion for a population entering a habitat or a pathogen invading a host: r must be positive.

But the model is also misleading: under severe conditions that make r negative, compounded by overcrowding so that x is greater than k, the equation tells us that the population would increase since both factors are negative! This nonsense result calls our attention to the fact that a model is applicable only within certain limits. This model is also misleading in more subtle ways: by labelling the parameters r and k we have turned them into objects. We then imagine that r refers to nonconsumable elements of the environment, such as temperature affecting fecundity, while the carrying capacity refers to limited resources, such as total available food or nesting sites. This commonsense inference is widespread in ecology. However, we can also pose the problem in a different way: suppose that R is some resource that enters the environment at some rate a, is consumed by the species of interest at rate p, and is removed by other processes at rate c. Suppose further that the population of x grows at a rate that depends on its food supply R and its death rate m. Then we might look at the pair of equations:

$$dR/dt = a - R(px + c), \tag{4.3}$$
$$dx/dt = x(pR - m). \tag{4.4}$$

When the population is very sparse, dR/dt approximates $(a - cR)$. It would reach an equilibrium value at $R^* = a/c$. Substitute this value in the equation for x:

$$dx/dt = x(pa/c - m). \tag{4.5}$$

Since this is the growth of the population when there is no density effect, $pa/c - m$ is equivalent to the r of the logistic model. When we remove this restriction and let R^* be $a/(px + c)$ using equation (4.3) and inserting in (4.4), we get:

$$dx/dt = x(pa/(px + c) - m), \tag{4.6}$$

from which we find that x reaches a population limit, k, at the equilibrium level:

$$x^* = a/m - c/p. \tag{4.7}$$

Now it appears that k and r depend on the same parameters and that these represent both consumable and nonconsumable aspects of the ecology. Furthermore, the resource level when x and R are at their equilibrium values is:

$$R^* = m/p, \tag{4.8}$$

the ratio of consumer mortality and harvesting rates, and quite independent of resource input or removal by other processes. Thus we cannot talk about "r" selection and "k" selection as if they were completely different processes (Pianka, 1970). Global change could affect "k" and "r" simultaneously. This discussion illustrates one way in which models can be used to check up on each other.

No model can satisfy all of our requirements: generality, realism, precision and manageability are all goals, but depend on somewhat different designs so that a research strategy should include several models. Ideally they agree on the important properties of the system studied and differ only in the details of the model that facilitate the analysis.

4.3 Models

The four main research approaches to complexity have been reduction, statistics, simulation and qualitative analysis.

Reduction assumes that to understand the whole situation it is sufficient to describe as completely as possible the smallest parts and their direct connections. This has been a highly successful strategy on the small scale, where detailed knowledge of the parts is sufficient, such as in the identification of molecules as having certain specific effects. However, it has led to disasters when the leap is made from physiological facts (e.g. pesticides kill bugs in bottles) to ecological or social claims (therefore, application of pesticides will control the bugs in the field or the pesticide–seed–fertilizer–mechanization package will improve the lives of Third World farmers and protect national economies). Reduction as a tactic is necessary; reduction as a philosophy of science gets us into more and more trouble as we tackle the complex ecosocial problems we now face. It leaves us with the paradox of growing sophistication on the small scale and irrationality at the level of the scientific enterprise as a whole. Such models are precise and manageable but lack realism and generality.

Statistical analysis attempts to reach for The Whole with a minimum of theoretical assumptions. The hope is that if we have enough information the numbers will speak to us. Statistical models are directed towards answering relatively few questions: are two or more populations different? What is the relative contribution to the observed data of the different factors that were considered as possibly relevant?

(a)

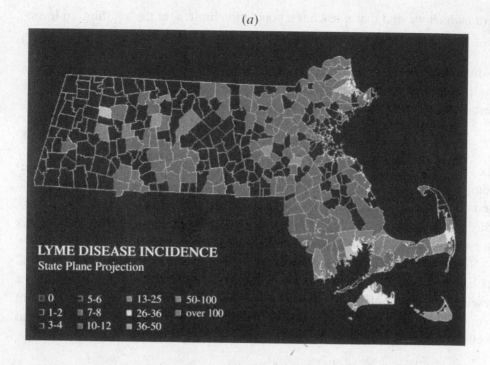

(b)

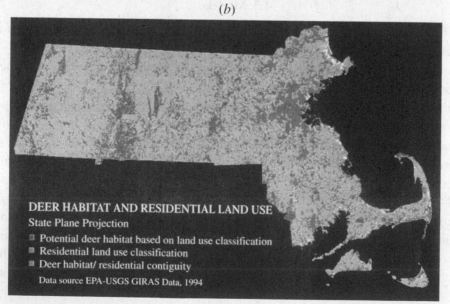

Fig. 4.3

For a colour version of this figure, see www.cambridge.org/9780521114028

(*c*)

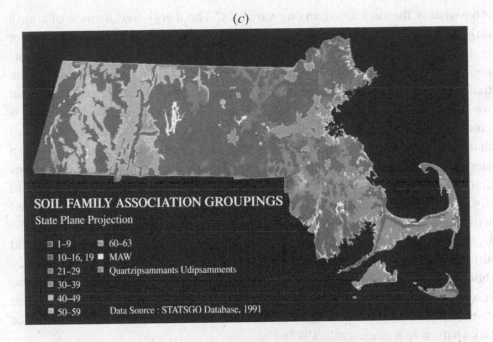

(*d*)

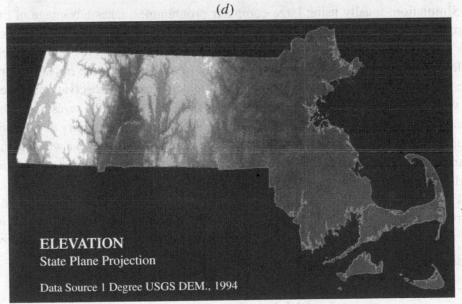

Fig. 4.3 (*cont.*)

For a colour version of this figure, see www.cambridge.org/9780521114028

And what is the association among variables? The theory-free status is of course only relative. The choice of relevant factors, the choice of appropriate measures (for instance, means or medians of income distributions), the choice of statistical tests and confidence levels for interpreting them and, underlying all this, the assumption that the partitioning of relative weights among factors is the understanding that we want, are all laden with theory. But given these caveats, statistical analysis is crucial for testing hypotheses and for scanning for possible causal relationships that have to be examined in other ways too. This approach in conjunction with a Geographic Information System has been useful for deriving spatial correlations among variables linked to the emergence of Lyme disease. It was surprising to discover that the distribution of Lyme disease cases in Massachusetts was associated with the distribution of the physical variables that support the development of the Lyme disease vector (the deer tick), such as soil type, proximity to water and altitude. There was no correlation with the spatial distribution of deer as previously thought. Satellite data were used to map these variables (Fig. 4.3(*a*)–(*d*)). These results are consistent with those obtained from the spatial analysis of Lyme disease distribution in Wisconsin, which was found to be correlated with that of the vector tick (Kitron & Kazmierczak, 1997).

Simulation, usually using large computer programmes, takes advantage of the capacity of computers to offer numerical solutions to simultaneous equations that are intractable analytically. The approach asks us to measure all the relevant variables and parameters, insert them into equations, and compute the trajectories of the variables. These predicted outcomes are then compared to observation. If they fit well, the model is supported. If they do not, we can go back to re-estimate the parameters or alter the model. The power of this approach is its ability to solve simultaneous nonlinear equations and display the results not only of what we think the situation is, but also possible alternative scenarios with different initial conditions and parameters.

From previous studies (Sandberg *et al.*, 1992), for example, we learn that the important features of the life cycle of the Lyme disease tick can be captured in terms of seasonal transitions, and its developmental progression in terms of three stages:

> Immature (egg and larvae) → Juvenile (1st year nymphs and 2nd
> year nymphs) → Adults.

It is possible to convert the system of three difference equations representing these transitions into an annual difference equation with delay. The stage of the adult tick is captured by:

$$Z(n+1) = a^* Z(n) + b^* Z(n-1)^* \exp[(-p^* Z(n)]$$
(4.9)

where $Z(n)$ is the abundance on the tick at its adult stage in year n, $Z(n-1)$ in the previous year, $Z(n+1)$ in the upcoming year. The model takes into account the biological characteristics of the Lyme disease tick: the moulting from immature to juveniles and from juveniles to adults (embedded in parameters a and b), and that they depend on feeding and surviving before moving to the next stage. Also included in the model are the two-year survival of the nymphs (Sandberg *et al.*, 1992) and the tick semelparaous reproductive behaviour (eggs are laid once and the adult female dies straight away). In addition, b includes a fecundity parameter and the probability that the adults will survive to lay eggs, which accounts for survival during harsh weather conditions, feeding opportunities and the successful mating necessary for laying eggs. Assuming that the adult ticks inhibit the immature from reproducing, through an inability of the adults to feed due to host immunity, we add density dependence represented by an exponential inhibition of growth.

Average parameter values estimated in a previous study (Sandberg *et al.*, 1992) are as follows: $b = 1.231$, $a = 0.084$, $p = 0.01$ and, starting with one adult as the initial condition, we get the pattern of growth depicted in Fig. 4.4(a).

However, if environmental conditions favour the survival of the immature (through a decline in the predation rates of this stage) and the adults that lay eggs we might get a chaotic pattern of population dynamics (Fig. 4.4(b)).

Simulation also has its limitations: the simulation requires a great amount of data, some of which are difficult to collect, so that the project is expensive and usually cannot be replicated if more data collection is required. Therefore, it tells us about a particular lake or forest as long as the situation remains more or less the same, but much less about lakes and forests in general since the numerical results do not reveal the reasons for the outcomes. The models also exclude poorly or nonmeasurable factors of qualitative nature, such as social pressures, policy implementation delays, or the level of panic in a population. This usually means that poorly measurable components are omitted from the model or consigned to footnotes. Systems simulation models are realistic and precise, but are not generalizable and are awkward to manage beyond the narrow purpose of projecting a trend.

An alternative simulation approach based on cellular automata (CA) replaces intractable sets of simultaneous equations with simple definitions of local interactions between discrete elements on a matrix (Toffoli, 1984). CA models employ a "bottom-up" approach that avoids homogeneous assumptions and allows for local variation in dynamics. Outcomes can be deterministic or stochastic. Qualitative factors can be depicted. Automata are particularly useful where spatial structures are too complex to be readily defined with systems of equations. Their greatest utility lies in situations where spatial structure and local interaction significantly affect the dynamics of a natural system. Fig. 4.5 illustrates an epidemiological application to a directly transmitted pathogen. Discontinuous spatial contact structures can

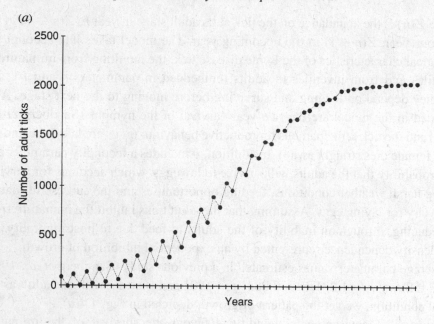

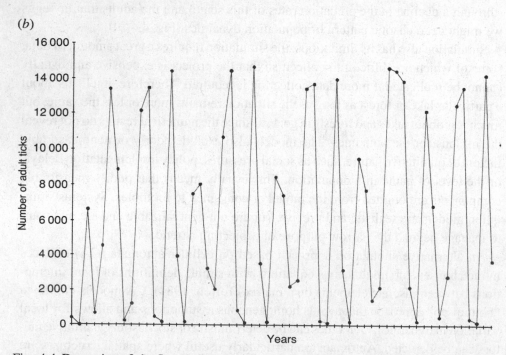

Fig. 4.4 Dynamics of the Lyme disease tick. (*a*) A simulation with average parameter values estimated in a site in Massachusetts (Sandberg *et al.*, 1992). (*b*) A simulation with increased parameter values due to higher survival rates at the various developmental stages.

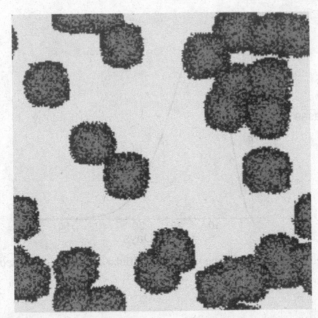

Fig. 4.5 For a colour version of this figure, see www.cambridge.org/9780521114028
A simple stochastic cellular automata model of directly transmitted pathogens. Introduction
of 25 pathogens into a homogeneous population of 50 000 hosts resulted in 25 foci; seen
after ten days. Uninfected organisms are indicated in yellow; purple represents infection
and pink represents recovry.

strongly affect the dynamics of an epidemic. Homogeneous spatial arrangements
produce a single steep incidence peak (Fig. 4.6), while a clustered host population (Fig. 4.7) displays an extended but shallower incidence curve that may
produce spatially induced pulses as new clusters acquire infection (Fig. 4.8). CA
models provide insights into special situations where homogeneous assumptions
do not apply and spatial structures are too complex to formulate deterministi-
cally.

Qualitative models emphasize realism and generality at the expense of precision.
Their strengths are that they depend on minimum assumptions, asking how much
we can get away with not knowing and still understand the processes of concern.
They allow for the inclusion of variables that are not readily measured and usually
require outcome data only of a rough sort: increase, decrease or no change; positive, negative or no correlation. This offers the opportunity for constructing a broad
model with variables spanning disciplines, and the final outcomes educate the in-
tuition to grasp at a glance those features of a system that determine the kinds of
behaviours that can happen. These models are inexpensive, and when in doubt we
can compare many of them for consistency and inconsistency. Their disadvantage
is the lack of precision: since they merely indicate the direction of change they do

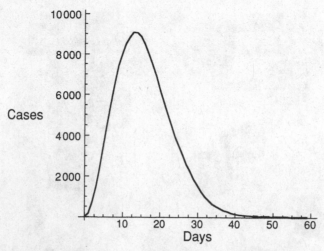

Fig. 4.6 Simulated epidemic using the stochastic cellular automata presented in Fig. 4.5.

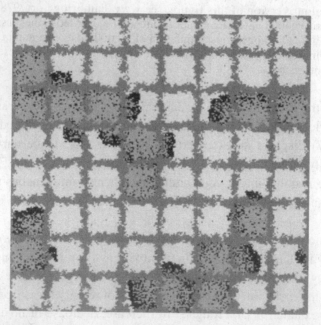

Fig. 4.7 Metapopulation configuration used for epidemic simulation in Fig. 4.8. 32 400 hosts are distributed into 64 metapopulations. The borders of each subpopulation are stochastically determined to allow limited contact between them. Blue represents infection and pink represents recovery. Unoccupied sites are red. For a colour version of this figure, see www.cambridge.org/9780521114028

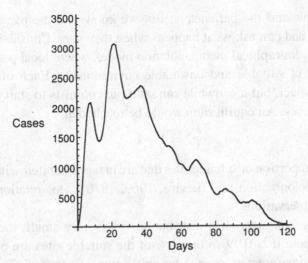

Fig. 4.8 The effect of spatial discontinuity on the dynamics of an epidemic. This simulation uses the same parameters as Fig. 4.7, except that the host population is divided into equivalent 64 subpopulations.

not tell managers and policy-makers how big an effect will be, or how soon it will occur.

Any complex research programme needs a mix of methods. A statistical scan may suggest possible relationships. A qualitative model may tell us what to expect and therefore what statistical analysis is needed. A reductionist exploration might identify the variables in the system and the possible ways they interact, namely in series or in parallel, and therefore they inform statistical and simulation models. A qualitative model can tell us what has to be measured and what structure the simulation model should have. Later the simulation model presents qualitative models with outcomes to explain. Thus, there are no "true" models – only models that reveal and conceal different aspects of reality. Our emphasis on qualitative models here is only because they are less familiar.

4.4 Qualitative models

Qualitative models attempt to grasp the behaviour of a system from some general system properties that we might catch at a glance. These include the shapes of curves and the structures of networks of interacting variables.

4.4.1 Concavity and convexity

Equations are often formulated in terms of constant parameters and variables. If the variables reach an equilibrium then we may find the relationship between the equilibrium value and the parameters. But we know that the parameters are not really constant, and can ask what happens when they vary. Consider, for example, the familiar biogeographical metapopulation model, where local populations exist in a patchwork of suitable and unsuitable environments. Each of them eventually becomes extinct, but meanwhile can send out colonists to start populations in unoccupied patches. An equilibrium would be reached at:

$$p^* = 1 - e/c, \tag{4.10}$$

where p is the proportion of suitable sites that are in fact inhabited, e is the extinction rate and c the colonization rate (Levins, 1969, 1970). This relationship gives us several useful observations.

First, near the boundary of a distribution, p^* is very small; therefore, e/c is close to 1. Suppose it is 0.99 so that 1 % of the suitable sites are occupied. Now let the biological parameters change by only some 1 % to 0.98. Then 2 % of the suitable sites are occupied. A very small change, perhaps not detectable at all by our methodology, can produce a very big change in the proportion of sites occupied. But near the centre of a distribution when e/c is say 0.5, half the sites are occupied and a 50 % decrease to 0.25 increases occupancy to 0.75, only a 50 % increase. The sites might be pools where mosquitoes can breed, people susceptible to malaria, or bits of woodland where white pine can grow. The extinction rate, e, can refer to pools drying, people being cured, or landslides, insects or developers wiping out a stand of trees. The changes may be in the distance between ponds, the time people spend outdoors or the temperatures that affect reproduction of herbivores. The general rule is that when an outcome is the small difference between large numbers, then small proportional changes in parameters can give very big changes in outcome. We may not be able to explain why one village has twice as much malaria as another, why Massachusetts supports more stands of white pine than Connecticut, or why the boundary of a species is in middle Vermont. But, we can understand that at the boundary small differences have big effects and that we can intervene by acting on those parameters.

If we plot $1 - e/c$ against e, we get the linear relationship shown in Fig. 4.9. Then if e varies, being above its mean value in some places and below in others, the proportion of sites occupied is not changed. In a linear relationship, the mean of a function is the function of the mean. But with c the situation is different. If c is below e then there is no occupation (p^* is negative). Then p^* increases at a diminishing rate with c and levels out at 1 when c becomes infinite (instant colonization, so all sites are occupied). The curve is convex upward (Fig. 4.10). Therefore, if there is some mean value of c with variation above and below, there is more decrease when c is below average than there is increase when c is above average, so that the average value of p^* is less than p^* at the average c.

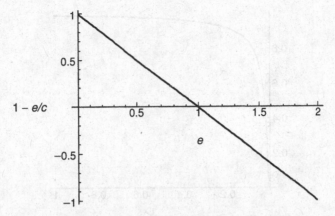

Fig. 4.9 The linear relationship between patch occupancy at equilibrium $(1 - e/c)$ and extinction rate (e) in a metapopulation, assuming that the relative rate of colonization (c) remains constant $(c = 1)$.

On the other hand, using an equation for the equilibrium level of the heartwater disease of cattle, Yonow *et al.* (1998) proposed a model that gives an equilibrium level of the proportion of cows that are infected:

$$x^* = (AC - mr)/A(C + r), \tag{4.11}$$

where x is the proportion of infected cows, r is the rate of removal of cows from a herd either by natural death or culling, m is the death rate of ticks, C the rate of infection of cows by the rickettsia and A the rate of acquisition of the infection by the ticks from an infected cow. Here we examine x^* as a function first of r and then of C (Levins, 2000). We can see from Fig. 4.11 that x^* is a concave upward decreasing function of r. Therefore, if r is not uniform but varies among farms, the average infection rate will be greater than the infection rate at average r. Thus, an

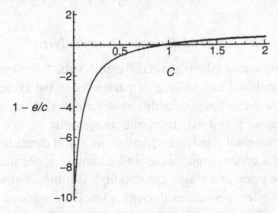

Fig. 4.10 The nonlinear relationship between patch occupancy at equilibrium $(1 - e/c)$ and colonization rate (c) in a metapopulation, assuming that the relative rate of extinction (e) remains constant $(e = 1)$.

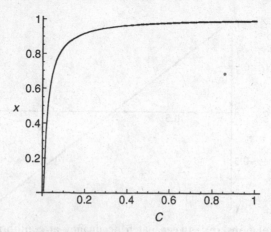

Fig. 4.11 The effect of varying the rate of bovine infection by ticks (C) on the equilibrium prevalence of heartwater disease in cattle (Yonow *et al.*, 1998, Levins, 2000) as determined by $x^* = (AC - mr)/A(C + r)$, where x is the proportion of infected cows, r is the rate of removal of cows from a herd either by death or culling, m is the death rate of ticks, C the rate of infection of cows by the rickettsia and A the rate of acquisition of the infection by the ticks from an infected cow ($m = 0.02$, $r = 0.01$, $A = 0.02$).

intervention programme to reduce regional infection will get best results if it acts to reduce both the mean and the variance of r by starting improvement where r is greatest (the worst off farms). But x^* is a convex upward function of C (Fig. 4.11), increasing from 0 when $AC = mr$ to 1 as C increases. Here the average x^* is less than x^* at the average C. Therefore, regional infection will decrease fastest if we reduce C but increase the variance, by building on the best. This result does not depend on the exact form of the equations but only on the concavity or convexity, the sign of the second derivative d^2x^*/dq^2 where q is any parameter.

4.4.2 Signed digraphs (loop analysis)

Environmental changes may take place at different levels. A minimum environmental change can be analysed as a change of parameter of the system. More radical changes may include the removal or addition of variables, changes in links among them, or the coupling of previously independent systems.

When any environmental condition changes, its initial direct impact may be on only one variable of a system (one species in a community, one molecule in a complex physiology, the price of a single commodity). But then that variable interacts with others, and the effect percolates through a whole network of interacting variables. Along some pathways the effect is nullified, along others enhanced, and along still others it may be reversed and we get counterintuitive outcomes. The method of signed digraphs is used to determine the direction of change of the equilibrium values of any variable in response to changes in any parameter (Puccia & Levins, 1985). It can be used as follows:

(i) To predict the outcomes if we know the network and the parameter change.
(ii) To identify the variable at which the parameter change enters the system, if we know the network. This is particularly useful in environmental assessment where some variable of interest has been altered (fish catch declines, a disease becomes more common, populations of algae explode). However, it is not obvious whether the change occurred because of a change of climate, reduced mineral run-off into the water, or over-fishing. Then we can analyse the outcomes under the assumptions of environmental inputs to each of the interconnected variables. It is not necessary to know the sign of the input: the table would have all predictions reversed if the sign of the input is reversed, but still tells us which variables change in the same or opposite direction. Each variable in the model increases, decreases or remains unchanged. This by itself is weak information, but for n variables there are n^2 outcomes predicted (the response of the equilibrium level of each variable to changed parameters entering the system at each variable). Even if only a fraction of these are observable, the set of outcomes as a whole supports strong statistical inference.
(iii) To compare the outcomes of alternative models, if we know only part of the network structure (e.g. the feeding relationships among species) but suspect other possible effects. Where the models agree on outcome, observation is a test of the whole set of models. Where they differ, the observations allow us to identify the best model, and if the outcomes are inconsistent with any of the models then the whole set of models is rejected and new ones are needed.
(iv) Experimental design: intervention in a system consists of providing an input to a variable, removing a variable (by holding its value constant or above some threshold so that it ceases to be a variable of the system even though it remains present physically) or changing the linkages among variables (inhibiting or severing a pathway, or adding a pathway through some monitor that then determines an input). The design gives us a new model or set of models that can then be studied for objectives (i)–(iii).

In Fig. 4.12, we present a model that helps us understand the counterintuitive result of a well-meant intervention to control malaria. The vector (**V**), its natural enemies (**N**) and pesticides (**P**) are linked indicating either a direct positive effect (right arrow) or a negative one (left arrow). The figure depicts how the existence of vectors prompts the application of pesticides, which in turn directly reduces mosquitoes. **V** and **P** are linked by negative feedback. The vectors and the natural enemies are linked by a similar loop. Note that the pesticide has a direct negative effect on both the vector and the natural enemies, and that the pesticide and the vector are both self-damped. An impact from the outside on a particular variable will percolate to the others through the network of interactions.

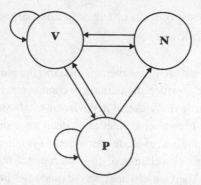

Fig. 4.12 The vector (V), its natural enemies (N) and pesticides (P) are linked with arrows indicating either a direct positive effect (→) or a negative one (←). V and P are self-damped. The figure shows how the existence of vectors prompts the application of pesticides which in turn reduce mosquitoes. V and P are linked with a negative feedback. The vectors and the natural enemies are linked by a similar loop. Note that the pesticide has a negative effect on both the vector and the natural enemies. An impact from the outside on a particular variable will percolate to the others through the network of interactions.

Carrying out a qualitative mathematical analysis of such a system will result in a table summarizing the perturbation that each of the variables will have on the others. The precise methodology of such an analysis and its mathematical basis are described in detail in Levins' seminal work (Levins, 1974, 1975), and its applications for the study of ecosystems in Levins (1998). Constructing a table (see below) for the pesticide example enabled us to examine the effect that adding pesticides to the system has on the vector, the natural enemies and the pesticide itself. We see from the last row that adding pesticides will result in an increase in the vector population, in a reduction of its natural enemies and in an increase in pesticides polluting the environment.

With the aid of such a table, we sometimes discover that the outcome can be zero at the point of impact, but strongest at some remote point in the system; thus we see from the first row that a direct positive impact on the vector actually has no effect on its equilibrium value, while an increase in its natural enemies will have the desirable effect of reducing the vector (second row). Also, a qualitative analysis of complex systems allows us to discover how positive feedback can reverse the expected outcome (Puccia & Levins, 1985).

	V	N	P
V	0	+, −	+
N	−	+	−
P	+	−	+

We learn that, in attempting to control infectious diseases, failures to consider complex interactions within an ecological context may lead to the resurgence of these diseases. An example is the malaria-control programme implemented in 1950s by WHO. Malaria came back with a vengeance following the worldwide application of pesticides, supposedly to eradicate the vectors that transmit the disease. A complex system analysis would have revealed the consequences of such a programme. Thus, in assessing the possible consequences of global change on a particular variable, the analysis should be carried out within an ecological context, also taking into account other variables that might affect the system.

4.4.3 Time averaging

The techniques of signed digraphs are used to study the behaviour of systems near equilibrium and to examine the changes in the equilibrium values. If the system is not at equilibrium then the methods of time averaging are more suitable. The time course of variables is summarized by measures analogous to the familiar statistical measures of mean, variance and covariance.

The basic principle is that if a variable is bounded then its derivative has a long-term average value that approaches zero (Puccia & Levins, 1985; Levins, 1998).

Time averages are used to predict the pattern of correlations among the variables of a system when the environmental change is not a shift of parameter to some new value, but a continuous perturbation. It can also determine whether a system reaches an equilibrium or permanently oscillates.

Time averaging is based on the following theorem. If the equation:

$$dx/dt = f(x, y, \ldots) \tag{4.12}$$

has only bounded solutions (or solutions that increase more slowly than t), then the average value of dx/dt approaches 0:

$$E\{dx/dt\} = 0. \tag{4.13}$$

This equation applies to any bounded function of x.

Then time averaging can be used to find the correlated responses of variables to variable parameters.

Consider a resource/consumer model in which there is a single resource, a single consumer and two predators of the consumer, one self-damped and the other not, as shown in Fig. 4.13. The equations might be:

$$dR/dt = a - R(pH + C), \tag{4.14}$$

where R is the resource, H the primary consumer, a the rate at which resource enters the system, p the rate of utilization by H, and C the rate of removal of the

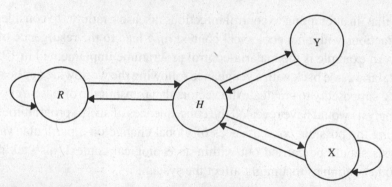

Fig. 4.13 A resource (R), a primary consumer (H) and two predators (X and Y) are linked with arrows indicating either a direct positive effect $(\rightarrow)$ or a negative one $(\leftarrow)$. R and X are self-damped. X does not influence the average level of H. See text for explanation of controlling factors using the method of time averaging.

resources by all other processes. "Removal" might mean a change of physical form, movement to somewhere else, or any other process that makes it unavailable to the primary consumer. If R is the product of photosynthesis made by a plant that is circulating in the sap, its removal may mean incorporation into the plant structure or consumption by herbivores not included in the model.

$$dH/dt = H(pR - m_1 - q_1x - q_2y), \qquad (4.15)$$

where q_1 and q_2 are mortality due to predators x and y and m_1 is the death rate due to all other causes.

$$dx/dt = x(q_1H - m_2 - x), \qquad (4.16)$$

and

$$dy/dt = y(q_2H - m_3). \qquad (4.17)$$

In this model, x is assumed to be self-damped and y not self-damped. The correlation pattern among the variables depends on which parameter varies. Suppose that m_3 varies and all other parameters are constant. Then we can begin the analysis at equation (4.14), the resource level. Taking expected values, we have:

$$a - E\{R\}E\{pH + C\} + p\text{Cov}(R, H) = 0. \qquad (4.18)$$

Dividing by $E\{R\}$ gives:

$$a/E\{R\} = E\{pH + C\} + p\text{Cov}(R, H). \qquad (4.19)$$

Since R is bounded away from 0 we can divide the equation by R. $(dR/dt)/R$ is $d(\log R)/dt$ so that it also has an expected value of zero.

Therefore:

$$aE\{1/R\} = E\{pH + C\}. \tag{4.20}$$

For any variable $R > 0$, $E\{1/R\} > 1/E\{R\}$.

Therefore:

$$E\{pH + C\} > E\{pH + C\} + p\mathrm{Cov}(R, H), \tag{4.21}$$

so that $\mathrm{Cov}\,(R, H) < 0$.

Thus, if the system is driven by environmental variation entering above the level of R, H will be negatively correlated with its resource.

R may either be the harvestable resource, or turn into the harvestable resource. Therefore, we are interested in a large average value for R. The average value of R is:

$$E\{R\} = [a - p\mathrm{Cov}(R, H)]/E\{pH + C\}. \tag{4.22}$$

The average value of H is found by dividing equation (4.17) by y and averaging:

$$E\{H\} = E\{m_3\}/q_2. \tag{4.23}$$

Note that $E\{H\}$ is determined completely by the parameters of y and that, surprisingly, variabilities do not affect the result. Since $\mathrm{Cov}(R, H) < 0$, $E\{R\}$ is increased by the covariance of R with H. Therefore, variation in m_3, a parameter of y, increases the variation of all variables, and in particular increases $E\{R\}$. x does not influence the average level of H. Equation (4.16) can be averaged to get:

$$q_1 E\{H\} = m_2 + E\{x\} \tag{4.24}$$

and

$$q_1 \mathrm{Cov}(H, x) = \mathrm{Var}\{x\} > 0. \tag{4.25}$$

4.5 Summary

As we asserted at the outset, surprises are inevitable in predicting the outcomes of scientific investigations. However, when we look at global environmental change we can minimize the hazards of surprises by making The Whole the focus of our investigation. Thus, this chapter is intended to promote methodologies that help us look at the effect of change on life complexities within an ecological context. It is not what global warming will do to the world as a whole but rather what it will do to each of the world's ecological systems, each with its particular characteristics.

Acknowledgements

Some of the ideas presented in this chapter were developed with the support of the Robert Wood Johnson Foundation Investigator Award.

References

Awerbuch T. E. (1994). Evolution of mathematical models of epidemics. In *Diseases in Evolution*, ed. M. Wilson, R. Levins & A. Spielman, pp. 225–31. New York City, NY; New York Academy of Sciences.

Awerbuch, T. E. & Sandberg, S. (1995). Trends and oscillations in tick population dynamics. *Journal of Theoretical Biology*, **175**, 511–16.

Awerbuch, T., Brinkmann, U., Eckardt, I., Epstein, P., Ford, T., Levins, R., Makhoul, N., Possas, C. A., Puccia, C., Spielman, A. & Wilson, M. (The Harvard Working Group on New and Resurgent Diseases) (1996). Globalization, development and the spread of disease. In *The Case Against the Global Economy*, ed. J. Mander & E. Goldsmith. San Francisco, CA: Sierra Club Books.

Coakley, S. M. & Scherm, H. (1996). Plant disease in a changing global environment. In *Implications of Global Environmental Change for Crops in Europe. Aspects of Applied Biology*, **45**, 227–38.

Doswell, C. A. III, Anderson, L. C. & Imy, D. A. (1991). *Basic Convection I: a Review of Atmospheric Thermodynamics*. Norman, OK: NOAA-NWS.

Kitron, U. & Kazmierczak, J. J. (1997). Spatial analysis of the distribution of Lyme disease in Wisconsin. *American Journal of Epidemiology*, **15**, 558–66.

Levins, R. (1969). Some demographic and genetic consequences of environmental heterogeneity for biological control. *Bulletin of the Entomological Society of America*, **15**, 237–40.

Levins, R. (1970). Extinction. In *Some Mathematical Problems in Biology*, ed. M. Gesternhaber, pp. 77–107. Providence, RI: American Mathematical Society.

Levins, R. (1973). The limits of complexity. In *Hierarchy Theory – the Challenge of Complex Systems*, ed. H. H. Pattee, pp. 109–27. New York City, NY: George Braziller.

Levins, R. (1974) The qualitative analysis of partially specified systems. *Annals of the New York Academy of Sciences*, **231**, 123–38.

Levins, R. (1975). Evolution in communities near equilibrium. In *Ecology and Evolution of Communities*, ed. M. Cody & J. Diamond. Harvard: Harvard University Press.

Levins, R. (1995a). Preparing for uncertainty. *Ecosystem Health*, **1**, 47–57.

Levins, R. (1995b). Rethinking the causes of disease. Review of Mark Lappé's "*Evolutionary Medicine: Rethinking the Origins of Disease*". Sierra Club Books. *Chemical and Engineering News*, 37–38, June 19.

Levins, R. (1998). Qualitative mathematics for understanding, prediction and intervention in complex ecosystems. In *Ecosystem Health: Principles and Practice*, ed. D. Rapport, R. Costanza, P. Epstein & R. Levins. Oxford: Blackwell.

Levins, R. (2000). Models for regional heartwater epidemiology in a variable environment. *Onderstepoort Journal of Veterinary Research*, **67**, 163–5.

Levins, R. & Lane, P. (1977). The dynamics of aquatic systems. 2. The effects of nutrient enrichment on model plankton communities. *Limnology & Oceanography*, **22**, 451–7.

Martens, P. (1998). *Health and Climate Change: Modeling the Impact of Global Warming and Ozone Depletion*. London, UK: Earthscan Publications.

McMichael, A. J. (1997). Integrated assessment of potential health impact of global environmental change: prospects and limitations. *Environmental Modeling and Assessment*, 1–9.

Pattee, H. H. (1973). *Hierarchy Theory. The Challenge of Complex Systems*. New York: George Braziller.

Pianka, E. R. (1970). On r- and K-selection. *American Naturalist*, **104**, 592–7.

Puccia, J. C. & Levins, R. (1985). *Qualitative Modeling of Complex Systems*. Cambridge, MA: Harvard University Press.

Sandberg, S., Awerbuch, T. E. & Spielman, A. (1992). A comprehensive multiple matrix model representing the life cycle of the tick that transmits the agent of Lyme disease. *Journal of Theoretical Biology*, **157**, 203–20.

Schmalhausen, I. I. (1949). *Factors of Evolution*, p. 276. Philadelphia, PA: Blakiston Company.

Toffoli, T. (1984). Cellular automata as an alternative to (rather than approximation of) differential equations in modeling physics. *Physica D*, **10**, 117–27.

Wagner, T. L., Wu, H., Sharpe, P. J., Schoolfield, R. M. & Coulson, R. N. (1984). Modeling insect development rates: a literature review and application of a biophysical model. *Annals of the Entomological Society of America*, **77**, 208–25.

Yonow, T., Brewster, C. C., Allen, J. C. & Meltzer, M. I. (1998). Models for heartwater epidemiology: practical implications and suggestions for future research. *Onderstepoort Journal* of *Veterinary Research*, **65**, 263–73.

5

Epidemiological and impacts assessment methods

KRISTIE L. EBI & JONATHAN A. PATZ

5.1 Introduction

Future global environmental exposures may be significantly different from those experienced in the past. Forecasting and preparing for the resultant potential ecological, social and population health impacts requires innovative and interdisciplinary research approaches, both to advance global change/health science and to contribute to informed policy decisions. These approaches include empirical analyses and scenario-based exposure modelling to achieve meaningful risk assessments of the potential impacts of climate and ecological changes. This chapter focuses on the application of epidemiology (an empirically based discipline) to understanding the potential health consequences of global environmental change. The empirical knowledge gained from epidemiological studies should be used iteratively with model development to strengthen the foundation of predictive models.

Epidemiological research can be used in the three domains introduced in Chapter 1: first, historical analogue studies to help understand current vulnerability to climate-sensitive diseases (including contributions to understanding the mechanisms of effects) and to forecast the health effects of exposures similar to those in the analogue situation; second, studies seeking early evidence of changes in health risk indicators or health status occurring in response to actual environmental change; and third, using existing empirical knowledge and theory to develop empirical-statistical or biophysical models of future health outcomes in relation to defined scenarios of change. This chapter discusses some standard epidemiological methods used to generate quantitative estimates of exposure–disease associations for studies in these three domains. The examples focus primarily on climate variability and change to maintain consistency throughout the discussion.

Climate change may be associated with many health outcomes; for example, Fig. 5.1 (Patz *et al.*, 2000). Most of these outcomes are expected to be adverse, although exceptions include decreased winter mortality in more northern latitudes.

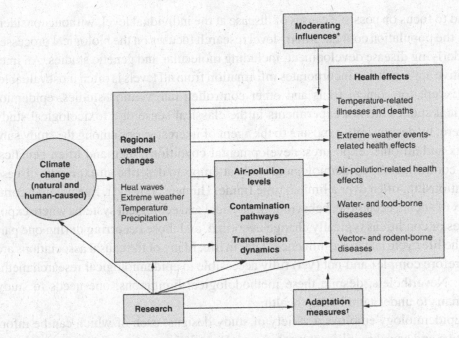

Fig. 5.1 Schematic of potential health effects (and moderating factors) of climate variability and change. Source: Patz *et al.* (2000). *Moderating influences include nonclimate factors that affect climate-related health outcomes, such as population growth and demographic change, standards of living, access to health-care, improvements in health-care and pub lic health infrastructure. †Adaptation measures include actions to reduce risks of adverse health outcomes, such as vaccination programmes, disease surveillance, monitoring, use of protective technologies, use of climate forecasts and development of weather warn ing systems, emergency management and disaster preparedness programmes, and public education.

5.2 Epidemiological study designs

Epidemiology is "the study of the distribution and determinants of health-related states or events in specified populations, and the application of this study to control of health problems" (Last, 1995). The purpose of epidemiology is to increase the health and well-being of populations through discovery of causative agents of ill-health and through discovery of interventions to minimize ill-health and suffering. As such, epidemiology is one of the core disciplines of public health.

Epidemiological studies can be undertaken at various levels: population, individual, or the molecular level, depending on how the research question is framed (Pearce, 1999). Population-level studies can involve comparisons between or within populations, or contain a mix of both population- and individual-level information. The key feature of population-level studies is that they start by considering the distribution and determinants of disease at a population level. Individual-level studies tend to focus on possible causes of disease at the individual level, without considering the population context. Micro-level research focuses on the biological processes underlying disease development, including molecular and genetic studies. An integrative approach that incorporates information from all levels is often most valuable.

Except for clinical trials and other controlled intervention studies, epidemiological studies are not experiments in the classical sense of a toxicological study where, by design, only exposure to the agent of interest varies among the study subjects and all other exposures, developmental conditions, diet, and often genetics, are controlled. An epidemiological study attempts to describe an exposure–disease relationship, often over a limited time frame. Humans, however, live within a complex of social, economic, behavioural, physical and ecological systems where exposures or conditions typically change over time, and those occurring during one part of the life cycle may have influences later in life. Many of the causal associations are therefore complex and not (yet) fully accessible to epidemiological research methods. Nevertheless, despite these methodological limitations, one needs to study humans to understand human health.

Epidemiology employs a variety of study designs, each of which can be informative and each has inherent weaknesses. Descriptive studies are the starting point for epidemiological analyses of a new adverse health event, whether a newly recognized disease or an existing disease in a new location. Such studies describe the distribution of a disease in time and space. Who gets the disease? When does the disease occur? Are there seasonal patterns suggestive of particular transmission dynamics? Where does the disease occur? What does the geographical pattern of disease occurrence suggest about possible causes? Are there other patterns that are informative for understanding the disease?

One approach to understanding the data gathered to answer these questions is to map any spatial and temporal variability. Since the days of John Snow (who, by mapping water distribution in London, discovered that drinking contaminated water caused cholera), epidemiologists have used maps to look for clues to the causes of a disease. In addition to simple maps, Geographical Information Systems are now a valuable tool for studying the distribution of a disease (see Chapter 9).

Study designs commonly employed to describe the distribution and determinants of disease include population-level or ecological studies, and individual-level studies including cross-sectional studies, surveys, and case–control studies.

5.3 Population-level or ecological studies

In ecological studies, the unit of study is the population (or some subdivision thereof). (The terminology is actually inappropriate as these studies have nothing to do with ecology or ecosystems.) In ecological studies, the frequency of disease is compared across different populations and its correlation with the population prevalence or level of some putative risk factor is determined. Ecological studies are often considered to be primarily hypothesis-generating studies because data are collected at the population level. The logical problem is that when individual-level factors, such as cholesterol levels, are collected at the population level there can be uncertainty as to whether any observed association (or lack of association) with some specified health outcome represents what actually happens at the level of the individual. Do the individuals with high or low blood cholesterol have heart attacks? One cannot tell from the population-level data. The extrapolation from population level to individual level is referred to as the "ecological fallacy".

Some exposures, however, are best measured at the population level. Examples include water fluoridation level, vaccine coverage, extreme weather events, or longer-term climate change. In addition, as noted by Pearce (2000), ecological studies are important because some risk factors operate essentially at the population level. For example, having a relatively low income in an area with high income neighbourhoods may be more detrimental than having the same income in an area with lower income neighbourhoods, possibly because of differential access to health-care and other resources (Yen & Kaplan, 1999).

An example of an ecological study is one by Ebi *et al.* (2001). The authors described and compared the associations between certain weather variables and hospitalizations for viral pneumonia, including influenza, during normal weather periods and El Niño events for girls and women in California over a 15.5-year period. One temperature variable could not describe the hospitalization patterns found during normal weather periods in the three geographical regions. In San Francisco and Los Angeles, a 5 °F decrease in the minimum temperature was associated with a large increase in hospitalizations (about 30–50 %). In Sacramento, a 5 °F decrease in the maximum temperature difference was associated with a large (about 25–40 %) increase in hospitalizations. Only in Sacramento did seasonal temperatures change during El Niño events, as did the pattern of viral pneumonia hospitalizations.

These results underscore the difficulties in trying to understand the potential health effects of climate variability. The associations between viral pneumonia hospitalizations and specific weather factors varied across the geographical regions. Developing a model based on either the inland or one of the coastal regions would not have been predictive for the other regions.

In another example, a community-based incidence survey assessed the impact of the construction of microdams on the incidence of malaria (90 % *Plasmodium falciparum*) among children in northern Ethiopia (Ghebreyesus *et al.*, 1999). Four quarterly cycles of malaria incidence surveys were undertaken in eight at-risk communities close to dams and in eight control villages at similar altitudes but beyond the flight range of mosquitoes (primarily *Anopheles arabiensis*). The results showed that this irrigation development programme led to increased malaria transmission over a range of altitudes and seasons. The overall incidence of malaria for villages close to dams was 14.0 episodes/1000 child months at risk, compared with 1.9 in the control villages. These results emphasize the need for health issues to be considered prior to the implementation of environmental development programmes.

Under climate change projections, increased extremes in the hydrological cycle are expected to affect the incidence of water-borne disease. Curriero *et al.* (2001) examined the association between extreme precipitation and water-borne disease outbreaks in the U.S. From 1948 to 1994, 548 reported outbreaks were analysed, along with total monthly precipitation. Analyses were conducted at the watershed level, including outbreaks due to both ground-water and surface-water contamination, and were further stratified by season and hydrological region. Of the total number of outbreaks, 51 % were preceded by a precipitation event above the 90 % percentile and 68 % above the 80 % percentile ($p < 0.001$). Outbreaks due to surface-water contamination were associated with extreme precipitation during the month of the outbreak, while those due to ground-water contamination had the strongest association with extreme precipitation two months prior to the outbreak.

Ecological studies are providing valuable information on how the distribution and rates of health outcomes vary in some geographical regions during El Niño events (discussed in more detail in Chapter 6). Studying what has happened during El Niño events provides information on the sensitivity of diseases to specific changes in weather patterns. El Niño events are "natural experiments" with worldwide impacts. Also, although there is uncertainty as to whether the frequency and magnitude of El Niño events will change with increasing global mean temperatures, understanding the potential consequences of today's climate variability would provide insights into future El Niño events. Until recently, there has been little research on the potential health impacts of El Niño events (Kovats *et al.*, 1999). As further outlined in Chapter 6, understanding these relationships is useful for developing current public health responses, for evaluating population vulnerability to current and future events, and for designing adaptation measures.

The study by Bouma and van der Kaay (1996) analysed historical malaria epidemics using an El Niño Southern Oscillation Index, then predicted high-risk years for malaria on the Indian subcontinent. There was an increased relative risk of 3.6 for an epidemic in El Niño years in Sri Lanka and a relative risk of 4.5 for a post El Niño year in the former Punjab province of India. Subsequently, associations were reported between El Niño events and malaria in other parts of the world, including South America (Bouma & Dye, 1997; Bouma *et al.*, 1997), thus improving the predictability of malaria epidemics.

The incidence of mosquito-borne Rift Valley fever occurs in association with El-Niño-driven flooding in East Africa, as demonstrated by the serious outbreak in Kenya during the strong El Niño of 1997–98 (Linthicum *et al.*, 1999). The mosquitoes that transmit Rift Valley fever lay their eggs at the tops of grasses. Only during periods of flooding are the eggs submersed, allowing for development; transovarian viral transmission can then render the emerging mosquitoes infectious. Therefore, potential epidemics can occur following flooding events.

5.4 Individual-level studies

Study designs used by epidemiologists to investigate individual-level determinants of disease include cross-sectional, case–control and cohort studies. Each of these designs examines the relationship between the occurrence of exposure to the occurrence of disease (cases). Because the situations are limited in which it is ethical to expose humans to suspected disease-causing agents, epidemiologists try to identify and take advantage of naturally occurring situations to study disease aetiology.

5.4.1 Cross-sectional studies and surveys

Cross-sectional studies collect data simultaneously on both the health outcome and the exposure(s) of interest at the level of the individual. Cross-sectional studies are designed to answer particular questions, while surveys describe the individual-level distribution of exposure, disease or other health-related index within a population. Because exposure and disease data are collected simultaneously in cross-sectional studies (thus taking a "snapshot" in time), it may be difficult to infer disease causality because of the lack of information on the temporal relationship between exposure and disease outcome. In addition, for studies of climate (or other environmental changes) there are few opportunities to differentiate exposures at the individual level.

5.4.2 Case–control studies

Case–control studies begin with individuals (cases) who have the disease. An appropriate comparison group of individuals who do not have the disease (controls) is then identified. The study determines whether the two groups differ in the proportion of individuals who were exposed to the factor of interest. Case–control studies are frequently used to study rare diseases or diseases with long latency periods. Advantages are that they are relatively inexpensive (compared with cohort studies) because generally fewer individuals are studied. In addition, results can be obtained relatively quickly. Case–control studies also are suitable when studying a disease with many possible causal exposures to be investigated.

Two potential disadvantages often arise in case–control studies, from uncertainty regarding the quality of exposure assessment and from whether the controls were appropriate. In the first instance, diseased individuals may have spent time thinking about which exposures could explain their condition. When asked about previous exposures, they may remember exposures more clearly than individuals without the disease, or may remember exposures as occurring at points in time different from when they actually happened. When such differences in recall arise between cases and controls, the resultant recall bias may explain at least part of any observed association. The second concern is about the choice of controls. The goal is to choose controls that are representative of the population from which the cases arose (Rothman & Greenland, 1998), after allowing for any matching deemed necessary to reduce/control confounding (see also below). When this happens, the results can be generalized to that population. If the controls are not representative, then the question arises as to whether any association observed was due to the exposure under study or was due to another difference between the cases and controls. For example, in the U.S., controls are often identified from a telephone solicitation process called random digit dialing. However, individuals who own a telephone, answer the telephone and agree to participate in a study are not representative of the total population (they tend to have higher socioeconomic status, among other differences) (Lasky & Stolley, 1994).

5.4.2.1 *Hantavirus and El Niño: a case–control design linking climate, ecology and infectious disease through the use of satellite remote sensing*

In the spring and summer of 1993, an outbreak of acute respiratory distress with a high fatality rate among previously healthy individuals was recognized in the Four Corners region of the Southwest United States. The disease, hantavirus pulmonary syndrome (HPS), was traced to infection with a previously unrecognized, directly transmissible hantavirus. The virus (Sin Nombre virus, SNV) was found to be maintained and transmitted primarily within populations of a common native field rodent, the deer mouse. Transmission to humans is thought to occur through contact with secretions and excretions of infected mice.

A case–control epidemiological study by Glass *et al.* (2000) tested the "trophic cascade" hypothesis for the potential link between El-Niño-driven heavy rainfall, regional ecological change and human disease. The hypothesis is:

(i) Changes associated with El Niño alter weather patterns in the Southwest U.S.
(ii) Increased winter–spring precipitation increases vegetation and insect populations.
(iii) Increases in food and shelter increase the size/density of rodent populations.
(iv) Increases in density alter the quantity/quality of SNV infection in *Peromyscus* (deer mouse) populations.
(v) Return of "normal weather patterns" leads to increased contact with humans.
(vi) Increased human/rodent contact leads to human infection and subsequent disease.

In this case–control study, the locations of 31 households with a case of HPS and 170 randomly selected control households were identified using a global positioning system. Landsat Thematic Mapper satellite imagery was obtained of the study area for the year prior to the outbreak. Satellite reflectance values at 17 case and

36 control households in a 12 200 km^2 training area were used to generate a landscape risk classification, which was used to test the predictability of the remaining case and control households. Three satellite wavelength bands and topographical elevation were found to be significantly correlated with the risk of HPS. When the probability of disease predicted by the model was 25 % risk or greater, 100 % of case households and 69 % of control households were correctly predicted. Case households tended to be at higher elevations with reflectance values indicative of locally higher levels of soil moisture and more green vegetation, which would be expected to support and benefit the rodent population as hypothesized.

This model was further validated with satellite images during the 1997–98 El Niño and interim non El Niño year of 1995. Predicted risk by satellite-derived risk classification and the number of human cases of HPS were both low in 1995, and elevated again in 1998. Further "ground truthing" of this risk model showed a very high predictability of rodent population (correlation factor = 0.93) following a rodent-trapping field validation study (unpublished data). This type of predictive modelling based on case–control studies and use of satellite remote sensing can apply to other diseases affected by changes in climate or ecology.

5.4.3 Cohort studies

Another study design for evaluating the individual-level determinants of a disease is the cohort study. This observational design best mirrors the logic of the classical experimental design in toxicology. Cohort studies begin with individuals free of disease, some of whom are exposed to the risk factor of interest. The cohort is followed over time to determine the incidence of disease both in the exposed and unexposed individuals. Occupational settings are particularly suitable for these types of studies, where exposures can often be readily identified and the workers can be systematically followed. Cohort studies are often the preferred study design because information about exposure is collected prior to the development and diagnosis of disease. Also, it may be easier to identify an appropriate comparison group within the cohort. Another advantage is the possibility for more accurate exposure and disease assessments. The major disadvantage is that these studies are generally much more expensive and lengthy to conduct. In addition, a cohort study is not efficient for studying rare diseases.

Cohort studies on the effects of lead exposure in children have had significant environmental policy implications worldwide. These studies revealed that low-level lead exposure can adversely affect the developing nervous system. A study by Needleman *et al.* (1979) on lead's impact on mean verbal Intelligence Quotient (IQ) in children showed an inverse relationship between lead concentration in dentine from collected deciduous teeth and IQ in later childhood. When followed to young adulthood, children with the highest dentine lead levels were much less likely to graduate from high school and suffered more reading disabilities, deficits in vocabulary, problems with attention and fine motor coordination, greater absenteeism, and lower class ranking compared with controls (Needleman *et al.*, 1990). In a prospective study from an Australian lead-smelting community, blood lead values obtained early in life (including prenatal) predicted IQ scores at the age of seven; a five-point difference in IQ associated with a 10 $\mu g/dl$ gradient in early-childhood blood lead level (Baghurst *et al.*, 1992). Lead levels between the ages of six months and 24 months proved most significant.

Several cohort studies have also been set up to follow children born to atomic bomb survivors in Hiroshima and Nagasaki, Japan. For example, persons who were exposed *in utero* at 8–15 weeks gestational age have shown a higher frequency of mental retardation (National Research Council, 1990).

5.5 Surveillance

Surveillance is not a formal research study design, but is a closely related function of public health. The goals of surveillance programmes are to monitor continuously the pattern of occurrence of specific health problems, to document their impact in defined populations, and to characterize affected individuals and those at greatest risk (Buehler, 1998).

The terms "active" and "passive" are used to describe the two approaches to surveillance. Active surveillance means that the public health agency initiates procedures to obtain case reports. Passive surveillance means that the agency leaves the initiation for reporting to others. Most surveillance systems are operated by public health agencies; they generally use the collected data to guide disease-prevention and control programmes. Surveillance can be important when seeking for early changes in the rate and range of weather-sensitive diseases. Targeting the most sensitive individuals and populations could increase the efficiency and effectiveness of such systems. For example, surveillance systems could be set up to monitor for changes in the range of malaria, dengue and other vector-borne diseases at the edges of their current ranges. Chapter 10 discusses this issue in detail. Surveillance also can aid in the detection of emerging and re-emerging diseases, such as West Nile virus in the eastern United States. Finally, surveillance programmes provide feedback on the success of interventions.

Surveillance alone is not sufficient to prevent illness, and continued efforts to develop predictive models and early warning systems should be a priority. Predictive models can be used to improve proactive health measures. As discussed later in this chapter and in depth in Chapter 6, while models cannot completely simulate real life, models are useful for conceptualizing dynamic processes and their outcomes. Modelling should iteratively combine the results from recent empirical studies to refine parameters. This two-pronged approach (empirical and modelling studies) is required to better understand the associations between climatological and ecological change as determinants of disease, and is a theme reinforced throughout this book.

5.6 Confounding and interaction

The epidemiological studies discussed above are not controlled experiments. Therefore, a reported association may not be due to the exposure, but instead may be due to another factor not accounted for in the analysis. This situation is termed "confounding". Confounding arises when such a factor is statistically associated with the exposure and is also an independent risk factor for the disease. In this situation, the causal influence attributed to the exposure may be partially or completely explained by the confounding factor. For example, if studying alcohol consumption, smoking should be considered as a potentially confounding factor for many chronic diseases because smoking is associated with alcohol and is a separate cause of a number of diseases. In another example, most studies of air pollutants control for the potential effects of weather in their models under the assumption that weather is a potential confounder. Studies by Hales *et al.* (2000) and Samet *et al.* (2000) suggest that weather and air pollutants are independent risk factors for daily mortality. Epidemiologists try to eliminate potential confounding, either in the way comparison populations are chosen or in the statistical analysis.

In addition to acting individually, a risk factor may modify the effect of another risk factor. Such "interaction" or "effect modification" can be either synergistic or antagonistic. For example, Sartor *et al.* (1995) reported that mortality during the Belgium heat wave of 1994 increased by 9.4 % in the ≤64 years age group (236 excess deaths) and by 13.2 % in the >65 years age group (1168 excess deaths). Daily death figures were mostly correlated with mean daily temperature and 24-hour ozone concentration from the previous day. A synergistic interaction between the effects of temperature and ozone on mortality was determined across age groups and explained 39.5 % of the variance for daily deaths in the >65 years age group. The authors concluded that elevated ambient temperatures combined with high ozone concentrations were likely to have been responsible for the unexpected excess mortality. A similar study by Piver *et al.* (1999) found that temperature and air pollution interactively affected the risk of heat stroke in Tokyo, Japan, although the results were not statistically significant.

5.7 Exposure assessment

Studying weather and climate variability offers unique challenges, in part because of the temporal and spatial variability in weather. Some methods used to analyse acutely varying exposures are discussed in the next section. Exposure assessment begins with incorporating a definition of the exposure of interest into the study

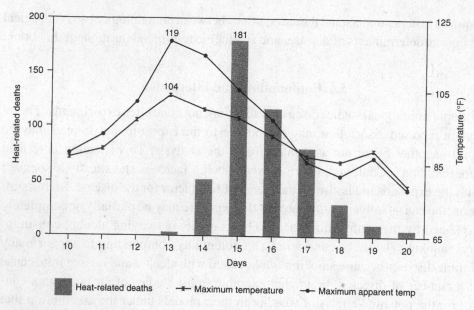

Fig. 5.2 Heat-related deaths and temperatures: Chicago residents, July 10 to July 20, 1995. Source: Whitman *et al.* (1997) reprinted with permission.

hypothesis. To use heat exposure as an example, "hot" needs to be defined and methods for measurement determined. Hot may be defined based on temperature alone or on a combination of temperature with other weather variables. For example, a study by Whitman *et al.* (1997) described the mortality in Chicago attributed to the July 1995 heat wave using the daily maximum temperature (Fig. 5.2). Another example is the previously mentioned study of the association between extreme precipitation and water-borne disease outbreaks (Curriero *et al.*, 2001). Extreme precipitation events were defined with Z-score thresholds; scores greater than 0.84, 1.28 and 1.65 corresponded to total monthly precipitation in the highest 20%, 10% and 5%, respectively.

A population's experience of weather is influenced by the physiological, behavioural and other adaptations of its members. For example, there are few standard definitions of a heat wave. The Netherlands Royal Meteorological Institute defines a heat wave as a period of at least five days, each of which has a maximum temperature of at least 25 °C, including at least three days with a maximum temperature of at least 30 °C (Huynen *et al.*, 2001). Many parts of the world would not consider such temperatures as a heat wave.

In addition, although weather varies on a day-to-day basis, there may be longer-term patterns that may affect human health. For example, a rainy spring followed the next year by a dry spring may lead to elevated pollen counts, resulting in more severe hay fever. When the goal is to seek for early health effects of climate change, the baseline climate has to be defined against which future changes will be measured. This is an important issue, as temperature and precipitation have been changing over at least the past 100 years (Patz *et al.*, 2000).

Once an exposure definition has been decided upon, questions may arise as to the appropriate lag between exposure and effect, and how long a time period after exposure deaths may be increased. Lag periods ranging from a few days to a year have been used, depending on the presumed underlying mechanism of effect. For example, the number of heat-related deaths in the July 1995 Chicago heat wave peaked about two days after temperatures peaked, and heat-related deaths continued to occur for almost a week (Whitman *et al.*, 1997). The medical examiner determined that a total of 514 heat-related deaths occurred during the month of July. The authors determined that there were 182 deaths in excess of those attributed by the medical examiner. This study encouraged medical examiners to adopt an explicit and standard definition of a heat-related death (Donoghue *et al.*, 1999).

In the viral pneumonia study of Ebi *et al.*, 2001, a 7-day lag was used, while the study of water-borne disease outbreaks used lags of 1 and 2 months (Curriero *et al.*, 2001), and the El Niño and malaria studies used 1-year lags (Bouma & Dye, 1997; Bouma *et al.*, 1997; Bouma & van der Kaay, 1996).

A related issue is mortality displacement. The length of time after an extreme temperature event for which the number of deaths may be increased is reportedly as long as a month for cold snaps (Huynen *et al.*, 2001). Knowing this time period can aid disentangling the proportion of deaths that would probably have occurred anyway and the proportion that were probably due to the weather event itself.

5.8 Analytical tools for epidemiological studies of global environmental change

A variety of analytical methods may be employed to study the relationship between the occurrence of disease or ill-health and the exposure of interest. A central research question is whether any differences among the study groups or any differences over time are statistically significant. The remainder of this chapter will focus on just a few methods of particular use for studying the health impacts of global environmental change.

5.8.1 Time series analysis

Time series analyses that are able to deal with acutely varying exposures have been used in a number of studies of short-term weather and climate variability. Time series analyses can take cyclical patterns into account when evaluating longitudinal trends in disease rates in one geographically defined population. This is important because many diseases follow seasonal patterns. However, seasonality alone does not prove a strong climate dependence. For example, calendar time (e.g. the school year) can confound the picture.

Two generally used approaches to time series analyses are generalized additive models (GAM) and generalized estimating equations (GEE). GAM have traditionally been used to evaluate associations between air pollution and hospital admissions (Zeger & Liang, 1986; Morgan *et al.*, 1988; Katsouyanni *et al.*, 1995). GAM entail the application of a series of semiparametric Poisson models that use smoothing functions to capture long-term patterns and seasonal trends from the data. The analyst has to choose the appropriate level of smoothing, which may introduce bias in the analysis by removing informative long-term or seasonal patterns from the original data. Normally, an optimization criterion is applied which determines the appropriate degree of smoothing.

The GEE approach is similar to GAM in the use of a Poisson regression model to estimate health events in relation to weather data. However, no *a priori* smoothing is performed for the time series. Instead, the GEE model allows for the removal of long-term patterns in the data by adjusting for overdispersion and autocorrelation. Autocorrelation needs to be controlled for in time series data of weather measurements because a particular day's weather is correlated with weather on the previous and subsequent days. Overdispersion may be present in count data (health outcomes) that are assumed to follow a Poisson distribution. When using GEE, the analyst does not have to choose the optimal degree of smoothing by running a number of smoothing applications to the data. The main disadvantage of the GEE approach is that it can be computationally prohibitive when dealing with large databases.

5.8.2 *Examples of other innovative statistical methods*

The unique natural complexities of weather and climate variability are leading to the development of innovative approaches to data analysis. The following are a few examples of approaches that are being used.

In the study of water-borne disease outbreaks (Curriero *et al.*, 2001), the goal of the analysis was to determine whether outbreaks clustered around extreme precipitation events as opposed to geographical clustering. A Monte Carlo version of Fisher's exact test was used to test for statistical significance of associations. The authors repeatedly generated sets of outbreaks in a random fashion, tabulating the percentage of these "artificial" outbreaks with extreme levels of precipitation at each step. This process produced a distribution of coincident percentages under the assumption of no association, which was then compared with the observed percentage to calculate a *p* value.

Samet *et al.* (2000) developed analyses to combine pollution/mortality relationships from many cities in the U.S. to gain a summary estimate of the health effects

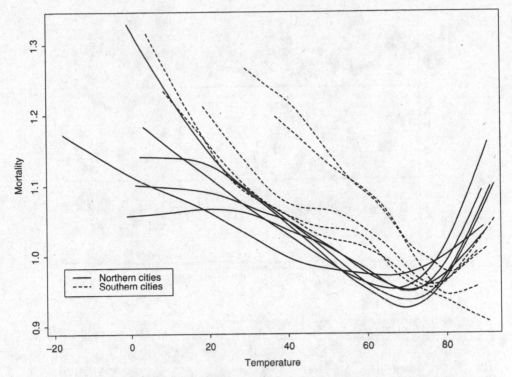

Fig. 5.3 The temperature–mortality relative risk functions for the 11 cities. Northern cities include Boston, Chicago, New York, Philadelphia, Baltimore and Washington DC. Southern cities include Charlotte, Atlanta, Taripa and Miami. Reprinted with permission from Oxford University Press (Curriero *et al.* (2002) *American Journal of Epidemiology*.

of air pollution. It was assumed that the estimated pollution effects for different cities would have a common distribution with a mean and variance structure. A hierarchical log-linear regression model was used with two stages. First, a log-linear regression of daily mortality was fitted on air pollution measurements and other confounders to estimate relative risks per city. The second stage used a Bayesian hierarchical model to combine relative risk across cities to determine an overall esti-mate, and to assess whether city population-specific characteristics were associated with the relative risk. In a similar study by Curriero *et al.* (2002) on temperature and mortality across 11 U.S. cities, a statistical bootstrapping resampling method was used for estimating variances across cities (Fig. 5.3).

Checkley *et al.* (2000) used the 1997–98 El Niño extreme event to assess the effects of unseasonable conditions on diarrhoeal disease in Peru (Fig. 5.4). The analytical methods of harmonic regression (to account for seasonality) and auto-regressive-moving average models (for Poisson time series data) were applied.

Hay *et al.* (2000) used spectral analyses to investigate periodicity in both climate data and in epidemiological time series data of dengue haemorrhagic fever (DHF) in Bangkok, Thailand. Dengue fever and DHF are mosquito-borne diseases caused

(a)

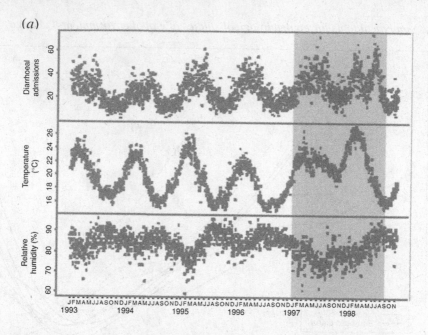

(b)

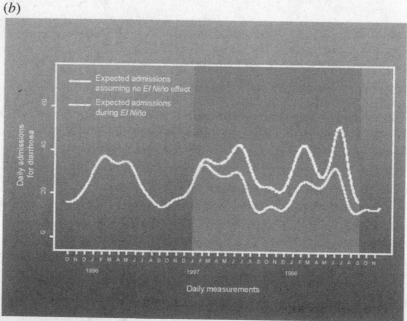

Fig. 5.4 (*a*) Daily time series between Jan 1, 1993, and Nov 15, 1998, for admissions for diarrhoea, mean ambient temperature, and relative humidity in Lima, Peru. Shaded area represents the 1997–98 El Niño event. (*b*) Observed versus expected daily admissions during the 1997–98 El Niño for childhood diarrhoeal disease to the Oral Rehydration Unit of the Instituto Nacional de Salud del Niño in Lima, Peru. Expected curve is based on five years of admission data prior to this El Niño event, utilizing methods of harmonic regression, semi-parametric curve fitting and autoregressive-moving averages. During the strong El Niño, observed hospital admissions more than doubled during the winter. Source: Checkley *et al.*, 2000. Reprinted with permission from Elsevier Science (*The Lancet* (2000) **355**, 442–50.)

by infection with dengue viruses of the family Flaviviridae. DHF exhibits strong seasonality, with peak incidences in Bangkok occurring during the months of July, August and September. This seasonality has been attributed to annual temperature variations. Spectral analysis (or Fourier analysis) uses stationary sinusoidal functions to deconstruct time series into separate periodic components. A broad band of two- to four-year periodic components was identified, as well as a large seasonal periodicity. One limitation of this method is that it is only applicable for stationary time series in which the periodic components do not change.

These methods provide a few examples of innovative statistical methods for analysing epidemiological studies of the health impacts of environmental variability and change. A general precaution is that the subtleties of such analyses can lead to spurious results if proper statistical tools are not used. Improved methods are still needed to fit time series data, as are methods to handle bivariate data (one series of counts and one of continuous data).

5.8.3 *Geographical Information Systems (GIS) and satellite remote sensing*

Geographic analysis of disease incidence is now recognized as an invaluable tool for the study of infectious diseases in the context of environment changes, as discussed in depth in Chapter 9. Risk prediction for many diseases has been enhanced using GIS. Satellite remote sensing also is useful in the analysis of diseases when their distribution or determinants depend on climate and landscape features. Spatial and temporal fluctuations in vector populations determine the distribution of vector-borne diseases (Hay *et al.*, 1996; Thomson *et al.*, 1996). Vegetation partly determines vector populations, with plant communities reflecting the aggregate effect of temperature, precipitation, humidity and other factors such as solar radiation and wind (Washino & Wood, 1994). For example, diurnal temperature differences obtained by satellite have been used as a surrogate for soil moisture to predict the prevalence of Bancroftian filariasis in the Nile delta (Thompson *et al.*, 1996). Remote sensing has been used to predict malaria transmission in several endemic regions (Beck *et al.*, 1997), hantavirus in the Southwest United States (Glass *et al.*, 2000), and Rift Valley fever in Kenya (Linthicum *et al.*, 1999).

5.9 Health impact assessment methods

Approaches to evaluating the potential impacts of global environmental change on human health vary depending on the extent of knowledge about the key variables of concern, such as exposure risk and susceptibility factors, and on the certainty and nature of the relationship between the risk and the potential outcome (Bernard & Ebi, 2001). Toxicological risk assessments are probably the most familiar method with their four-step paradigm of hazard identification, dose–response estimation, exposure assessment and risk characterization. The evaluation of information on the hazardous properties of environmental agents using this paradigm results in a qualitative or quantitative statement about the probability and degree of harm to exposed populations.

Although this approach is informative, there are several distinctions when trying to understand the potential health affects associated with future climate scenarios. One is that the environmental exposures are large-scale, they act on a decadal time frame, and are likely to affect health via complex pathways. A second is that the exposure of concern usually lies in the future. A third is the inapplicability of the primary assumptions underlying traditional risk assessment – that a defined exposure to a specific agent causes an adverse health outcome to identifiable exposed populations (Bernard & Ebi, 2001). The health outcome for some toxic exposures is distinctive, and the association between immediate cause and health effect can be fairly clearly determined (e.g. asbestos and mesothelioma). Even for less well-defined situations, data may be available from epidemiological and animal studies.

Many of the potential health impacts of global environmental change are neither linear nor direct, and complex modifying and/or interacting factors with feedback mechanisms may influence whether an adverse health outcome occurs. Most diseases associated with environmental exposures have many causal factors, making it difficult to isolate the role of one factor in population disease patterns. There are many unresolved questions about the sensitivity of any given health outcome to weather, climate and climate-induced changes in environmental conditions critical to health, e.g. water resources. In addition, although by definition measurements of the impact of climate change entail identifying effects in response to "exposure" above a baseline, the baseline is often difficult to determine. As many of the questions focus on potential future health risks, there is the added uncertainty of future population density, level of economic and technological development, local environmental conditions, and the quality and availability of health care and of a public health infrastructure.

There are many uncertainties as to the magnitude, timing and nature of future changes in the climate system. These necessitate an estimate of the impacts of a range of possible climate scenarios on health. Thus, there are uncertainties in projections of both future climate and future potential health consequences of climate change.

Because of the nature of the problem and its inherent uncertainties, the standard four-step risk assessment paradigm may not be the most appropriate approach for assessing future health risks associated with climate variability and change. The assessment methods used to date include expert judgement, historical or analogue studies, and modelling.

Expert judgement has traditionally been used to rapidly assess the state of knowledge through extensive literature review in combination with the collective experience and judgement of a group of experts. Examples include the IPCC (McMichael *et al.*, 2001) and the health sector of the U.S. National Assessment of the Potential Consequences of Climate Variability and Change (Patz *et al.*, 2000).

Historical studies analyse spatial or temporal variations in either exposure or disease as an analogue for future climate. Although informative for future directions in research and adaptation, the predictive value of these analogues may be limited. For example, as mentioned above, experiences from past El Niño events suggest areas of concern for future health effects. However, the extent of vulnerability of a population changes over time. The 1995 heat wave caused considerably more loss of life in Midwestern states in the U.S. than did the 1999 heat wave, in part because of programmes established in the interim (Palecki *et al.*, 2001). There are similar issues for geographical analogues, such as using the current experience of heat waves in a more southern region to predict what might happen in a more northern region. Cities differ on a number of important factors, such as living standards and behaviours, which could explain the cross-city differences in mortality/temperature curves in Fig. 5.3 (Curriero *et al.*, 2002). A more complete discussion of analogue studies can be found in Chapter 6.

5.9.1 Modelling future risks to health

Modelling (discussed in detail in Chapter 8) can take a number of approaches. An empirical-statistical approach begins with quantification of current associations between risk factors and disease outcomes. Estimates of future populations and potential exposures are used to project quantitative relationships and to estimate statistical uncertainties. An example is a retrospective study of the relationships between climatic extremes and mortality in the five largest Australian cities during the period 1979–90 (Guest *et al.*, 1999). Expected numbers of deaths per day in each city were calculated. Observed daily deaths were compared with expected rates according to temporal synoptic indices of climate. The climate-attributable mortality in the five cities was approximately 160 deaths per year, with the numbers evenly distributed across summer and winter. The authors then applied these relationships to scenarios for climate and demographic change to predict potential impacts on public health in the same cities in 2030. After allowing for increases in population, the analyses showed a net 10 % reduction in mortality (Table 5.1).

A process-based approach begins with integration of what is known about the system of interest. Theoretical and empirical information about the disease of interest is derived from epidemiological, clinical and micro-level (molecular and genetic) studies. A model is developed to estimate the association between climate and disease under a range of potential scenarios. A benefit of these models is that they allow for integration across disciplines of multiple factors that could influence a climate–disease association. Limitations of these models include the lack

Table 5.1. *Mortality associated with climate (summer + winter): summary of analyses by temporal synoptic indices, with current total mortality and projected changes in mortality.*

Age group (years)	1979–90 annual mean	Projected total[a]	2030	
			Change[b]	
			Low	High
0–4	11	11	3	4
5–19	4	5	4	3
20–49	3	16	1	1
50–64	6	17	−2	−1
≥65	136	469	−47	−69
Total	160	518	−41	−62

[a] With allowance for aging and population growth.
[b] Increase or decrease to projected total, according to low or high climate change scenarios.
Source: Guest CS *et al.* (1999); Reprinted with permission.

of quantitative information about some of the links in the models, and that uncertainties accumulate through the model. Examples include models for dengue fever (Focks *et al.*, 1995) and malaria (Martens *et al.*, 1999).

Although not frequently used by epidemiologists, integrated assessment modelling (discussed in detail in Chapter 8) is a valuable tool for putting a risk factor into perspective with other key drivers.

Integrated assessment is a multidisciplinary process of synthesizing knowledge across scientific disciplines to inform policy and decision-making rather than advance knowledge for its intrinsic value. Integrated assessment is a systems-based approach to understanding complex problems. Its main advantage is that it can facilitate gaining insights that typically are difficult or impossible to achieve from traditional, single disciplinary research. The outcome of an integrated assessment can be used to understand the interactions and feedbacks within an entire system, including estimating the potential impact for each of multiple drivers within a system. Specifically for climate change and health, it allows climate change to be put into perspective with other key drivers of health status (Chan *et al.*, 1999; Martens *et al.*, 1999; Casman *et al.*, 2000). It also allows us to evaluate how adaptation measures could change the system response. Finally, the outcome can be used to prioritize decision-relevant uncertainties and research needs. Such information is necessary for decision-makers when prioritizing where and when to focus resources.

Integrated assessment modelling is one method employed to conduct an integrated assessment. One approach to integrated assessment modelling is to develop

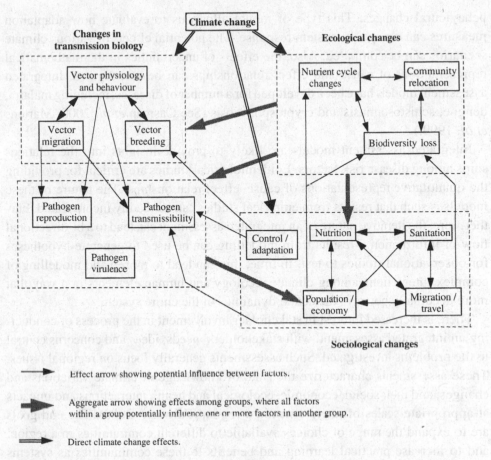

Fig. 5.5 Details of integrated assessment framework for evaluating research on the association between climate change and infectious diseases. Source: Chan *et al.* (1999) reprinted with permission.

a conceptual model of all the relevant aspects of an issue. An array of component modules is developed, each with mathematical representations of cause and effect relationships. Each of the input modules is a simplified, parametric representation of more detailed and complex processes and feedbacks contained in the system. When knowledge is limited, qualitative estimations of relationships may be used. The modules are linked to show the inter-relationships and feedback mechanisms among the key components. This method is particularly useful for modelling potentially indirect health effects of climate change. For example, Chan *et al.* (1999) proposed an integrated assessment model for climate change and infectious diseases. Included in the model are pathogen transmission dynamics, ecological forces (such as changing land use) and sociological forces (such as population movement) (Fig. 5.5).

Sensitivity analyses can be used to better understand the system's sensitivity to current or future changes in each relationship, including socioeconomic and behavioural changes. This type of model allows us to evaluate how adaptation measures can change the system response. The potential effects of various climate scenarios can be compared. Also, the effects of uncertainties in the mathematical representations of cause and effect relationships can be investigated. Integrated assessment models have been developed for a number of diseases, including malaria, dengue, schistosomiasis and cryptosporidiosis (See Casman *et al.* 2000; Martens *et al.* 1999.).

Integrated assessment models are likely to provide insights into the relationships among disease risk factors. Epidemiological studies are critical for providing the quantitative representations of cause–effect relationships. The nature of these models is such that results from empirical studies can be readily incorporated. Furthermore, the iterative nature of an integrated assessment can lead to a bi-directional flow of information – results from modelling can be used to generate hypotheses for observational studies to test. In time, this can lead to more valid modelling of complex interactions among climate, ecology and human elements in a way that more accurately portrays the actual dynamics of the entire system.

There is increased interest in stakeholder involvement in the process of conducting an integrated assessment, with stakeholders' needs, ideas and concerns central to the problems investigated. Such assessments generally focus on regional issues. These assessments characterize the state of knowledge of climate variations and changes, and their social, economic, ecological and health interactions and impacts at appropriate scales of interest within a region (Pulwarty, 2000). The end goals are to expand the range of choices available to different communities in a region, and to increase practical learning and benefits to these communities as systems evolve and new knowledge and information arise. The major tasks include: characterizing the current state of knowledge of climate variability, including social and environmental impacts; assessing vulnerability to climate on various time scales; developing awareness with respect to climate impacts on regional systems; and iteratively refining mechanisms of interaction and learning among the research and user communities, including the understanding of public goals and expectations. This allows research to be focused on realizing the benefits of integrated knowledge and on providing an informed basis for decision-making.

5.10 Conclusion

Empirical epidemiological approaches to describing the distribution of and risk factors associated with a disease or health outcome are useful for: developing quantitative estimates of current vulnerability to climate-sensitive diseases (including contributions to understanding the mechanisms of effects); determining early health effects of climate change; estimating potential health risks associated with global environmental change via empirical–statistical or biophysical models; and understanding what adaptation measures might be desirable given various future projections.

However, these standard methods are not enough. The convergence of expertise, methods and databases from multiple disciplines is required to understand and prepare for markedly different future global environmental exposures and their potential ecological, social and population health impacts. Additional tools and methods are discussed in Chapter 11. Capacity building to improve human and ecological data quality, and the development of innovative interdisciplinary, perhaps transdisciplinary, methods remain high priorities as we face the challenges of assessing actual and potential risks from global environmental change.

Acknowledgements

Funding support for Dr. Ebi is from the Electric Power Research Institute, EPRI. Partial funding support for Dr. Patz is provided by the U.S. Environmental Protection Agency, Global Change Research Program.

References

Baghurst, P. A., McMichael, A. J., Wigg, N. R. *et al.* (1992). Environmental exposure to lead and children's intelligence at the age of seven years: the Port Pirie cohort study. *New England Journal of Medicine*, **327**, 1279–84.

Beck, L. R., Rodriguez, M. H., Dister, S. W. *et al.* (1997). Assessment of a remote sensing-based model for predicting malaria transmission risk in villages of Chiapas, Mexico. *American Journal of Tropical Medicine and Hygiene*, **56**, 99–106.

Bernard, S. M. & Ebi, K. L. (2001). Comments on the process and product of the health impacts assessment component of the National Assessment of the Potential Consequences of Climate Variability and Change for the United States. *Environmental Health Perspectives*, **109** [Suppl. 2], 177–84.

Bouma, M. J. & van der Kaay, H. J. (1996). El Niño Southern Oscillation and the historic malaria epidemics on the Indian subcontinent and Sri Lanka: an early warning system for future epidemics? *Tropical Medicine and International Health*, **1**, 86–96.

Bouma, M. J. & Dye, C. (1997). Cycles of malaria associated with El Niño in Venezuela. *Journal of the American Medical Association*, **278**, 1772–4.

Bouma, M. J., Poveda, G., Rojas, W. *et al.* (1997). Predicting high-risk years for malaria in Colombia using parameters of El Niño Southern Oscillation. *Tropical Medicine and International Health*, **2**, 1122–7.

Buehler, J. W. (1998). Surveillance. In *Modern Epidemiology*, 2nd edn. ed. K. J. Rothman & S. Greenland, pp. 435–58, Philadelphia: Lippincott-Raven Publishers.

Casman, E. A., Fischhoff, B., Palmgren, C. *et al.* (2000). An integrated risk model of a drinking-water-borne cryptosporidiosis outbreak. *Risk Analysis*, **20**, 495–511.

Chan, N. Y., Ebi, K. L., Smith, F. *et al.* (1999). An integrated assessment framework for climate change and infectious diseases. *Environmental Health Perspectives*, **107**, 329–37.

Checkley, W., Epstein, L. D., Gilman, R. H. *et al.* (2000). Effects of the El Niño and ambient temperature on hospital admissions for diarrhoeal diseases in Peruvian children. *Lancet*, **355**, 442–50.

Curriero, F. C., Patz, J. A., Rose, J. B. *et al.* (2001). Analysis of the association between extreme precipitation and waterborne disease outbreaks in the United States, 1948–1994. *American Journal of Public Health*, **91**, 1194–9.

Curriero, F. C., Heiner, K., Zeger, S. *et al.* (2002). Temperature and mortality in eleven cities of the Eastern United States. *American Journal of Epidemiology*, **155**, 80–7.

Donoghue, E. R., Graham, M. A., Jentzen, J. M. *et al.* (1999). Criteria for the diagnosis of heat-related deaths: National Association of Medical Examiners. *American Journal of Forensic Medicine and Pathology*, **18**, 11–14.

Ebi, K. L., Exuzides, K. A., Lau, E. *et al.* (2001). Association of normal weather periods and El Niño events with hospitalization for viral pneumonia in females: California 1983–1998. *American Journal of Public Health*, **91**, 1200–8.

Focks, D. A., Daniels, E., Haile, D. G. *et al.* (1995). A simulation model of the epidemiology of urban dengue fever: literature analysis, model development, preliminary validation, and samples of simulation results. *American Journal of Tropical Medicine and Hygiene*, **53**, 489–506.

Ghebreyesus, T. A., Haile, M., Witten, K. H. *et al.* (1999). Incidence of malaria among children living near dams in northern Ethiopia: community based incidence survey. *British Medical Journal*, **319**, 663–6.

Glass, G., Cheek, J., Patz, J. A. *et al.* (2000). Predicting high risk areas for Hantavirus Pulmonary Syndrome with remotely sensed data: the Four Corners outbreak, 1993. *Journal of Emerging Infectious Diseases*, **6**, 239–46.

Guest, C. S., Willson, K., Woodward, A. J. *et al.* (1999). Climate and mortality in Australia: retrospective study, 1979–1990, and predicted impacts in five major cities in 2030. *Climate Research*, **13**, 1–15.

Hales, S., Salmond, C., Town, G. I. *et al.* (2000). Daily mortality in relation to weather and air pollution in Christchurch, New Zealand. *Ausralia and New Zealand Journal of Public Health*, **24**, 89–91.

Hay, S. I., Tucker, C. J., Rogers, D. J. *et al.* (1996). Remotely sensed surrogates of meteorological data for the study of the distribution and abundance of arthropod vectors of disease. *Annals of Tropical Medicine and Parasitology*, **90**, 1–19.

Hay, S. I., Meyers, M. F., Burke, D. S. *et al.* (2000). Etiology of interepidemic periods of mosquito-borne disease. *Proceeding of the Naional Academy of Sciences*, **97**, 9335–9.

Huynen, M. M. T. E., Martens, P., Schram, D. *et al.* (2001). The impact of cold spells and heat waves on mortality rates in the Dutch population. *Environmental Health Perspectives*, **109**, 463–70.

Katsouyanni, K., Schwartz, J., Spix, C. *et al.* (1995). Short term effects of air pollution on health: a European approach using epidemiological time-series data. The APHEA protocol. *Journal of Epidemiologic Community Health*, **50**, [Suppl. 1], S12–S18.

Kovats, R. S., Bouma, M. & Haines, A. (1999). *El Niño and Health*. (WHO/SDE/PHE /99.4). Geneva: WHO.

Lasky, T. & Stolley, P. D. (1994). Selection of cases and controls. *Epidemiologic Reviews*, **16**, 6–17.

Last, J. M. (ed). (1995). *A Dictionary of Epidemiology*, 3rd edn. New York City, NY: Oxford University Press.

Linthicum, K. J., Anyamba, A., Tucker, C. J. *et al.* (1999). Climate and satellite indicators to forecast Rift Valley fever epidemics in Kenya. *Science*, **285**, 397–400.

Martens, P., Kovats, R. S., Nijhof, S. *et al.* (1999). Climate change and future populations at risk of malaria. *Global Environmental Change*, **9**, S89–S107.

McMichael, A. J., Githeko, A., Akhtar, R. *et al.* (2001). Health. In *The Third Assessment Report of the Intergovernmental Panel on Climate Change*. Cambridge: Cambridge University Press.

Morgan, G., Corbett, S. & Wlodarczyk, J. (1988). Air pollution and hospital admissions in Sydney, Australia, 1990 to 1994. *American Journal of Public Health*, **88**, 1761–6.

National Research Council. (1990). *Health Effects of Exposure to Low Levels of Ionizing Radiation*. (BEIR V). Washington DC: National Academy Press.

Needleman, H. L., Gunnoe, C., Leviton, A. *et al.* (1979). Deficits in psychologic and classroom performance of children with elevated dentine lead levels. *New England Journal of Medicine*, **300**, 689–95.

Needleman, H. L., Schell, A., Bellinger, D. *et al.* (1990). The long-term effects of exposure to low dose of lead in childhood: an 11-year follow-up report. *New England Journal of Medicine*, **322**, 83–8.

Palecki, M. A., Changnon, S. A. & Kunkel, K. E. (2001). The nature and impacts of the July 1999 heat wave in the midwestern U.S.: learning from the lessons of 1995. *Bulletin of the American Meteorological Society*, **82**, 1353–67.

Patz, J. A., McGeehin, M. A., Bernard, S. M. *et al.* (2000). The potential health impacts of climate variability and change for the United States: executive summary of the report of the health sector of the U.S. National Assessment. *Environmental Health Perspectives*, **108**, 367–76.

Pearce, N. (1999). Epidemiology as a population science. *International Journal of Epidemiology*, **28**, S1015–S1018.

Pearce, N. (2000). The ecological fallacy strikes back. *Journal of Epidemiological Community Health*, **54**, 326–7.

Piver, W. T., Ando, M., Ye, F. *et al.* (1999). Temperature and air pollution as risk factors for heat stroke in Tokyo, July and August 1980–1995. *Environmental Health Perspectives*, **107**, 911–16.

Pulwarty, R. (2000). *The NOAA-OGP Regional Integrated Sciences and Assessments Program*. Silver Spring: NOAA Office of Global Programs.

Rothman, K. J., Greenland, S. (eds). (1998). *Modern Epidemiology*, 2nd edn. Philadelphia: Lippincott-Raven Publishers.

Samet, J. M., Dominici, F., Curriero, F. C. *et al.* (2000). Fine particulate air pollution and mortality in 20 U.S. cities, 1987–1994. *New England Journal of Medicine*, **343**, 1742–9.

Sartor, F., Snacken, R., Demuth, C. *et al.* (1995). Temperature, ambient ozone levels, and mortality during summer 1994, in Belgium. *Environmental Research*, **70**, 105–13.

Thomson, M. C., Connor, S. J., Milligan, P. J. M. *et al.* (1996). The ecology of malaria as seen from Earth observation satellites. *Annals of Tropical Medicine and Parasitology*, 243–64.

Thompson, D. F., Malone, J. B., Harb, M. *et al.* (1996). Bancroftian filariasis distribution in the southern Nile delta: correlation with diurnal temperature differences from satellite imagery. *Emerging Infectious Diseases*, **3**, 234–5.

Yen, I. H., Kaplan, G. A. (1999). Neighborhood social environment and risk of death: multilevel evidence from the Alameda County Study. *American Journal of Epidemiology*, **149**, 898–907.

Washino, R. K., Wood, B. L. (1994). Application of remote sensing to arthropod vector surveillance and control. *American Journal of Tropical Medicine and Hygiene*, *Supplement*, **50**, 134–44.

Whitman, S., Good, G., Donoghue, E. R. *et al.* (1997). Mortality in Chicago attributed to the July 1995 heat wave. *American Journal of Public Health*, **87**, 1515–18.

Zeger, S. L., Liang, K. (1986). Longitudinal data analysis using generalized linear models. *Biometrika*, **73**, 13–22.

6

Retrospective studies: analogue approaches to describing climate variability and health

R. SARI KOVATS & MENNO BOUMA

6.1 Introduction

There is a long history of man's awareness of both climate and environmental influences on his health and well-being. Hippocrates wrote his *"On Airs, Waters and Places"* to educate his students on the climate and geographical risk factors that could aid in the prediction and diagnosis of diseases. Early climatologists defined climate in terms of the effects on the organs of the human body (such as von Humboldt in the early nineteenth century), and weather prediction has long been based on aching corns and squeaky joints.

Extensive literature has accumulated over the last few hundred years on the associations between climate, environment and disease. However, the consistency of these associations was often found to depend on geographical location, and seldom resulted in a scientific consensus on the causative pathway. Malaria obtained its name from its presumed cause – the bad odours emanating from marshes in Italy – but this olfactorial connection is absent from many other malarious regions. Our biological and ecological knowledge regarding the dynamics of malaria has greatly improved over the last century. However, the plethora of factors that determine the outcome of disease still causes major disputes over assessing the contribution of a single factor.

Medical geography and medical climatology have existed as scientific disciplines for a long time. However, they are largely descriptive and have been of little practical significance. The current need to assess the impact of global environmental change on disease intensity and distribution has given a new relevance to these disciplines. Attributing risk to specific environmental variables requires extensive information, in time and/or space, about the disease of interest, the environmental exposures and the many relevant covariates. Prospective studies typically take a long time and are therefore unappealing. As an alternative, retrospectives studies seek analogues of climate change in climate variability, which may provide guidance to defining the environmental influences on disease. The paucity of available time series data on disease poses considerable limitations upon this retrospective research strategy, and uncovering suitable historical data for retrospective analysis appears a worthwhile task. The recent great increase in computational capacity and new developments in Geographical Information Systems offer unprecedented opportunities to re-interpret historical data.

Variability is an inherent characteristic of climate, whether the climate system is subject to change or not. Spatial and temporal variability in climate thus offer some opportunity to look at climate–disease interactions. This chapter examines what studies of past or current climate–health relationships can tell us about the impacts of future climate change.

6.2 Climate change, climate variability and epidemiology

Epidemiologists have traditionally viewed the influences of weather and climate on health as part of the natural backdrop to life. Not only are natural climate variations not amenable to our control, but their relationships with health can usually only be studied at the level of whole communities or populations. In popular terms, weather (meteorology) is what we experience on a day-to-day basis. Climate means the "average weather" and its longer-term variability over a particular period.

It is important to distinguish between *climate change* and *climate variability*. *Climate change* is defined as a statistically significant variation in either the mean state of the climate or in its variability, persisting for an extended period (typically decades or longer) (IPCC, 2001). Climate varies at many geographical and temporal scales and can therefore be described in various ways. Many large-scale climate phenomena occur, however, that are difficult to characterize at the local scale. The Asian summer monsoon is a dramatic seasonal phenomenon that is an important component of the global atmospheric circulation and, at the local level, can also cause major flooding in the region around the Bay of Bengal. Each year's monsoon is a little different – the amount of rainfall and the onset of the monsoon both vary, for example. The North Atlantic Oscillation (NAO) has recently been found to be a strong determinant of interannual variability in Western Europe. The El Niño Southern Oscillation (ENSO) is a strong determinant of interannual variability in countries bordering the Pacific. El Niño is not a new phenomenon: it has been occurring for millennia and individual events have been described back to around 1500. However, since the major event of 1997 there has been increasing scientific interest in the effect of El Niño on epidemic disease.

Retrospective studies on climate and health can be categorized using the following types of "exposures":

(i) Long-term changes in mean temperatures, and other climate "norms"
  • climate change requires changes over decades or longer.
(ii) Interannual climate variability
  • including indicators of recurring climate phenomena such as El Niño.
(iii) Short-term variability
  • including monthly, weekly or daily meteorological variables.
(iv) Isolated extreme events
  • simple extremes, such as temperature or precipitation extremes.
  • complex events, such as tropical cyclones, floods or droughts.

Current epidemiological research methods are best able to deal with the health impacts of short-term (daily, weekly, monthly) variability, which require only a few years worth of health data. For most diseases we lack sufficiently accurate and long-term disease data to study the effect of gradual changes in mean climate. Even if we possessed such information, identifying climate as a driver of gradual change in disease is technically difficult in the presence of usually many other explanatory variables that have changed in the same direction. For gradual changes in disease, similar trends are usually found in possible "independent" variables, e.g. the decrease in use of insecticides and local environmental changes that could account for the change in malaria distribution or incidence. The temptation to utilize monthly data to lengthen the time series also introduces seasonality, which can seriously complicate the interpretation of results.

The epidemiological methods used to describe the impact of short-term variations in climate are discussed in Chapter 5. These methods are used to describe the impact of day-to-day variations in weather on daily counts of mortality or morbidity and are not useful as analogues of climate change. In such studies, the seasonal components and any indication of a trend are explicitly removed.

Population health is often sensitive to isolated extreme events (e.g. heavy rainfall and flooding, high temperatures, tropical cyclones) directly, or indirectly through damage to the public health infrastructure. Such impacts are not likely to be significantly affected by long-term, incremental, climate change, unless these same meteorological extremes also change in frequency or character. There is limited evidence that changes in the frequency of extremes over recent decades are attributable to anthropogenic climate change (Easterling *et al.*, 2000; IPCC, 2001). Inferring causal relationships from a single weather event is usually not possible unless reliable data are sufficient, and spatial and temporal resolution are available.

Our understanding of the climate determinants of disease need to be derived from studies of the interannual variability of weather, including the role of extreme

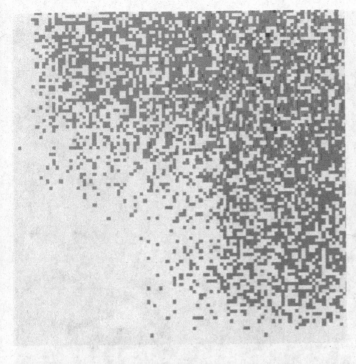

Fig. 4.1
Simulation of habitat competition across two orthogonal resource gradients. Conditions for parameter 1 vary stochastically from left (lowest concentration) to right (highest concentration). Conditions for parameter 2 vary from north to south. Organisms (red and green) populate sites that fall within their range of tolerance. Competitive outcomes are determined by least divergence from optimum conditions.

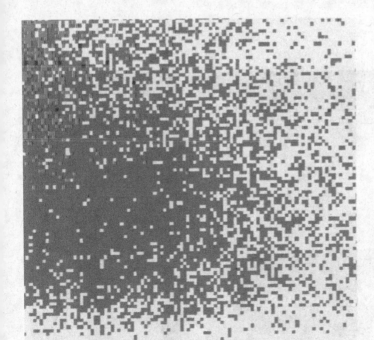

Fig. 4.2
Simulation of two-species competition for habitat across two orthogonal gradients. Parameter 1 remains unchanged from Fig. 4.1. Parameter 2 is increased uniformly across the entire matrix.

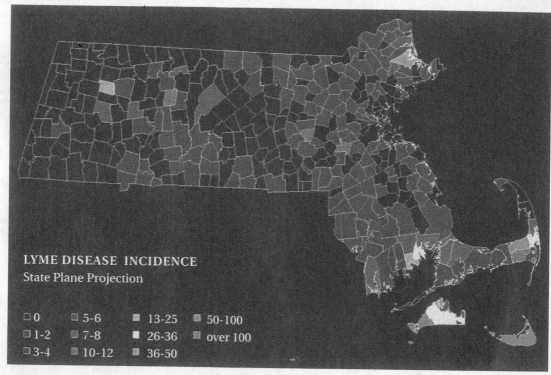

LYME DISEASE INCIDENCE
State Plane Projection

☐ 0   ☐ 5-6   ◼ 13-25   ◼ 50-100
☐ 1-2   ☐ 7-8   ☐ 26-36   ◼ over 100
☐ 3-4   ◼ 10-12   ◼ 36-50

Fig. 4.3a

Fig. 4.3b

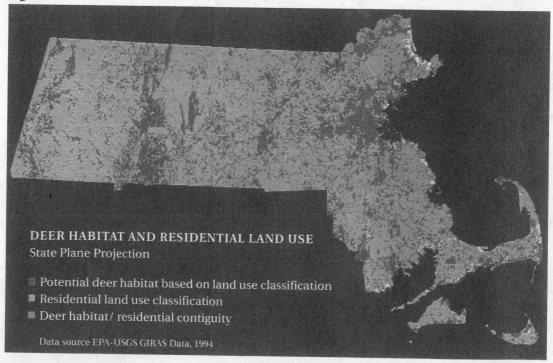

DEER HABITAT AND RESIDENTIAL LAND USE
State Plane Projection

◼ Potential deer habitat based on land use classification
◼ Residential land use classification
◼ Deer habitat/ residential contiguity

Data source EPA-USGS GIRAS Data, 1994

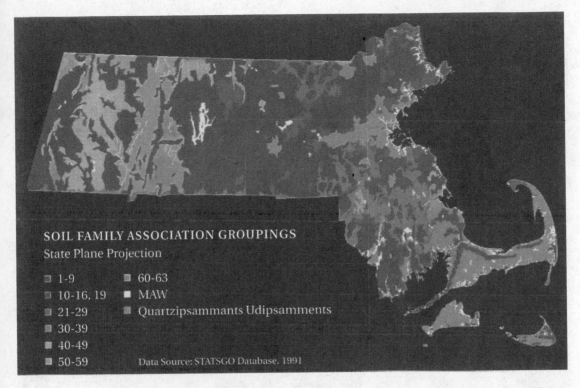

SOIL FAMILY ASSOCIATION GROUPINGS

State Plane Projection

- 1-9
- 10-16, 19
- 21-29
- 30-39
- 40-49
- 50-59
- 60-63
- MAW
- Quartzipsammants Udipsamments

Data Source: STATSGO Database, 1991

Fig. 4.3c

Fig. 4.3d

ELEVATION

State Plane Projection

Data source 1 Degree USGS DEM,, 1994

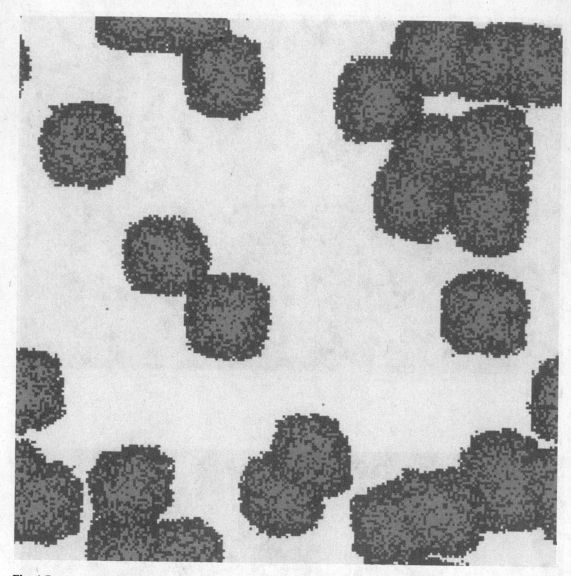

Fig. 4.5

A simple stochastic cellular automata model of directly transmitted pathogens. Introduction of 25 pathogens into a homogeneous population of 50 000 hosts resulted in 25 foci; seen after ten days. Uninfected organisms are indicated in yellow; purple represents infection and pink represents recovery.

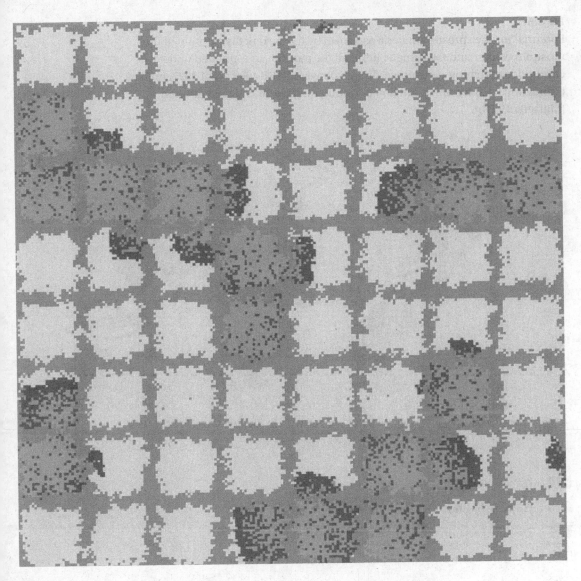

Fig. 4.7

Metapopulation configuration used for epidemic simulation in Fig.4.8. 32 400
hosts are distributed into 64 metapopulations. The borders of each subpopulation
are stochastically determined to allow limited contact between them. Blue
represents infection and pink represents recovery. Unoccupied sites are red.

Fig 8.6

Potential for the spread of malaria as shown by the current distribution of the mosquito vector and the climatic limits of the parasite, as estimated under conditions matching the average climate between 1961-1990. As of 1997, malaria is endemic (red) for only a fraction of its potential world-wide distribution [Martens,1999b].

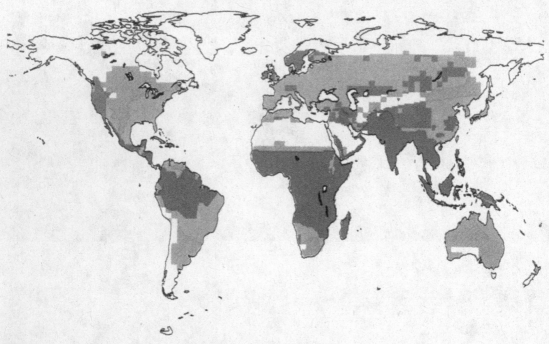

■ Vectors present, transmitting the parasite

■ Vectors present, currently not transmitting the parasite

■ Vectors present, but current climate too cold for parasite

■ Climate warm enough for parasite but no vectors present

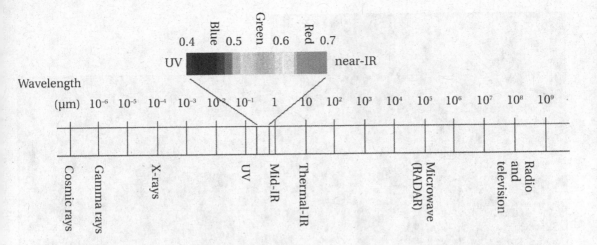

Fig 9.1

The electromagnetic spectrum.

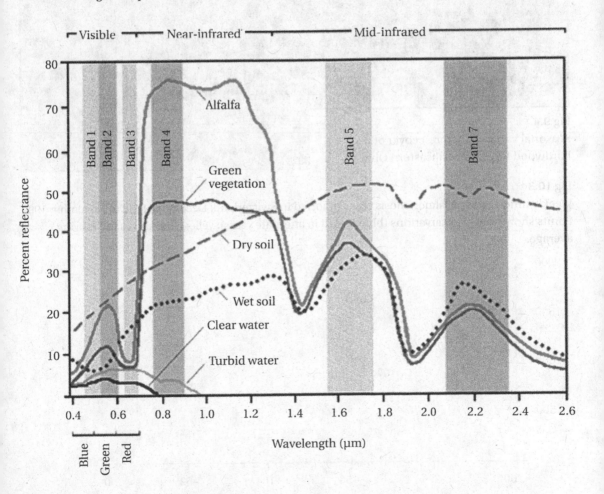

Fig 9.2

Spectrum response functions of different land-cover types.

Fig 9.4
Seasonal variation in forest cover of a
hardwood forest in northeastern Ohio.

Fig 10.3
Weekly time-series of salmonellosis cases reported in Switzerland: December 1998 to November 1999.
Points show weekly observations (blue and rd in alternate years); yellow line shows the moving
average.

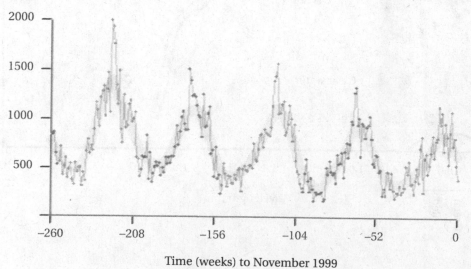

Time (weeks) to November 1999

events. Interannual variability is often considerable, and may equal many decades of gradual change as anticipated under global climate change. Some of the interannual variability in climate is attributed to cycles such as the Quasi-Biennial Oscillation (QBO), ENSO, NAO and sunspot cycles, often in geographically distinct regions. The impact of ENSO – which has often been described as a partial analogue for climate change (e.g. McMichael *et al.*, 2000) – is discussed in more detail below.

6.3 Analogues: a definition

In philosophy, analogy is the process of reasoning from parallel cases. Within the context of the emerging discipline of climate change research, more formal methods have been developed that use current or past climates (or factors of that climate) to predict the impact of future climates. Unfortunately, the term "analogue" has been used imprecisely in health impact assessment. It has sometimes been used to describe any study of the relationship between climate variability and a health outcome.

The Intergovernmental Panel on Climate Change (IPCC) has produced technical guidelines that describe analogue methods for generic climate impact assessment (Carter *et al.*, 1994; Parry & Carter, 1998; Table 6.1). We shall quickly review these methods and discuss their applicability to health impact assessments. In the climate literature, there are two approaches to the use of analogues. The first approach uses

Table 6.1. *Analogue methods used to forecast future health impacts of climate change*

Type of analogue	Nonhealth example	Health example
Historical trends		Association of changes in malaria incidence in a highland area with a trend in temperature
Events	Agriculture studies have used the dry 1930s period in central North America as an analogue of future climate	Assess mortality impact of an individual heatwave, or very hot summer
Spatial analogue	One current location, X, is compared with another current location, Y, the present climate conditions of which are deemed analogous to the future conditions of X	Comparison of vector activity in two locations, the second location having a climate today that is similar to that forecast for the first location; must have similarity in nonclimate factors, e.g. land use

situations that simulate anticipated aspects of future climate change and which can therefore be considered as a plausible climate change "scenario". The second approach looks at impacts associated with an "analogue" situation.

6.3.1 Analogue climate scenarios

Analogue scenarios are constructed by identifying recorded climate regimes that may be used as analogues for plausible scenarios of future climate in a given region. These records can be obtained from the past (temporal analogues) or from another region (spatial analogues). Then the interaction between climate and outcome is observed or assessed. An advantage of this climate scenario is that it is internally consistent and representative of a real climate, even at high spatial or temporal resolution.

Climate scenarios that are derived from general circulation models (GCM) need to be downscaled if climate exposures are to be satisfactorily modelled at the local level. That is, additional information needs to be included before the output can approach a realistic, observed climate. This is particularly true for health impact studies. There is a major mismatch between the spatial and temporal scales of environmental factors that affect health (e.g. local concentrations of air pollutants, focal vector distributions) and GCM-based scenarios that are supplied on grids of several hundred kilometres.

An alternative method identifies a distinct climate period from the instrumental record (i.e. the last 100 years). Several studies have looked at the dry period in the U.S. in the 1930s. The MINK study of the regional impacts of climate change on agriculture, water resources and economy in Minnesota, Iowa, Nebraska and Kansas used the climate of 1931–40 as the analogue of climate change (Rosenberg, 1993). It is also possible to select planning scenarios, representing not the most extreme events but those having sufficient impact and frequency to be of concern. For example, one could describe the impacts of a one-in-ten-year drought or the impact of consecutive events for which the combined effect is greater than the sum of individual events.

6.3.2 Analogues using long-term changes in mean climate variables

The climate is always changing and it is difficult to define absolutely the climate for a particular place. The average over a 30-year period is therefore used to define the climate in a location. The convention for describing the current climate, as specified by the World Meteorological Organization, is the average for the period 1961–90. The last Ice Age ended about 10 000 years ago. Since then, when humans began to develop into settled communities, there have been some "second-order" climate changes such as the Medieval Warm Period and then the Little Ice Age in Europe.

Detecting the impact on health of long-term changes in mean temperatures or other climate "norms" is difficult for reasons that have already been discussed (see also Kovats *et al.*, 2001). However, such an observed association would provide a good analogue for global climate change, providing other factors could be excluded. Long-term data are required for a proper analogue of time-trends in climate. Time series of 15–20 years are required in order for moderate-strength correlations to pass the test of statistical significance of an association. As described above, health data are often not available beyond the very recent past. There is very little literature looking at historical trends in climate and disease as an analogue for climate change impacts.

Evidence on rapid responses to climate change in insect populations have been demonstrated using fossil data (Elias, 1991). The distributions of fossil beetles in high latitudes of the northern hemisphere are used as indicators of environmental and climate change throughout the Quaternary period (1.7 millions years ago to the present time). This provides some limited information about the potential response of insect vectors of human disease to future climate change.

With respect of climate change and human health, there has been some discussion about the historical causes of vector-borne disease transmission. "Tropical" diseases such as malaria, dengue and yellow fever were present in temperate countries in Europe and in the U.S. in the nineteenth century. Medical historians have described these diseases from many contemporary sources. Malaria declined significantly at beginning of twentieth century in most temperate countries. There is some debate about the main causes. However, the general consensus is that social changes, particularly changes in agricultural and domestic habitats, plus the introduction of treatments for malaria were the main trigger for the decline of the disease. The malaria agent also seems to have become less virulent (Dobson, 1994).

Reiter (2001) proposes to go back in human history to provide analogues of past climate and the effects on vector-borne diseases such as malaria to "assess the significance of climate variables in the context of many other factors that affect transmission". Malaria was present in England during the coldest period of the Little Ice Age (1560s to 1700s) and therefore it was suggested that climate was not a significant factor determining transmission. However, Lindsay and Joyce (2001) argue that the cold summers experienced in the 1800s in the UK may have contributed to the disappearance of malaria. However, it is unlikely that information about malaria from such a long time ago could have any relevance to the climate change issue today.

Earth has now experienced a period of recent rapid sustained warming – mean global temperatures have increased 0.2 °C per decade since the 1970s (IPCC, 2001).

The IPCC Third Assessment Report states that "most of the warming observed over the last 50 years is likely to be attributable to human activities" (IPCC, 2001). Therefore, effects attributed to this recent warming would constitute the first impacts of global climate change on human population health. However, several methodological problems persist in demonstrating that vectors and diseases are responding to climate trends (Kovats *et al.*, 2001, Chapter 7 in this volume).

6.3.3 Event analogues

Human society and human individuals are adapted to dealing with the average climate conditions. It is difficult to be adapted to extreme climate events – which, by definition, are rare events. Many health outcomes are sensitive to extreme events. It is important to note that the impacts of extreme events are not a component of climate change unless these same meteorological extremes will alter in frequency or character under changes in climate associated with human activities.

Extreme climate events can trigger a disaster. Some parts of the world are especially vulnerable to natural disasters. For disasters triggered by an extreme weather event, the principle risks zones are tropical Asia and tropical America. In developing countries, populations are becoming more, rather than less, vulnerable to disasters. To date, the greatest number of people affected by natural disasters has been in Asia, principally because of its larger population and higher population density. Droughts, famines and floods account for more than half of all people affected globally (International Federation of Red Cross and Red Crescent Societies, 2000). Very large numbers of people have been killed or affected by famines associated with drought, such as the Sahelian famines in Africa of the early 1970s and mid-1980s. Studies of the impacts of extreme weather events on health have primarily focused on mortality. The IPCC has reviewed the health impacts of weather disasters (McMichael & Githeko, 2001), which include:

- Physical injury.
- Decreases in nutritional status, especially in children.
- Increases in respiratory and diarrhoeal diseases due to crowding of survivors, often with limited shelter and access to potable water.
- Impacts on mental health, which may in some cases be long-lasting.
- Increased risk of water-related diseases due to disruption of water supply or sewage systems.
- Release and dissemination of dangerous chemicals from storage sites and waste-disposal sites into flood waters.

It is a reasonable assumption, and implicit within the IPCC assessment, that the impacts of weather extremes in the future will be very similar to the impacts in the recent past. However, there are two important caveats that are often overlooked:

(i) The validity of scenarios of future climate variability. Are more heatwaves and tropical cyclones forecast in the future, either globally or in the region of interest?
(ii) How will vulnerability of the target population change in the future?

The impacts of unusually warm years or seasons may also be considered as analogues for future climate change. A report on the impact of mild winters and warm summers (1989–90) in the U.K. described many important impacts on terrestrial and freshwater ecosystems. Impacts that were relevant to health included an increase in insect abundance and activity, particularly indoor pests, and an increase in freshwater algal blooms (Cannell & Pitcairn, 1993). The impacts of the unusually warm year of 1995 in the UK were addressed more formally using time series statistical methods to estimate the attributable effect on mortality and food-poisoning incidence (Palutikof *et al.*, 1997). An analogue of the hot summer of 1995 has been used to look at people's behaviour with respect to domestic tourism (Giles & Perry, 1998). However, as with all event analogues, it is necessary to make assumptions about a future world and the other influences on behaviour that affect health outcomes.

There have been several attempts to investigate the usefulness of the impact of current weather variability in climate change research (Changnon & Changnon, 1998). The ACACIA project (Parry, 2000) used an expert-judgement-based approach, sometimes assisted by predictive models where these were available. However, instead of using future socioeconomic development scenarios, a study of the response to current-day weather extremes was used to illuminate the understanding of climate change impacts. This approach was thought most useful to decision-makers. An analysis based on analogue extremes can reveal real sensitivities and vulnerabilities to weather events that can be expected to become more common in the future as a result of climate change. These studies greatly contributed to clarifying the issues and helped to establish methods for studying the impact of changes in weather variability and extremes.

6.3.4 Spatial analogues

Spatial analogues rely on assumptions that an observed relationship between health and climate over space can applied to other geographical areas to represent the effects of climate change at some point in the future. Comparison between areas in health end-points is frequently undertaken in epidemiology. So-called ecological studies have been used to describe the impacts of environmental exposures such as air pollution (e.g. Dockery *et al.*, 1993).

Spatial analogue methods identify the contemporary climate of a geographically distinct location that can be used as an analogue for the future climate of the target location. This must be done with sufficient care and not based on a simplistic assumption about annual mean temperature; for example, the climate of southwest England by 2050 may be said to resemble the current climate of the Bordeaux region of France following a 2 °C warming (Hulme *et al.*, 2002). Climate cannot be transposed in space on the basis of mean annual temperature. The seasonal and multivariate nature of climate is extremely important, as are the impact of local topography and land surface characteristics.

Furthermore spatial analogues must consider the nonclimate differences between areas. This may not be important for nonhuman systems such as crop selection. It is likely that differences in socioeconomic factors are particularly important for health outcomes of interest. Therefore, there are very few examples of spatial analogues for health outcomes, although some work has been done on the vectors of human disease. For example, a geographical analogue has been used to explore the effect of a change in temperature on the mosquito vector of St. Louis encephalitis (Reeves *et al.*, 1994). Vector surveys were undertaken in two adjacent valleys in California, U.S., but one valley was, on average, 5 °C warmer than the other. Although other factors such as the timing and quantity of agricultural irrigation influence vector abundance, temperature was found to be an important determining factor with respect to the transmission of a viral agent by a vector (Reisen *et al.*, 1993).

Some methods used in mapping the distribution of vectors or vector-borne diseases are called "pattern matching". Multivariate techniques have been developed to describe the current area at risk using climate and other environmental variables (sometimes derived from remote sensing). These methods are described more fully in Chapter 9. Rogers *et al.* (2002) used a "Climate Envelope" approach to estimate the risk of *new* vector-borne disease in the U.K. for a variety of climate scenarios. The projected U.K. climates were matched to other current climates in the world. The diseases prevalent in those areas were then identified. This method may be used where there are not sufficient data to use empirical statistical mapping methods. Similar (analogue) climates to the U.K. were found in northern Europe, Asia west of the Caspian sea, Japan and neighbouring regions of China and New Zealand. It is not clear how useful these methods are for estimating the impacts of climate change.

6.3.4.1 Analogues for temperature-related mortality and acclimatization

Human beings, like all other species, are adapted to the climate in which they live. Unlike other species, human populations, over many thousands of years, have strayed outside their original climatic zones, and are able to deploy culture and technology to ensure adaptation to otherwise unfamiliar climates. There are interesting differences in the responses of populations to direct effects of temperature on mortality which are thought to reflect both the local climate regime and other (as yet unidentified) factors.

Studies of the direct effects of temperature on mortality have been undertaken mainly on populations in temperate countries. The statistical methods used are different in each study and this makes comparison between the results difficult to interpret. However, there is some evidence that populations at mid to high latitudes are more responsive to hotter weather than populations at lower latitudes. (However, the attributable burden is not greater because the number of "hot" days is much lower (Keatinge *et al.*, 2000; Curriero *et al.*, 2002).) It may be assumed that the population-specific response will change with climate warming to reflect the response of warmer countries. However, this ignores the rate of change of climate and whether adaptation can occur fast enough to reduce an additional burden of climate change on weather-related mortality. It is reasonable to assume that populations are able acclimatize physiologically to a warmer climate regime. Individuals are able to acclimatize within a single season. However, acclimatization is also dependent upon social behaviour and on the physical infrastructure (e.g. building design). Changes in these factors are likely to take longer. Spatial analogues may be able to provide important information on adaptation to warmer climates.

6.3.5 *Situation analogues*

Climate change is also likely to affect health through indirect effects. For example, populations may be displaced by sea level rise, food insecurity or water stress. It is difficult to describe the health implications of population displacement. However, a case study approach may be the only method for describing such potential impacts of global climate change.

The depopulation pattern of coastal zones affected by subsidence provides an analogue for predicting population movement resulting from sea level rise. In fringe areas of densely populated land areas especially, coastal populations will tend – for cultural and economic reasons – to resist moving over long distances. Adaptive measures may be accepted relatively well. Inhabitants of low-lying islands, however, may not have that choice, and, unless managed transmigration schemes are put in place in good time, island populations could become environmental refugee populations. Thus, the health implications of past population displacements should be used to inform predictions and guide preventive actions. In this case, the analogue is not at all climate related; rather the "trigger" of population movement is analogous to the sea level rise.

6.4 El Niño Southern Oscillation and interannual climate variability

El Niño is often considered to be an analogue for aspects of global climate change. A growing number of studies have shown that the El Niño cycle is associated with changes in the risk of diseases transmitted by mosquitoes, such as malaria and dengue and other arboviruses. Some of these studies are reviewed below in order to determine whether El Niño can be used as an analogue for the health impacts of global climate change.

6.4.1 The El Niño climate phenomenon

The El Niño Southern Oscillation (ENSO) is a complex climate phenomenon that affects the likelihood of anomalous weather in certain regions around the Pacific. Weather patterns associated with ENSO include regional land and sea surface warming, changes in storm tracks and changes in rainfall patterns (particularly heavy rain or prolonged drought) (Ropelewski & Halpert, 1987, 1989). For example, countries in Southern Africa and Indonesia often experience drought during El Niño. The effect of El Niño along the western coast of South America is very significant, particularly in Peru and Ecuador. Nearly every El Niño event, whether weak or strong, is associated with heavy rainfall in this region. ENSO also has an important effect on the character of the annual monsoon in Asia.

"El Niño" describes the exceptionally strong and prolonged warm periods that appear in the Pacific Ocean around the equator. These warm periods are linked to weather changes all over the world. The cycle of warming (and cooling) of the eastern Pacific waters is closely mirrored by air pressure deviations in the East and West Pacific – the Southern Oscillation. In the 1960s, the link was made between the atmospheric Southern Oscillation and the oceanic El Niño – now referred to as El Niño Southern Oscillation (ENSO). The two extremes of ENSO are El Niño (warm event) and La Niña (cold event). Global weather patterns associated with La Niña are generally less pronounced and, in some areas, tend to be the opposite of those that are associated with El Niño.

El Niño and La Niña events usually last for 12–18 months, and occur every 2–7 years. Some changes in frequency have been reported over the last 200 years. Events vary in frequency, time of onset, duration and intensity. Scientists sometimes describe events as very weak, weak, moderate, strong and very strong, depending on their impacts (Glantz, 1996). *Individual* El Niño events should not be confused with long-term climate change. In reality, these fluctuations introduce more noise into the long-term trends, which makes it more difficult to detect the climate change signal.

6.4.2 *El Niño and health studies*

There has been growing interest in the effects of El Niño on health (PAHO, 1998; Kovats *et al.*, 1999; Kovats, 2000). Two main approaches have been used to describe the relationship between health outcomes and El Niño:

(i) Description of individual outbreaks of disease associated with extreme weather events.
(ii) Time series studies that look at the effect of several El Niño events over a long data series.

Individual outbreaks should only be attributed to El Niño with extreme caution. It is not possible to attribute an individual extreme weather event with any certainty to El Niño. El Niño affects the likelihood of an extreme event occurring (Webster & Palmer, 1997). Regional factors can mitigate or amplify the influences of ENSO events on local conditions. ENSO-related anomalies are quite localized and inappropriate aggregation of data can easily mask an ENSO signal.

Statistical analyses are needed to assess how disease incidence or epidemics vary over time with the ENSO cycle. This requires a long (at least 20 years) data series due to the low frequency of ENSO events. Such analyses should demonstrate that disease incidence or epidemics vary over time with the weather pattern (e.g. rainfall) that is associated with ENSO. The analyses should take into account potential confounding factors over the time period which may account for the observed association; for example, changes in land use which may affect vector abundance. Confounding factors are unlikely to vary coincidentally with the ENSO cycle over long periods and therefore the likelihood of confounders explaining the observed relationship is greater for short time series or single-event case studies. Changes in population vulnerability to climate variability may also occur. For example, changes in public health infrastructure, including changes in vector control, may change vulnerability and therefore enhance or reduce the magnitude of the relationship. Again, these are more likely to be important to the interpretation of single-event case studies.

Box 6.1 describes some of the parameters that are used in time series studies. Additional considerations for the quantitative analyses include (Kovats *et al.*, 2001) the following:

- Variability in the climate series (e.g. year to year) should correspond to variability in the health time series.
- Both time series and spatial analyses of correlations between climate and health outcomes should make adjustments for autocorrelation. Failure to do so will tend to overestimate the effect of climate variables.
- Quoted statistical significance values for the association between temporal variation in climate and health outcomes should clearly distinguish between the effects of (in increasing order or relevance) seasonal variation, interannual variation, and long-term trends in climate.

• Analyses should take into account, as far as possible, other changes that have occurred over the same time period and which could plausibly account for any observed association with climate.

BOX 6.1
ENSO parameters

Time series studies of the impacts of ENSO use a variety of parameters, both raw meteorological station data (i.e. temperature, precipitation, humidity) and derived meteorological indices:

• *El Niño or La Niña years.* El Niño does not run to the calendar year, and some events go on for more than 12 months. There are differences between climatologists in the definition of El Niño and in the designation of individual events.
• *Sea surface temperatures (SST) in the Pacific.* Four geographical regions have been defined in the Pacific (named NIÑO1 to NIÑO4). Each region gives different kinds of information about the ENSO phenomenon. SST anomalies are defined as deviations for a specified region from the averaged climate for 1961–90 (a standard recommended by the World Meteorological Organization).
• *The Southern Oscillation Index (SOI).* A simple index, defined as the normalized pressure difference between Tahiti, French Polynesia and Darwin, Australia. There are several slight variations in the SOI values calculated at various centres. El Niño events are associated with large negative values and La Niña events are associated with large positive values. The relationship between El Niño year and SOI values is not perfectly correlated but it is very strong.
• *The Multivariate ENSO Index* (MEI) derived by Wolter (Wolter & Timlin, 1998).

6.4.3 El Niño and mosquito-borne disease

Several mechanisms can explain an association between rainfall anomalies (drought, heavy rain, flood) and malaria incidence. Rainfall is known to affect diseases spread by mosquitoes that breed in surface water. The ecology of the local vector species needs to be understood in order to determine the epidemiology of the disease. Vector-borne disease transmission is also sensitive to temperature fluctuations. Increases in temperature decrease the intrinsic incubation period of the malaria parasite and vectors become infectious more quickly. Increases in temperature also accelerate vector life cycles or allow the vector to colonize areas that

Table 6.2. *High risk years for malaria in relation to the ENSO cycle*

Country/District	Time period	Relative risk	Reference
Sri Lanka (SW)	1870–1945	3.6 El Niño year	Bouma & van der Kaay (1996)
India + Pakistan (Punjab)	1867–1943 pre-DDT	4.5 post-El-Niño year	Bouma & van der Kaay (1996)
India (Rajasthan)	1982–92	* El Niño year	Bouma *et al.* (1994)
Venezuela (Carabobo)	1910–35 deaths	5.0 post-El-Niño year	Bouma, unpublished
Venezuela	1975–90 cases	* post-El-Niño year	Bouma & Dyc (1997)
Colombia	1960–1992	* post-El-Niño year	Bouma *et al.* (1997)
Colombia	1959–1993	* post-El-Niño year	Poveda *et al.* (1999)

* Relative risk not calculated.
Source: Kovats *et al.* (2000).

were previously too cold. Temperature may also affect the behaviour of the human population with regard to exposure.

6.4.3.1 Malaria

The majority of El Niño studies have addressed interannual changes in malaria. Table 6.2 summarizes the findings of several studies that have found a relationship between El Niño and malaria incidence.

At the fringes of its distribution, malaria is an epidemic disease, subject to periodic outbreaks. Najera *et al.* (1998) describe three types of epidemics:

Type 1. Temporary disturbances of a stable hypoendemic equilibrium; for example, due to abnormal meteorological conditions.
Type 2. Major changes in the eco-epidemiological systems, shifting towards a new equilibrium of higher endemicity. For example, due to different kinds of environmental modification such as the introduction of irrigation systems or colonization of sparsely populated areas, or the establishment of a new and more efficient vector.
Type 3. Interruptions of anti-malarials which have kept malaria under control but in an unstable equilibrium in areas with all the epidemiological characteristics of high endemicity. The resulting epidemics are caused by resurgence or failures of control.

Climate change would be expected to trigger a Type 2 epidemic. Weather patterns associated with El Niño would be expected to trigger a Type 1 epidemic.

The periodic nature of epidemics may be due to many nonclimate factors, the most important being the waxing and waning of herd immunity. Thus, the size of the susceptible (nonimmune) population increases as children (who have no immunity)

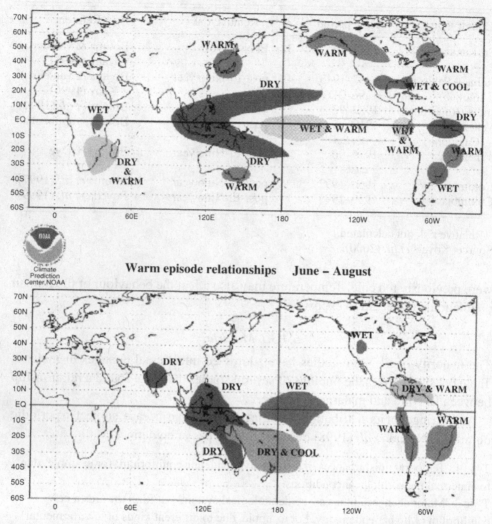

Fig. 6.1 Climate anomalies associated with El Niño. These are averaged effects and are likely to differ from the impacts of individual events. These maps should not be used for planning purposes. Source: U.S. National Oceanic and Atmospheric Administration.

are born into the population. These changes in population immune status are an important factor for the periodic epidemics of measles and many other diseases (Anderson & May, 1979a, b).

Choice of the geographical area of analysis is important. Figure 6.1 indicates the locations that are affected by ENSO at different stages of the cycle. Health researchers must necessarily rely on the assessment by climatologists of the teleconnections in the region of interest. Teleconnections do not respect national boundaries. Aggregation of data at the national level may lead to a loss of information. In most countries that are affected, local data are necessary to evaluate the impact of El Niño. Unfortunately, good-quality long time series of surveillance data at such resolution may not be available. Additional factors that should also be addressed in studies investigating links between ENSO and health include (PAHO, 1998): disease ecology (existing vector reservoirs, host/parasite interactions); magnitude of the El Niño/La Niña events; other climate influences; social change (such as lack of access to health-care services); and increased case detection.

Some highland malaria epidemics have been caused by short-term increases in precipitation associated with ENSO events. It is often assumed that highland malaria transmission is very sensitive to temperature changes. However, precipitation is an important limiting factor of transmission in many highland areas. Kilian *et al.* (1999) observed a short-term increase in malaria transmission in highland areas of southwest Uganda during January to February 1998. This was attributed to an increase in rainfall associated with the El Niño event. The peak in rainfall occurred about 2–3 months in advance of the peak in malaria incidence. Lindsay *et al.* (2000) compared malaria transmission in highland areas of Tanzania with unstable transmission before and after the 1997–98 El Niño. This event was associated with rainfall two to four times heavier than normal, as well as small increases in temperature. Malaria was found to decrease following the El Niño. One suggested explanation is that the heavy rainfall flushed out the vectors' breeding sites. Unfortunately, no entomological surveys were carried out to support this hypothesis.

In desert fringe areas, conditions are normally too dry for malaria transmission. The semi-arid region of northwest India experiences malaria epidemics following excessive monsoon rainfall. Historical analysis (1868–1943) of malaria in the Punjab region shows that the risk of a malaria epidemic increased fivefold during the year following an El Niño (Bouma & van der Kaay, 1996). The sequence of a dry year followed by a wet year appears to be significant for the genesis of epidemics. Although there is some speculation about the mechanisms – historically famine may have increased susceptibility to infection – these have not been determined. The Punjab no longer experiences malaria epidemics, because of the influence of economic and ecological changes. However, epidemic malaria remains a serious problem in the more arid areas in Gujarat in India. There is a significant relationship between annual rainfall and malaria incidence in most districts of Rajasthan and some districts in Gujarat (Akhtar & McMichael, 1996). Years with a high risk of malaria and excessive monsoon rainfall can be expected in years following El Niño and during La Niña years.

The relationship between ENSO and malaria has been examined in detail for Venezuela (Bouma & Dye, 1997) and Colombia (Poveda & Rojas, 1996, 1997;

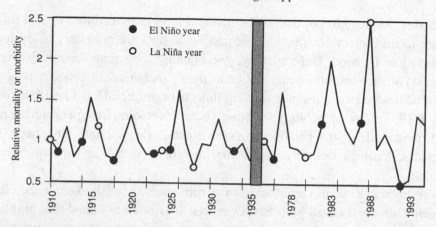

Fig. 6.2 Malaria mortality and morbidity in Venezuela. Malaria increases by 36.5 % on average in years following recognized El Niño events. Relative change in mortality (deaths 1910–35) and morbidity (cases 1975–95), calculated as cases in year *n* divided by cases in year *n* − 1. Source: Bouma & Dye, 1997.

Bouma *et al.*, 1997). These countries usually have below average rainfall during El Niño. In Venezuela, malaria increased on average 37 % in the post-Niño year (see Fig. 6.2). However, the analysis used aggregated data and does not reflect the dramatic focal surges of malaria during these years. In Colombia, malaria cases increased by 17.3 % during the El Niño and by 35.1 % in the post-Niño year (Bouma *et al.*, 1997). El Niño is associated with a reduction of the normal high rainfall regime in much of the country (Poveda & Mesa, 1997). Reduced runoff and stream-flow may increase mosquito abundance by increasing the number of breeding sites (Poveda *et al.*, 1999). Higher temperatures during El Niño episodes may also favour malaria transmission. The reasons why malaria increases after a dry period are not completely understood. Further studies are required to elucidate the mechanisms that underlie these associations. Other countries in the region with similar ENSO–related rainfall anomalies appear to have similar malaria–ENSO relationships.

6.4.3.2 Murray Valley encephalitis and Ross River virus disease in Australia

ENSO usually has a strong effect on the weather in parts of Australia. Murray Valley encephalitis (MVE), also known as Australian encephalitis, is an arboviral disease reported, to date, only in Australia. Interannual rainfall variability in eastern and northern Australia are closely related to the Southern Oscillation. Frequent small epidemics of MVE occur in tropical Australia. However, infrequent but severe epidemics of MVE have occurred in temperate southeast Australia, after well above average rainfall and flooding associated with La Niña episodes (Nicholls, 1986).

Epidemic polyarthritis is caused by infection with Ross River virus, which is transmitted by a wide range of mosquito species in complex transmission cycles with several intermediate hosts. The disease is distributed throughout Australia and

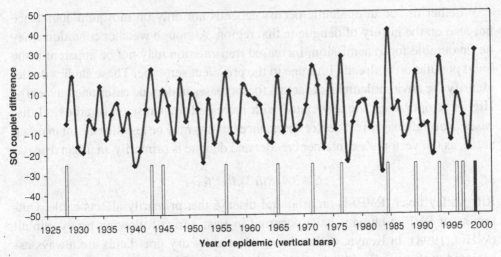

Fig. 6.3 Ross River virus epidemics and the Southern Oscillation Index: Southeast Australia, 1928–99.

elsewhere in the South Pacific. Although the relationship with ENSO is less certain than for MVE, the public health impact of Ross River virus infection is greater in terms of the number of people affected. In arid areas of Australia, the virus is thought to persist in mosquito eggs for a considerable time. When environmental conditions become favourable, such as with heavy rain or flooding, the eggs hatch into infected mosquitoes and a localized outbreak of the disease may occur. Epidemics of Ross River virus in southern Australia (1928–98) were associated with the Southern Oscillation index (Maelzer *et al.*, 1999, Fig. 6.3). Some outbreaks of Ross River virus disease may be linked to weather patterns associated with ENSO but a strong relationship has not been proven (Harley & Weinstein, 1996; Tong *et al.*, 1998).

6.4.3.3 Dengue

Dengue incidence is seasonal and is usually associated with warmer, more humid weather. There is some evidence to suggest that increased rainfall in many locations can affect the vector density and transmission potential (Foo *et al.*, 1985). ENSO may also act indirectly by causing changes in water-storage practices brought about by the disruption of regular supplies (WHO, 1998).

Epidemics of dengue in islands in the South Pacific (aggregated data, 1970–95) were positively correlated with the Southern Oscillation Index (SOI), a parameter of ENSO (Hales *et al.*, 1996). Hales *et al.* (1999) examined the relationship between ENSO and monthly reports of dengue cases in 14 island nations in the Pacific. There were positive correlations between SOI and dengue in ten countries. In five of these (American Samoa, Nauru, Tokelau, Wallis and Western Samoa) there were positive correlations between SOI and local temperature and/or rainfall. During La Niña, the five islands are likely to experience wetter and warmer than normal conditions.

Whether or not an epidemic occurs depends not only on mosquito abundance but also on the history of dengue in that region. Although weather conditions may be favourable for transmission, increased transmission may not be apparent if the local population is already immune to the prevalent serotype. These studies do not identify the environmental risk factors for increases in dengue cases unequivocally. Higher temperatures are associated with increased transmission of arboviral diseases. Rainfall may affect vector abundance but this may be less significant in urban areas, as the vector is a container breeder and dengue is primarily an urban disease.

6.4.3.4 Rift Valley fever

Rift Valley fever (RVF) is an arboviral disease that primarily affects cattle. Outbreaks of RVF in humans have occurred in East Africa following heavy rainfall (WHO, 1998). In Kenya, outbreaks in the usually dry grasslands are always associated with periods of heavy rain. An analysis of RVF outbreaks between 1950 and 1982 in Kenya found a strong and consistent relationship with an aggregated measure of rainfall, i.e. persistent rainfall at a number of sites (Davies *et al.*, 1985). It is thought that the eggs of the mosquito vector are present in large numbers in the grassland depressions (Fontenille *et al.*, 1995). These eggs are already infected with the virus. Flooding in this habitat enables the mosquitoes to develop and appear in high enough densities to cause an epidemic.

The 1997–98 El Niño event has been linked to very heavy rainfall in northeastern Kenya and Southern Somalia, from October 1997 to January 1998. The rain was 60- to 100-fold heavier than normal. In December 1997, there was a large outbreak of RVF in the North Eastern Province of Kenya and Southern Somalia. Linthincum *et al.* (1999) found an association between RVF activity (1950–98) monthly SOI, and sea surface temperatures anomalies in the Pacific and Indian Oceans. However, these relationships were not found to be sufficiently predictive for operational use. The authors recommended the additional use of satellite data for the development of an early warning system. A vegetation index (the Normalized Difference Vegetation Index or NDVI) could identify more localized areas where heavy rainfall had occurred and RVF activity was likely to be present.

6.5 Is El Niño a useful analogue for climate change?

The large-scale pollution of the lower atmosphere with greenhouse gases will significantly affect the global climate over the next few decades. Climatologists are becoming more confident that climate change caused by human activities has already begun. This climate change "signal" will be superimposed on the background "noise" of natural climate variability. There is considerable concern about how climate change will affect the frequency or intensity of ENSO events in the *future*. If the recurring frequency and/or intensity of El Niño and La Niña were to increase there would be profound changes to many local and regional climate extremes, with consequent health impacts.

Hales *et al.* (2000) have argued that El Niño may be used as an analogue for climate change in two ways:

(i) Assessment of the impacts of extreme events.
(ii) Extrapolation based on the climate parameter as a continuous variable, based on a quantitative assessment of the climate–health association.

As discussed above, the impacts of individual events can provide an analogue for the impacts of climate change. However, the inherent characteristics of a flood determine its impact, not whether it is attributed to ENSO or not. As discussed above, such attribution is not straightforward. The majority of deaths and disease that are associated with El Niño can be attributed to weather-related disasters. This is illustrated by the impacts of torrential rain, landslides and flooding in South America during the major El Niño event of 1997–98. In Peru, the impacts of El Niño in 1983 increased total mortality by 39.79 % and infant mortality by 103 % compared to previous years (Toledo Tito, 1997). Many health consequences were recorded after flooding in South America associated with the major El Niño of 1982–83. Increases in the incidence of acute diarrhoeal diseases and acute respiratory diseases were recorded in Bolivia (Telleria, 1986) and in Peru (Gueri *et al.*, 1986).

The extrapolation of a dose–response relationship between climate variable and health outcome in relation to ENSO has not been undertaken in the literature, as far as we are aware. Such empirical–statistical models have been derived from daily time series for temperature and mortality (e.g. Langford & Bentham, 1995) and for temperature and food-borne disease (Benthan & Langford, 1995). Many studies have shown that the impact of ENSO is greater than would be expected from the effect of the climate parameters alone (e.g. Checkley *et al.*, 2000). Until the mechanisms for such relationships are clarified, it would be unwise to derive quantitative estimates for the impacts of climate change on such models.

There are many important differences between the climate "exposures" associated with El Niño and those projected for global climate change (Table 6.3) (Hales *et al.*, 2000). The rate of warming (and cooling) associated with ENSO is much greater than that forecast for global climate change. El Niño events typically wax and wane over a year. However, climate change is likely to be irreversible and therefore will become increasingly important in the coming decades.

Most of the effects of El Niño on mosquito-borne disease are mediated by precipitation extremes. Precipitation is a complex factor to describe and project under global climate change. One would need to look at regional climate models to estimate the change in precipitation patterns forecast under climate change.

Table 6.3. *Comparison of ENSO and global climate change*

Factor	ENSO (Observed/attributed)	Global climate change (Observed and projected)
Time scale	Semi-periodic, with a return period of two to seven years	Progressive over decades to centuries
Predictions – spatial scale	Regional – distribution of climate anomalies is fairly well described, e.g. Indonesia, NW Australia, southern Africa	Global – spatial patterns determined by impact of climate signal on local climatology. Warming greater at higher latitudes
Temperature change	Rapid warming or cooling, superimposed on seasonal changes	Gradual warming, superimposed on interannual variations
Sea surface temperatures	SST increase rapidly in eastern equatorial Pacific	Gradual global increase in SST
Precipitation	Increased or decreased – superimposed on seasonal changes	Increased or decreased – superimposed on interannual changes
Sea level	Increased or decreased, in areas around the Pacific	Increased in all areas (except those with compensatory land uplift)
Frequency of climate extremes – event extremes (tropical cyclones, floods, droughts)	Increased or decreased	Increased or decreased
Frequency of simple climate extremes	Increased or decreased	Increased or decreased (based on current definitions)

Based on: Hales *et al.* (2000).

For areas with desert fringe malaria, it may be that the epidemic zone increases as decreases in precipitation result in a reduction in seasonal transmission (see Kovats *et al.*, 2001). El Niño provides an opportunity to illustrate the importance of the ecological basis of many diseases. These linkages need to be more fully appreciated by many health professionals, policy-makers and the general public.

There has been a tendency to oversimplify the mechanisms by which climate change may affect disease transmission. In particular, discussions of highland malaria have relied on assumptions of shifts in mean temperatures and a simple threshold effect or "altitude limit". Vector-borne disease transmission often does not have a clear threshold or geographical boundary. Variations in transmission

Table 6.4. *The effects of El Niño and long-term climate change on malaria*

Effects	Long-term climate change – warming	Long-term climate change – decreases in precipitation	ENSO – short-term increase/decrease in precipitation
Mosquito vectors	• Changes in vector abundance in any one site, as new climate conditions alter reproduction and survival rates • Changes in geographical distribution, as vectors expand into newly suitable areas • Changes in seasonality, wherever temperature and rainfall restrict vector activity to part of the year	• Changes in vector abundance in any one site, as new climate conditions alter reproduction and survival rates • Changes in geographical distribution, as vectors retreat from unsuitable areas • Changes in seasonality, wherever rainfall restricts vector activity to part of the year	• Temporary changes in vector abundance in any one site • Temporary changes in distribution
Human effects	• Changes in the incidence of disease in particular sites • Changes in the geographical distribution of disease transmission • Changes in the seasonality of disease transmission	• Changes in the incidence of disease in particular sites • Changes in the geographical distribution of disease transmission • Changes in the seasonality of disease transmission	• Temporary change in the incidence of disease in particular sites • Temporary changes in the geographical distribution of disease transmission

intensity due to altitude become increasing unstable as altitude increases and within this "epidemic zone" there will be a significant amount of seasonal and interannual variation. It is therefore very difficult to extrapolate from the effects of El Niño to the impacts of long-term climate change (Table 6.4).

The IPCC has recognized that the 1997–98 El Niño event and other natural hazards have provided global opportunities to identify lessons in impact assessment and adaptation options (UNEP/NCAR/UNU/WMO/ISDR, 2000). This could be achieved through an assessment of the literature on a worldwide series of case studies of specific ENSO-related events. Such science assessments would not only evaluate all aspects of the event, but also address the question of how one would respond to exactly the same event should one know it was going to occur again (MacIver & Klein, 1999). Part of the reason for the increased interest in El Niño and infectious disease has been the potential to use seasonal forecasting in epidemic prediction and control. Seasonal forecasts – also called medium-range climate forecasts – are used to predict major climate trends for a single season up to 12 months in advance. In many areas, especially within or near the tropics, the reliability of such forecasts is greatest when El Niño or La Niña events are occurring. Farmers have been successfully using them for several years to mitigate the effects of drought in countries such as Brazil and Australia.

6.6 Discussion and conclusions

There is good epidemiological evidence that El Niño is associated with an increased risk of certain diseases in specific geographical areas where climate anomalies are linked with the ENSO cycle. The associations are particularly strong for malaria but suggestive for other mosquito-borne and rodent-borne diseases. More research is needed to determine the nature of the ecological mechanisms of these relationships. El Niño provides an opportunity to illustrate the importance of the ecological basis for many diseases. These linkages need to be more fully appreciated by many health professionals, policy-makers and the general public. The recent advances in seasonal forecasting also provide the opportunity to develop longer-term forecasting of epidemic risk in some vulnerable areas (Fig. 6.4). The benefits of current research

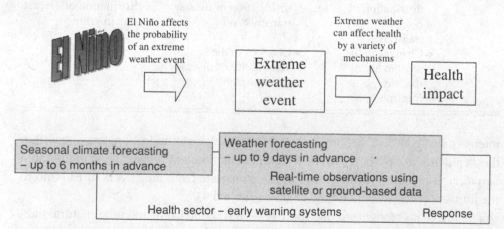

Fig. 6.4 Time scale for epidemic early warning systems that incorporate seasonal forecasts. Source: Kovats, 2000.

on incorporating climate forecasts into epidemic early warning systems may be seen soon. Reducing vulnerability to weather extremes is an important way to reduce the vulnerability to global climate change.

Climate change, as an inter-decadal process, presents us with many research challenges. It is not enough to know that weather affects patterns of disease. The central new task lies in estimating the health impact of future changes in climate on a current and future world. Climate impact assessment requires a variety of tools and methods. The analogue methods described in this chapter are but one component of an approach that will assess the impact of future climate change.

The "analogue" methods have several advantages compared to the other methods of climate impact assessment. One advantage is that it looks at "real world" situations whereas models of the health impacts of future climate scenarios necessarily include many, sometimes unrealistic, assumptions. The scale used (local/national) will also be more appropriate to assessing health impacts than those models that use highly aggregated information. The suitability of the analogue situation should be assessed in relation to the conventional GCM forecasts for future climate change. The climate change that is projected over the coming century (and beyond) exceeds the limits of documented human experience – and may also exceed the limits of adaptation.

References

Akhtar, R. & McMichael, A. J. (1996). Rainfall and malaria outbreaks in western Rajasthan. *Lancet*, **348**, 1457–8.

Anderson, R. M. & May, R. M. (1979a). Population biology of infections diseases I. *Nature*, **280**, 361–7.

Anderson, R. M. & May, R. M. (1979b). Population biology of infections diseases II. *Nature*, **280**, 455–61.

Bentham, G. & Langford, I. H. (1995). Climate change and the incidence of food poisoning in England and Wales. *International Journal of Biometeorology*. **39**, 81–6.

Bouma, M. J. & Dye, C. (1997). Cycles of malaria associated with El Niño in Venezuela. *Journal of the American Medical Association*, **278**, 1772–4.

Bouma, M. J. & van der Kaay, H. J. (1996). The El Niño Southern Oscillation and the historic malaria epidemics on the Indian subcontinent and Sri Lanka: an early warning system. *Tropical Medicine and International Health*, **1**: 86–96.

Bouma, M. J., Sondorp, H. J. & van der, Kaay. H. J. (1994). Health and climate change. *Lancet*, **343**, 302.

Bouma, M. J., Poveda, G., Rojas, W., Chavasse, D., Quinones, M., Cox, J. & Patz, J. (1997). Predicting high-risk years for malaria in Colombia using parameters of El Niño Southern Oscillation. *Tropical Medicine and International Health*, **2**, 1122–7.

Cannell, M. G. R. & Pitcairn, C. E. R. (1993). *Impacts of the Mild Winters and Hot Summers in the United Kingdom in* 1988–1990. London: HMSO.

Carter, T. R., Parry, M. L., Harasawa, H. & Nishioka, S. (1994). *IPCC Technical Guidelines for Assessing Climate Change Impacts and Adaptations*. London: University College London.

Changnon, D. & Changnon, S. A. Jr. (1998). Evaluation of weather catastophe data for use in climate change investigations. *Climate Change*, **38**, 435–45.

Checkley, W., Epstein, L. D., Gilman, R. H., Figueroa, D., Cama, R. I., Patz, J. A., Black, R. E. (2000). Effects of El Niño and ambient temperature on hospital admissions for diarrhoeal diseases in Peruvian children. *Lancet*, **355**, 442–50.

Curriero, F. C., Heiner, K., Zeger, S., Somet, J. & Patz, J. A. (2002). Analysis of heat-mortality in the Eastern United States. *American Journal of Epidemiology* (in press).

Davies, F. G., Linthicum, K. J. & James, A. D. (1985). Rainfall and epizootic Rift Valley fever. *Bulletin of the World Health Organization*, **63**, 941–3.

Dobson, M. J. (1994). Malaria in England: a geographical and historical perspective. *Parassitologia*, **36**, 35–60.

Dockery, D. W., Pope, C. A., & Xu, X. (1993). An association between air pollution and mortality in six cities. *New England Journal of Medicine*, **329**, 1573–9.

Easterling, D. R., Meehl, G. A., Parmesan, C., Changnon, S. A., Karl, T. R. & Mearns, L. (2000). Climate extremes: observations, modelling and impacts. *Science*, **289**, 2068–70.

Elias, S. A. (1991). Insects and climate change. *BioScience*, **41**, 552–9.

Fontenille, D., Traore-Lamizana, M., Zeller, H., Mordo, M., Diallo, M. & Digontte, J. P. (1995). Short report: Rift Valley Fever in western Africa: isolations from *Aedes* mosquitoes during an inter-epizootic period. *American Journal of Tropical Medicine and Hygiene*, **52**, 403–4.

Foo, L. C., Lim, T. W. & Fang, R. (1985). Rainfall, abundance of *Aedes aegypti* and dengue infection in Selangor, Malaysia. *South East Asian Journal of Tropical Medicine and Public Health*, **16**, 560–8.

Giles, A. R. & Perry, A. H. (1998). The use of a temporal analogue to investigate the possible impact of projected global warming on the UK tourist industry. *Tourism Management*, **19**, 75–80.

Glantz, M. H. (1996). *Currents of Change: El Niño's Impact on Climate and Society*. Cambridge: Cambridge University Press.

Gueri, M., Gonzalez, C. & Morin, V. (1986). The effect of the floods caused by "El Niño" on health. *Disasters*, **10**, 118–24.

Hales, S., Weinstein, P. & Woodward, A. (1996). Dengue fever epidemics in the South Pacific: driven by El Niño Southern Oscillation? *Lancet*, **348**, 1664–5.

Hales, S., Weinstein, P., Souares, Y. & Woodward, A. (1999). El Niño and the dynamics of vector-borne disease transmission. *Environmental Health Perspectives*, **107**, 99–102.

Hales, S., Kovats, R. S. & Woodward, A. (2000). What El Niño can tell us about human health and climate change. *Global Change and Human Health*, **1**, 66–77.

Harley, D. O. & Weinstein, P. (1996). The Southern Oscillation Index and Ross River virus outbreaks [letter]. *Medical Journal of Australia*, **165**, 531–2.

Hulme, M., Jenkins, G., Brooks, N., Cresswell, D., Doherty, R., Durman, C., Gregory, J., Lowe, J. & Osborn, T. (2002). Climate change in the UK. In: *Health Effects of Climate Change in the UK*. London: Department of Health in press.

International Federation of Red Cross and Red Crescent Societies. (2000). *World Disaster Report 2000*. New York: Oxford University Press.

IPCC (2001). *Summary for Policy Makers Climate Change 2001: The Scientific Basis*. Cambridge: Cambridge University Press.

Keatinge, W. R., Donaldson, G. C., Cordioli, E., Martinelli, M., Kunst, A. E., Mackenbach, J. P., Nayha, S. & Vuori, I. (2000). Heat related mortality in warm and cold regions of Europe: observational study. *British Medical Journal*, **81**, 795–800.

Kilian, A. H. D., Langi, P., Talisuna, A., & Kabagambe, G. (1999). Rainfall pattern, El Niño and malaria in Uganda. *Transactions of the Royal Society of Tropical Medicine and Hygiene*, **93**, 22–3.

Kovats, R. S. (2000). El Niño and human health. *WHO Bulletin*, **78**, 1127–35.

Kovats, R. S., Bouma, M. & Haines, A. (1999). *El Niño and Health*. Geneva: WHO (WHO/SDE/PHE/99.4).

Kovats, R. S., Campbell-Lendrum, D., Reid, C. & Martens, P. (2000). Climate and vector-borne disease: an assessment of the role of climate in changing disease patterns. Maastricht: International Centre for Integrative Studies, Maastricht University, Maastricht.

Kovats, R. S., Campbell-Lendrum, D., McMichael, A. J., Woodward, A. & Cox, J. (2001). Early effects of climate change: do they include changes in vector-borne disease? *Philosophical Transactions of the Royal Society B* [Theme Issue: Population Biology of Emerging and Re-emerging Pathogens.] **356**, 1057–68.

Langford, I. H. & Bentham, G. (1995). The potential effects of climate change on winter mortality in England and Wales. *International Journal of Biometeorology*, **38**, 141–7.

Lindsay, S. W. & Joyce, A. (2001). Climate change and the disappearence of malaria from England. *Global Change and Human Health*, 1, 184–7.

Lindsay, S. W., Bodker, R., Malima, R., Msangeni. H. A. & Kisinzia, W. (2000). Effect of 1997–98 El Niño on Highland malaria in Tanzania. *Lancet*, **355**, 989–90.

Linthincum, K. J., Anyamba, A., Tucker, C. J., Kelley, P. W., Myers, M. F. & Peters, C. J. (1999). Climate and satellite indicators to forecast Rift Valley Fever epidemics in Kenya. *Science*, **285**, 397–400.

Loevinsohn, M. E. (1994). Climate warming and increased malaria incidence in Rwanda. *Lancet*, **343**, 714–18.

MacIver, D. C. & Klein, R. J. T. (eds.) (1999). IPCC Workshop on Adaptation to Climate Varability and Change: Methodological issues. *Mitigation and Adaptation Strategies for Global Change*, **4**, 189–361.

Maelzer, D., Hales, S., Weinstein, P., Zalucki, M. & Woodward, A. (1999). El Niño and arboviral disease prediction. *Environmental Health Perspectives*, **107**, 817–18.

McMichael, A. J. & Githeko, A. (2001). Human Health. In *Climate Change 2001: Impacts, Adaptation and Vulnerability*. Contribution of Working Group II to the Third Assessment Report of the Intergovernmental Panel on Climate Change, ed. J. J. McCarthy. O. F. Canziani, N. A. Leary, D. J. Dokkery & K. S. White. New York: Cambridge University Press.

McMichael, A. J., Martens, W. J. M., Kovats, R. S. Lele, S. (2000). Climate change and human health: mapping and modelling future impacts. In *Disease Exposure and Mapping*, ed. P. Elliot, D. Briggs *et al.*, pp. 444–61. Oxford: Oxford University.

Najera, J. A., Kousnetzsov, R. L. & Delcollette, C. (1998). *Malaria Epidemics: Detection, Control, Forecasting and Prevention*. Geneva: WHO (WHO/MAL/98.1084).

Nicholls, N. (1986). A method for predicting Murray Valley encephalitis epidemics in southeast Australia. *Australian Journal of Experimental Biology and Medical Science*, **64**, 587–94.

PAHO (1998). *El Niño and its Impact on Health*. Report presented to 122nd Executive Assembly of PAHO May 1998 (PAHO Document CE122/10).

Palutikof, J. P., Subak, S. & Agnew, M. D. ed. (1997). *Economic Impacts of the Hot Summer and Unusually Warm Year of 1995*. Norwich: University of East Anglia.

Parry, M. L. ed. (2000). *Assessment of Potential Effects and Adaptations for Climate Change in Europe: The Europe ACACIA Project*. Norwich: Jackson Environment Institute, University of East Anglia.

Parry, M. L. & Carter, T. (1998). *Climate Impact and Adaptation Assessment*. London: EarthScan.

Poveda, G. & Mesa, O. J. (1997). Feedbacks between hydrological processes in tropical South America and large scale-atmospheric phenomena. *Journal of Climate*, **10**, 2690–702.

Poveda, G. & Rojas, W. (1996). Impact of El Niño phenomenon on malaria outbreaks in Colombia (in Spanish). In *Proceedings of the XII Colombian Hydrological Meeting*, pp. 647–54. Bogotá: Colombian Society of Engineers.

Poveda, G. & Rojas, W. (1997). Evidences of the association between malaria outbreaks in Colombia and the El Niño Southern Oscillation [in Spanish]. *Revista Academia Colombiana de Ciencias*, **XXI (81)**, 421–9.

Poveda, G., Graham, N. E., Epstein, P. R., Rojas, W., Quiñonez, M. L., Vélez, I. D. & Martens, W. J. M. (1999). Climate and ENSO variability associated with vector-borne diseases in Colombia. In *El Niño and the Southern Oscillation, Multiscale Variability and Global and Regional Impacts*, ed. H. F. Diaz & V. Markgraf. Cambridge: Cambridge University Press.

Reeves, W. C., Hardy, J. L., Reisen, W. K. & Milby, M. M. (1994). Potential effect of global warming on mosquito-borne arboviruses. *Journal of Medical Entomology*, **31**, 323–32.

Reisen, W. K., Meyer, R. P., Presser, S. B., & Hardy, J. L. (1993). Effect of temperature on the transmission of Western Equine Encephalomyelitis and St Louis Encephalitis viruses by *Culex tarsalis* (Diptera: Culicidae). *Journal of Medical Entomology*, **30**,151–60.

Reiter, P. (2001). Climate change and mosquito-borne disease. *Environmental Health Perspectives*, **109** [Suppl. 1], 141–61.

Rogers, D. J., Randolph, S., Lindsay, S. W. & Thomas, C. (2002). Vector borne diseases and climate change. In *Health Effects of Climate Change in the UK*. Department of Health Report (in press).

Ropelewski, C. F. & Halpert, M. S. (1987). Global and regional scale precipitation patterns associated with the El Niño/Southern Oscillation. *Monthly Weather Review*, **115**, 1606–25.

Ropelewski, C. F. & Halpert, M. S. (1989). Precipitation patterns associated with the high index phase of the Southern Oscillation. *Journal of Climate*, **2**, 268–83.

Rosenberg, N. J. (1993). A methodology called' 'MINK' for study of climate change impacts and responses on the regional scale. *Climatic Change*, **24**, 1–6.

Telleria, A. V. (1986). Health consequences of floods in Bolivia in 1982. *Disasters*, **10**, 88–106.

Toledo Tito, J. (1997). *Impacto en la Salud del Fenomeno d'El Niño 1982–83 en el Peru*. Presented at the "The health impact of the El Niño phenomenon" Central American workshop held in San Jose, Costa Rica, 3–5 November 1997. Geneva: WHO/PAHO.

Tong, S., Peng, B., Parton, K., Hobbs, J. & McMichael, A. J. (1998). Climate variability and transmission of epidemic polyarthritis. *Lancet*, **351**, 1100.

UNEP/NCAR/UNU/WMO/ISDR (2000). *Lessons Learned from the 1997–1998 El Niño: Once Burned, Twice Shy?*. Tokyo: United Nations University.

Webster, P. J. & Palmer, T. N. (1997). The past and future of El Niño. *Nature*, **390**, 562–4.

WHO (1998). Dengue in the WHO Western Pacific Region. *Weekly Epidemiological Record*, **73**, 273–7.

Wolter, K. & Timlin, M. S. (1998). Measuring the strength of ENSO events: how does 1997/98 rank? *Weather*, **53**, 315–23.

7

Detecting the infectious disease consequences of climate change and extreme weather events

PAUL R. EPSTEIN

7.1 Introduction

According to the World Health Organization (WHO, 1996) 30 infectious diseases new to medicine emerged between 1976 and 1996. Included are HIV/AIDS, Ebola, Lyme disease, Legionnaires' disease, toxic *Escherichia coli*, a new hantavirus, a new strain of cholera and a rash of rapidly evolving antibiotic-resistant organisms. In addition, there has been a resurgence and redistribution of several old diseases on a global scale; for example, malaria and dengue ("breakbone") fever carried by (vectored by) mosquitoes. The factors influencing this lability of infectious diseases are many and varied. They include urbanization, increased human mobility, long-distance trade, changing land-use patterns, drug abuse and sexual behaviours, the rise of antibiotic resistance, the decline of public health infrastructure in many countries, and a quarter century of predominantly anthropogenic climate change. This complex mix of potential influences means, of course, that the scientific task of attributing causation is difficult. This chapter discusses the types of evidence relevant to the detection of changes in infectious disease occurrence in response to climatic variations and trends.

Arthropods such as mosquitoes and ticks are extremely sensitive to climate. Throughout the past century public health researchers have understood that climate circumscribes the distribution of mosquito-borne diseases, while weather affects the timing and intensity of outbreaks (Gill, 1920, 1921; Dobson & Carper, 1993). Paleo-climatic data (Elias, 1994) demonstrate that geographical shifts of beetles have been closely associated with changes in climate. Their distribution – using the mutual climatic range (MCR) method to map fossilized species assemblages – is particularly sensitive to changes in minimum temperatures. Dramatic shifts in species distributions accompanied the beginning and the end of an abrupt cool period called the Younger Dryas, near the end of the Last Glacial Maximum 12 000 years ago. Conversely, the battle between insects and plants may have been a central influence on climatic conditions during the carboniferous period several hundred million years ago. By developing and coevolving multiple means of defence against herbivory, woody terrestrial plants may have thrived, thus drawing down atmospheric carbon and helping to cool the biosphere (Retallack, 1997).

A growing number of investigators propose that vector-borne diseases (VBDs – e.g. involving insects and snails as carriers) could shift their range in response to climate change (Leaf, 1989; Shope, 1991; Patz *et al.*, 1996; McMichael *et al.*, 1996; Haines *et al.*, 2000). Models, linking climate scenarios to vectorial capacity (which incorporates temperature-dependent insect reproductive and biting rates, and micro-organism reproductive rates), indicate the potential for spread of the geographical areas that could sustain VBD transmission to higher elevations and higher latitudes under conditions of global warming (Maskell *et al.*, 1993; Matsuoka & Kai, 1994; Focks *et al.*, 1995; Martin & Lefebvre, 1995; Martens *et al.*, 1997). The transmission season may also be extended. Studies in the U.S. indicate a potential for the northern movement of mosquito-borne encephalitides (e.g. western equine encephalomyelitis and St. Louis encephalitis) within the continental U.S. and Canada in response to regional warming (Reisen *et al.*, 1993; Reeves *et al.*, 1994).

Two components of climate change can significantly influence the pattern of infectious diseases. Warming affects their range, while extreme weather (e.g. excessive rains) affects the timing and intensity of outbreaks. Warming alters the boundary conditions for transmission potential, while atmospheric, land surface and ocean warming also alter the intensity, frequency and temporal/spatial distribution of extreme weather events that are associated with disease outbreaks.

In addition, increased climatic variability and anomalous weather can precipitate "clusters" of water-borne, mosquito- and rodent-borne diseases. Moreover, long-term changes (e.g. winter warming), combined with greater extremes, can destabilize the predator/prey relationships that hold opportunistic species and pests in check. Hence, the impacts of long-term climate change may interact with the impacts of increased variability and weather extremes affecting the incidence, prevalence, seasonality and distribution of infectious diseases. Furthermore, the numerous nonclimatic confounding factors that concurrently influence infectious disease patterns – such as deforestation – often act as *compounding* ecological stresses.

The geographical distributions of several important vector-borne infections have shifted in recent decades. While local environmental changes and population migrations have contributed, these observed changes in disease occurrence are consistent with model projections of shifts associated with warming, *and* are consistent with other data involving plants, mountain glaciers and temperature isotherms. These findings, taken together as a pattern, suggest that the *conditions conducive to* the transmission of VBDs in montane areas have shifted upward, supporting the conclusions that changes in insect ranges within sensitive montane areas are related to

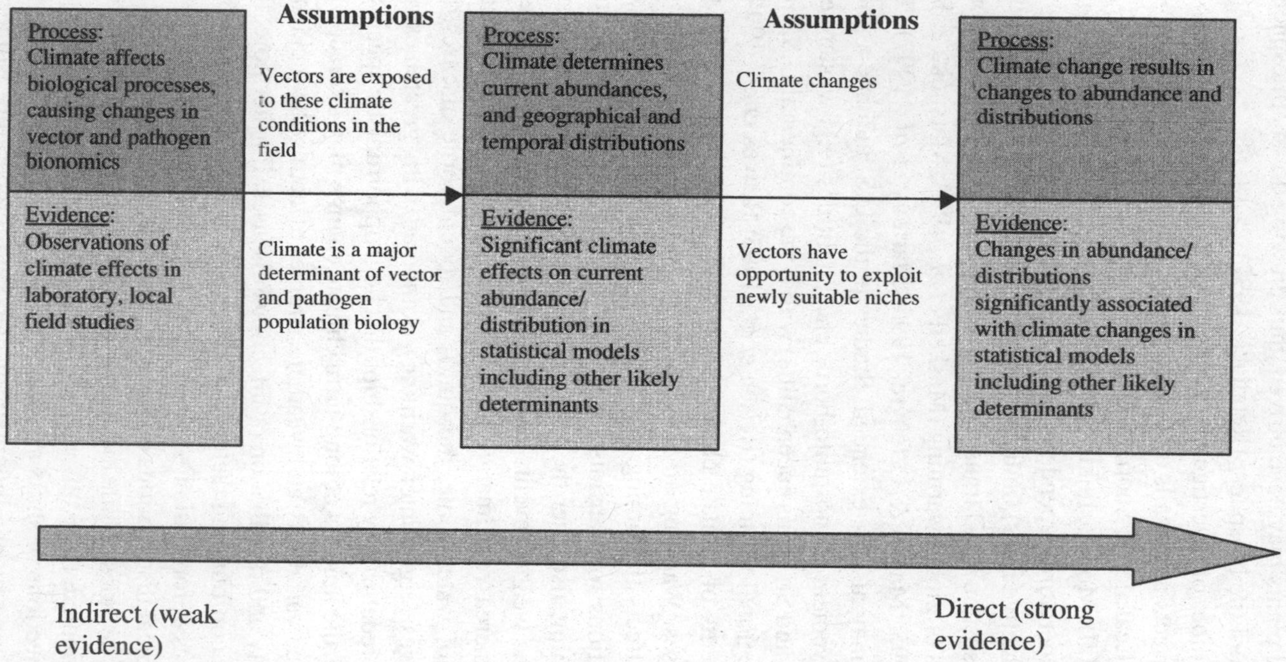
Process:
Climate affects biological processes, causing changes in vector and pathogen bionomics
Evidence:
Observations of climate effects in laboratory, local field studies
Assumptions
Vectors are exposed to these climate conditions in the field
Climate is a major determinant of vector and pathogen population biology
Process:
Climate determines current abundances, and geographical and temporal distributions
Evidence:
Significant climate effects on current abundance/ distribution in statistical models including other likely determinants
Assumptions
Climate changes
Vectors have opportunity to exploit newly suitable niches
Process:
Climate change results in changes to abundance and distributions
Evidence:
Changes in abundance/ distributions significantly associated with climate changes in statistical models including other likely determinants
Indirect (weak evidence)
Direct (strong evidence)

Fig. 7.1. Scheme of climate-driven processes resulting in gradual changes in patterns of disease vectors (Kovats *et al.*, 2000).

warming. Indeed, this is one of the most consistent signals of the influence of climate change on VBD distribution. A second approach to exploring such a signal is to work backwards, beginning with extreme weather events and documenting the various health outcomes (with attention to plausible connections). A third approach, covered in other chapters of this book (e.g. Chapter 6), is to examine long-term data series on infectious disease outbreaks, looking for a climate driver such as ENSO events. This chapter focuses on the first two approaches to identifying a climate signal for mosquito-borne diseases. It also considers the implications for: (i) surveillance and public health, (ii) an interdisciplinary approach to climate change detection, and (iii) environmental policies.

7.2 Assessing the evidence for climate-change impacts on disease

Three factors must be considered when assessing the "causality" of an empirically observed relationship between climate change and changes in human health outcomes, including infectious diseases (Kovats *et al.*, 2000):

(i) *Evidence of biological sensitivity to climate.* Although components of the transmission cycles of most infectious diseases are sensitive to climate conditions in the laboratory, these effects might not necessarily occur in field conditions. Evidence that climate has an influence on current distributions and seasonal patterns of vector and pathogen (the first step, lower section, in Fig.7.1) can be taken as supportive evidence that the disease may respond to climate change.

(ii) *Meteorological evidence of climate change.* Meteorological measurements must be carried out for a sufficiently long time, and in enough sites, to determine whether a long-term trend in meteorological conditions has occurred. Data from a single site or from a short time period should be interpreted with caution.

(iii) *Evidence of entomological and/or epidemiological response to climate change.* In order to show that vectors and diseases are indeed responding to any observed change in meteorological conditions (the second step in Fig.7.1), it is necessary to carry out standardized monitoring of changes in vector populations and disease patterns (see also Chapter 10).

Little research effort has yet been directed at measuring effects of long-term, interdecadal climate change on vectors and disease. Consequently, the criteria for assessing the evidence for climate change effects in the recent past have not been well defined. A framework for assessing climate change effects has been proposed by Kovats and colleagues (2000), depicting the processes through which climate change may be expected to affect vector-borne disease transmission (Fig.7.1).

Earlier processes in the chain are necessary for later processes to occur. It is clear, however, that they may not always be sufficient, as one or more of the intervening assumptions may not hold true. For example, vectors that are highly sensitive to climate conditions in the laboratory may be able to survive outside of the ambient climatic range by exploiting microhabitats with more amenable conditions. This is likely to be particularly true of vectors that live in close association with humans, whose houses are usually more climatically stable than the outside environment. Alternatively, factors other than climate (e.g. control programmes, availability of suitable habitats) may prevent vectors from exploiting all of their climatically suitable range.

At the next stage, even if climate is important in determining current distributions of vectors or diseases, climate change will only be reflected in changing distributions if vectors and pathogens are both able to expand into the newly suitable areas. This will depend on the ability of the vector to disperse into newly suitable ranges before their current range becomes unsuitable, against the challenges of interspecies competition or other possible limiting factors. Recent work indicates that mosquito evolution can be influenced by global warming. Over the last 30 years, the genetically controlled photoperiodic response of the pitcher-plant mosquito, *Wyeomyia Smithii* (not a disease vector), has shifter towards shorter, more southern, daylengths as growing seasons have become longer (Bradshaw & Holzapfel, 2001). This effect is more pronounced at high latitudes.

7.3 Detecting climatic changes, changes in disease vectors and human health

Ideally, climate effects on VBDs should be analysed as a whole, combining climate data with concurrent measurements of the vectorial capacity and infection rate of vectors, the abundance and infection rate of reservoir hosts (if any), and the infection rate and eventual health impacts on humans. It is often more practical, however, to carry out separate analyses of climate effects on the vectors themselves, for the following reasons (Kovats *et al.*, 2001):

- Climate effects on vector populations must occur before they are reflected in changes in health impacts in human populations, and are therefore potentially more sensitive early indicators of climate-change effects on health.
- Climate-driven changes in vector populations are themselves of relevance to public health. For example, it may be important to be aware of the threat caused by the invasion of vectors into new areas, even if pathogen transmission is not (yet) endemic.
- Good quality data on vector populations may not necessarily correspond to the same times and places as good quality health data, as they are often collected by different bodies (e.g. research entomologists or ecologists rather than public health personnel).

- Although affected by many nonclimatic factors, analyses of climate effects on vectors may avoid some confounding variables that affect later stages of the climate–vector–health chain of causation (e.g. changes in treatment regimes or public awareness of diseases).

7.3.1 Climatic changes

In 1995 the Intergovernmental Panel on Climate Change (1996) concluded that there is "discernible evidence" that humans – through changes in multiple forcing factors – have begun to alter the Earth's climate regime. These conclusions were based primarily upon so-called fingerprint studies, studies where data match the projections of general circulation models driven with increased greenhouse gases. The three most prominent fingerprint studies at that time were:

(i) The warming pattern in the mid-troposphere in the southern hemisphere (Santer *et al.*, 1996).
(ii) The disproportionate rise in night-time and winter temperatures (Easterling *et al.*, 1997).
(iii) The statistical increase in extreme weather events occurring in many nations (Karl *et al.*, 1995a,b).

Those three characteristics have been confirmed over the subsequent half-decade (McCarthy *et al.*, 2001). Global warming is not uniform: warming is occurring disproportionately at high latitudes, high elevations and during winter and night-time. All these aspects of climate change and climate variability have implications for biological systems and disease patterns. In particular, the disproportionate rise in minimum temperature (Karl *et al.*, 1993) accompanying climate change favours insect overwintering and activity, and pathogen maturation rates. More recently, the detection of human-induced climate change has drawn upon a long time series of multiple "paleothermometers" (Mann *et al.*, 1998) that indicate that the twentieth century was the warmest in over 1000 years. Data from recent years also demonstrate a significant increase in the *rate* at which warming is occurring (McCarthy *et al.*, 2001).

Ocean warming may have contributed to the northward shift in marine flora and fauna reported since the 1930s along the California coast (Barry *et al.*, 1995), to the drop in zooplankton in the same region (Roemmich & McGowan, 1995) and to the increase in marine-related diseases being reported globally (Harvell *et al.*, 1999). Ocean warming – along with increased terrestrial evapotranspiration – contributes to the accelerated hydrological cycle. This, in turn, increases temperatures through-out most of the tropics (Graham, 1995), thus altering precipitation patterns that affect the timing of outbreaks of VBDs. Warming sea surface temperatures appear related to the isotherm shift (Diaz & Graham, 1996) and the world's oceans may indeed be the main repository for the past century's global warming. Now deep ocean warming has been reported from subtropical transects in the Atlantic (Parrilla *et al.*, 1994), Pacific (Thwaites, 1994) and Indian Oceans (Bindoff & Church, 1992), and near the poles (Travis, 1994; Regaldo, 1995). Indian Ocean waters tested down to 900 metres warmed by up to 0.5 °C between 1962 and 1987, and the Indian Ocean – between 500 and 1500 metres deep – contains more fresh water than in the past. Recent publication of basin-wide data for the Pacific, Atlantic and Indian Oceans corroborate earlier findings that warming is extensive, and is occurring down to a depth of three kilometres (Levitus *et al.*, 2000).

As the oceans warm, atmospheric water vapour is rising. These changes in the hydrological cycle underlie the disproportionate warming and intensification of extremes. A warmer atmosphere holds more moisture (6 % more for every 1°C) and increased evaporation produces more atmospheric water vapour, higher humidity and increased cloudiness, reducing daytime warming and retarding night-time cooling (Karl *et al.*, 1997). Meanwhile warming oceans fuel storms and alter ocean and air currents (e.g. the Jet Stream), affecting weather patterns globally. Large systems may be shifting due to land mass warming. The Asian monsoons appear to be intensifying and losing their previous association with El Niño events, due to warming of the Eurasian landmass (Rajagopalan *et al.*, 1999).

Several types of extreme weather events measurably increased in intensity in the twentieth century and many are projected to increase in frequency during this century (Easterling *et al.*, 2000). Extremes of weather – heatwaves, droughts, floods, windstorms and ice storms – can certainly have direct impacts on public health (e.g. trauma) as well as impacts on the social infrastructures underlying public health in general. They can also influence the populations of pests.

7.3.2 Detection of climate (change) effects on vectors

Demonstrations of climate effects on the abundance and distribution of vectors, either now or in the recent past, constitute indirect evidence that they have been, or will be, affected by climate change. The recent increase in affordable computing power and the advent of Geographic Information Systems (GIS) software has facilitated the mapping of available data on vector abundance and distributions. New ground- and satellite-based sensors (Easterling *et al.*, 2000) have allowed these data to be matched against increasingly accurate climate measurements (see also Chapter 9).

In general, the mapping studies that have so far been carried out confirm the importance of climate as a limiting factor in the distribution of many important insect vectors. Such confirmations are suggestive (though not conclusive) tests that the distributions are likely to change as climate change progresses.

Care should be taken to differentiate the effects of long-term climate change from those of extreme climate events (Kovats *et al.*, 2001). Individuals that live at the edge of the distribution therefore tend to be under more climatic stress and are also more likely to respond to extreme climate events (Parmesan *et al.*, 1999). Assuming that the statistical distribution of the weather event does not change, a shift in the mean will entail a disproportionate increase in the frequency of extreme events. Local population extinctions of certain insect species have been associated with particular climate events (e.g. Edith's Checkerspot butterfly, see Parmesan *et al.*, 1999). Droughts are often associated with widespread reductions in insect populations. However, they can also lead to increases in the abundance of some vector species (e.g. rivers turned into pools, which led to an increase in breeding sites, Bouma *et al.*, 1994a,b). Precipitation extremes (especially heavy rainfall) have long been known to increase mosquito vector abundance. However, it should be noted that biological responses to individual extreme events are not necessarily related to changes in mean climate conditions. Such events (and the associated health outcomes) should not be attributed to long-term climate change unless it can be demonstrated that the frequency of events is changing over time.

In summary, direct tests of climate change-driven effects on vectors continue to face great methodological difficulties (Kovats *et al.*, 2001).

- The lack of standardization and the temporary nature of most vector-monitoring efforts hamper the detection of changes in distribution and abundance.
- The high sensitivity of many vectors to climate parameters means that, unless monitoring is carried out frequently, interobservation changes caused solely by interannual climate variability may be misinterpreted as long-term trends.
- Many nonclimatic factors may also change over time, so that it is often difficult to attribute observed changes to climate changes.
- Changes in vector abundance or distribution that correlate with climate change are more likely to be reported than those that do not, creating a sampling bias. This is a particular problem given the increasing interest in climate change and its impacts in recent years.

7.3.3 Detection of climate (change) effects on human disease

Many of the same principles and problems (variability in monitoring and surveillance, necessity for long data sets to overcome statistical "noise", confounding variables, publication bias, etc.) that apply to the detection of changes in vector distributions hold for the detection of changes in human disease. However, there are significant additional challenges, mainly involved with differentiating climate influences from the other multiple determinants of VBD transmission. Climate-change impacts on VBD occur against a background of rapidly evolving patterns of emergence of new health threats (human immunodeficiency virus (HIV) and human variant Creutzfeldt-Jakob Disease (CJD)), and the resurgence of previously recognized problems (e.g. tuberculosis, malaria). Many of the factors are linked to global environmental changes (travel, population movement, trade). The widespread breakdown in public health measures, such as vector control, has also been a major factor behind the resurgence of VBDs such as malaria and dengue.

BOX 7.1

Causes of recent emergence and resurgence of infectious diseases

Factor	Examples of specific factors	Examples of diseases
Ecological changes including those due to economic development and land use	Agriculture; dams; changes in water ecosystems; deforestation/reforestation; irrigation; climate change	Schistosomiasis (dams); Rift Valley fever (irrigation); Argentine haemorrhagic fever (agriculture); hantavirus pulmonary syndrome (seasonal climate anomalies–ENSO)
Human demographics and behaviour	Societal events: population growth and migration (movement from rural areas to cities); war or civil conflict; urban decay; sexual behaviour; intravenous drug use; use of high-density facilitates	Introduction of HIV; spread of dengue; spread of HIV and other sexually transmitted diseases
International travel and commerce	Worldwide movement of goods and people; air travel	"Airport" malaria; dissemination of mosquito vectors; rodent-borne hantaviruses; introduction of cholera into South America; dissemination of O139 *Vibrio cholerae*
Technology and industry	Globalization of food supplies; changes in food processing and packaging; organ or tissue transplantation; drugs causing widespread immunosuppression; widespread use of antibiotics	Haemolytic uraemic syndrome (*Escherichia coli* contamination of meat), bovine spongiform encephalopathy and human variant CJD; transfusion-associated hepatitis (hepatitis B and C); opportunistic infections in immunosuppressed patients
Microbial adaptation and change	Microbial evolution, response to selection in environment	Antibiotic-resistant bacteria, "antigenic drift" in influenza virus

BOX 7.1 (*cont.*)

Factor	Examples of specific factors	Examples of diseases
Breakdown in public health measures	Curtailment or reduction in prevention programs; inadequate sanitation and vector control measures	Resurgence of tuberculosis in the U.S.; cholera in refugee camps in Africa; resurgence of diphtheria in the former Soviet Union; malaria in South America, eastern Europe, Africa

Source: Morse, 1995.

Emerging diseases are defined as infections that have newly appeared in a population, or have existed but are rapidly increasing in incidence or geographical range (Morse, 1995). Resurging infections (a subset of emerging infections) are diseases such as malaria that have previously declined in incidence and distribution and are now rapidly increasing. Many of these factors may apply to a single disease. The main determinants of malaria may be:

- human migration due to travel
- migration of refugees
- parasite resistance to antimalarial medications
- ecological disturbances due to water resource development, e.g. irrigation, tree plantations
- natural disasters
- changes in health care, particularly vector control programs

7.3.4 Changes in disease distribution, transmission intensity and seasonality

Changes in the distribution of diseases are clearly important, as new populations are exposed to risk. Such populations lack immunity and therefore morbidity and mortality can be significantly higher. Populations may also lack health expertise and mechanisms for responding to the disease. For example, the spread of malaria from areas of constant endemic transmission (and therefore high levels of immunity) to regions where the disease has not been experienced before is likely to lead to epidemic transmission with large impacts on public health.

To detect an attributable change, one needs to define a "baseline" distribution before climate change. Mapping efforts have recently received new impetus from the advent of remote sensing and GIS technology (Chapter 9). Even so, it is very difficult to obtain accurate and reliable information on current or previous distributions of disease. Maps can sometimes be misleading where they are based on insufficient data, use artificially drawn boundaries, or are used for purposes other than originally intended. Another consideration in interpreting data is that meta-analyses of

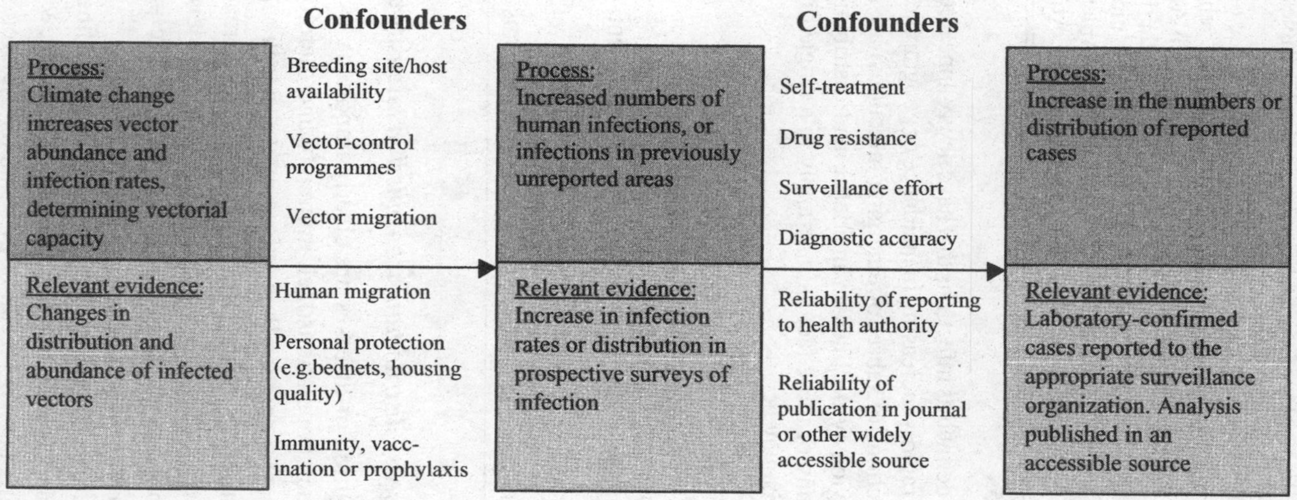

Confounders

Breeding site/host availability

Vector-control programmes

Vector migration

Confounders

Human migration

Personal protection (e.g. bednets, housing quality)

Immunity, vaccination or prophylaxis

Confounders

Self-treatment

Drug resistance

Surveillance effort

Diagnostic accuracy

Reliability of reporting to health authority

Reliability of publication in journal or other widely accessible source

---

Process:
Climate change increases vector abundance and infection rates, determining vectorial capacity

Relevant evidence:
Changes in distribution and abundance of infected vectors

→

Process:
Increased numbers of human infections, or infections in previously unreported areas

Relevant evidence:
Increase in infection rates or distribution in prospective surveys of infection

→

Process:
Increase in the numbers or distribution of reported cases

Relevant evidence:
Laboratory-confirmed cases reported to the appropriate surveillance organization. Analysis published in an accessible source

Fig. 7.2 Processes leading from climate change to reported changes in the burden of vector-borne disease (Kovats et al., 2000).

multiple studies can reflect a bias toward the publication of positive results and multiple publications with the same data. Fig. 7.2 (Kovats *et al.*, 2000) outlines the relevant evidence, as well as the potential confounding influences and biases, in using published information to identify an effect of climate change on VBD transmission.

It is also potentially possible to detect climate effects on the intensity of transmission or the length of the transmission season. Again, it is important to address confounders – particularly for transmission intensity, which is influenced by a large number of other factors. When considering short-term changes, it is important to distinguish between "climate–health" relationships and "weather–health" relationships. Weather events occur at the synoptic time-scale (two to six days) and are associated with many health impacts, such as outbreaks of mosquito-borne disease triggered by heavy precipitation and flooding. Such individual events should not be considered as climate–health relationships. However, a long-term trend of increased frequency or intensity of these events (and associated health impacts) would be considered a climate-change impact.

The frequent lack of data on potential confounders can be partially overcome by careful study design (see Kovats *et al.*, 2001). The confounding influence of other factors such as land use may be reduced by conducting studies in which nonclimate variables have stayed relatively constant. Secular trends and seasonality may be removed by appropriate use of time-series analysis (although over-controlling for seasonality may result in the loss of information). Even with careful monitoring and analysis, however, it will remain difficult to provide strong direct evidence of climate-change effects on disease transmission.

7.4 Examples of physical and biological signs of climate change

Various simple physical and biotic systems, unbuffered by human intervention, have undergone changes plausibly caused by climatic change over the past quarter-century (McCarthy *et al.*, 2001). Viewed together, these changes comprise a compelling and coherent pattern.

7.4.1 Shifting isotherms and shrinking glaciers

Data from the physical and biological sciences indicate a significant warming trend during the twentieth century (IPCC WGII, 2001). Moreover, Diaz & Graham (1996) report that, between 1970 and 1989, the freezing level (0 °C isotherm) in the mountains within the 30 °N to 30 °S latitude range shifted upward approximately 160 metres.

As tropospheric temperatures rise, most of the glaciers and ice caps from the tropics to mid-latitudes – South American Andes, African highlands, European

Alps, Asian Highlands, Indonesia and New Zealand – are retreating, many at rates that continue to accelerate (Kaser & Noggler 1991; Hastenrath & Kruss, 1992; Thompson *et al.*, 1993; Haeberli, 1995; IPCC, 1996; Parkinson *et al.* 1999). For example, the edge of the Qori Kalis glacier that flows off the Quelccaya ice cap high in the Peruvian Andes Mountains was retreating at a rate of four metres annually between 1963 and 1978. By 1995, that rate had grown to 30 metres each year (Mosley-Thompson, 1997). Many of the smaller ice fields may soon disappear, potentially jeopardizing local water supplies that are critical for human consumption, regional agriculture and the generation of hydroelectric power. As the Arctic ice cap thins, recent evidence indicates that the west Antarctic glacier is thinning and retreating at an accelerating rate (Krabill *et al.*, 1999; Rothrock *et al.*, 1999; Shepherd *et al.*, 2001).

7.4.2 Plant migrations

Past and potential displacements of plant distributions in response to climate change have received considerable attention (Jacobson *et al.*, 1987; Davis, 1989; Overpeck *et al.*, 1991; Davis & Zabinski, 1992; Billings, 1995). From a climatic perspective, a small elevational displacement of plant distribution corresponds to a much larger latitudinal displacement. To accommodate to a 2°C rise in temperature in the Northern Hemisphere, for example, plant distributions may rise 300 metres in elevation or shift 18 kilometres poleward (MacArthur, 1972; Peters, 1991). Upward displacements of plant distributions have been documented on 30 Alpine peaks (Grabherr *et al.*, 1994; Pauli *et al.*, 1996), along with a mean air temperature rise in Austria of 1 °C since 1850, and elsewhere in Europe (Diaz *et al.*, 1997). Maximum upward displacement rates of plant distributions approach four metres per decade. Similar upward shifts have been observed in Alaska, the U.S. Sierra Nevada and New Zealand (Yoon, 1994). Meanwhile, for reasons of latitude, aspect, etc., some researchers report examples not indicative of the overall trend.

7.4.3 Insects and ticks

The distribution of Edith's Checkerspot butterflies (order Lepidoptera) has shifted northward in recent decades (Parmesan, 1996, 1999; Parmesan *et al.*, 1999). Population extinctions are four times greater at the far southern end of its range (in Mexico) than at the far northern end of its range (in Canada); and about 2.5 times as great at lower elevations as compared to populations above 2700 metres.

Europeans colonizing Africa were well aware of the benefits afforded by living at higher, cooler elevations (Lindsay & Martens, 1998), separated from *mal* areas. Mosquitoes, in particular, are highly sensitive to climatic factors (Gill, 1920,

1921; MacArthur, 1972; Billett, 1974; Burgos, 1990; Burgos *et al.*, 1994). *Anopheline* spp. and *Aedes aegypti* mosquitoes have established temperature thresholds for survival, and there are temperature-dependent incubation periods for the parasites and viruses within them (the extrinsic incubation period or EIP). Provided sufficient moisture, warmer temperatures – within the survivable ranges – increase mosquito populations, biting rates (blood meals), mosquito activity and abundance, and decrease the extrinsic incubation period or duration of sporogony (MacArthur, 1972; Patz *et al.*, 1996; Martens *et al.*, 1997). Mosquito survival drops as ambient temperatures approach 40 °C.

Bioclimatographs of temperature and humidity levels, and of the geographical distribution of areas permitting mosquito and pathogen development, can be constructed (Dobson & Carper, 1993). The aquatic stages of *Anopheline* mosquitoes in the tropics do not develop (Leeson, 1939) or breed (DeMeillon, 1934) below approximately 16 °C, and *Plasmodium falciparum* malaria parasites do not develop within a mosquito lifetime below 18–19 °C (Molineaux, 1988). In general, isotherms present boundary conditions, and transmission is limited by the 16 °C winter isotherm. (Mosquitoes may avoid these restrictions by resting in houses or shaded areas; and biogeographical factors, land use, dams and irrigation ditches, control measures plus population movements and the "immunological history" of inhabitants all contribute to the precise areas where disease outbreaks occur.)

The northern limit of the distribution of important tick vectors moved north in Sweden between 1980 and 1994 (Tälleklint & Jaenson, 1998). Further analysis shows that changes in the distribution and density of tick species over time have been correlated with changes in seasonal daily minimum temperatures (Lindgren *et al.*, 2000; Lindgren & Gustafson, 2001).

The distribution of agricultural pests (many cold-blooded *stenotherms*) can also shift, for they – and the pathogens some transport – also require specific temperatures and conditions for survival (Dahlstein & Garcia, 1989; Sutherst, 1990; Rosenzweig & Hillel, 1998). Extreme weather also influences crop pests and pathogens: drought favours aphids (Homopteran family Aphididae), locusts (Orthopteran family Acrididae) and whiteflies (Homopteran family Aleyrodidae) (see Rosenzweig *et al.*, 2000, 2001). *Bemisia tabaci* whiteflies, for example, are responsible for injecting numerous *geminiviruses* into the leaves of bean, squash, tomato and other staple crops in Latin America (Anderson & Morales, 1993). A change in the distribution of plant pests has implications for food security (Rosenzweig & Hillel, 1998).

At high elevations, the overall trends regarding glaciers, plants, insect ranges and shifting isotherms show remarkable internal consistency, and there is consistency between model projections and the ongoing changes (Epstein *et al.*, 1998). Such emerging patterns consistent with model projections may be considered biological "fingerprints" of climate change and suggest that the climate regime has already affected the geographical distribution of biota on Earth (see also McCarthy *et al.*, 2001).

7.4.4 Mosquito-borne diseases

Mosquito-borne diseases have recently been reported at unusually high elevations in the highlands of Asia, Central Africa and Latin America. *Plasmodium falci- parum* malaria is a growing public health threat in the eastern, southern, western and Chimbu highlands of Papua New Guinea (PNG) (Rozendaal, 1996), and in 1997 malaria was reported up to 2100 metres in the highlands of Irian Jaya and PNG (ProMED, 1997). A steady rise in annual temperatures has been associated with expanding malaria transmission in the Usamabara Mountains in Tanzania (Matola *et al.*, 1987), and highland malaria has been reported in Kenya (Some, 1994). In association with the warm El Niño Southern Oscillation (ENSO) event of 1987 – and an increase in night-time temperatures – *Plasmodium falciparum* in- creased significantly in the highlands of Rwanda (Loevinsohn, 1994). Satellite temperature profile data show that temperature anomalies over tropical land are in phase with the ENSO (Susskind *et al.*, 1997).

Dengue fever – previously limited to about 1000 metres in elevation in the tropics by the 10 °C winter isotherm – has appeared at 1700 metres in Mexico (Koopman *et al.*, 1991); and *Aedes aegypti* – the mosquito that can carry dengue and yellow fever viruses – has been reported at an elevation of 2200 metres in Colombia (Suarez & Nelson, 1981). There is therefore a growing potential for the resurgence of urban yellow fever in highland regions of Latin America.

In the 1990s, outbreaks of *locally transmitted* (as opposed to "airport" or imported cases of) malaria have occurred in North America: New Jersey, 1991 and Queens, NY, 1993, during hot wet summers (Zucker, 1996); Michigan (CDC, 1996); Texas (CDC, 1995); Florida (ProMED, 1996); Georgia (CDC, 1997); Long Island, NY, 1999 (ProMED, 1999); Toronto, 1998 (see Epstein *et al.*, 1998); and California (Maldonado *et al.*, 1990). In the 1980s, only California reported local transmission of malaria in the continental U.S. Malaria occurred within the U.S. earlier this century, and these reports lack the supporting documented changes occurring at high elevations in plants, glaciers and isotherm shift. But these reports are consistent with model projections that warmer, wetter *conditions conducive to* greater transmission potential can be expected at higher latitudes.

Montane changes in distribution are not only of academic interest. Malaria threat- ens elevated urban centres such as Nairobi, Kenya. Mountain ranges that have been barriers to spread may no longer be so. After the warm wet year of 1993, dengue fever blanketed the large western plains of Costa Rica, later appeared in mountain ranges above the capital, San José, then appeared on the Caribbean coast (Pan American Health Organization, unpublished data 1996).

7.4.5 Extreme weather events and infectious diseases

While distributional changes are occurring at high elevations, altered weather patterns and more extreme events (Karl *et al.*, 1995a,b) may be the chief characteristics of climate change, and weather extremes also contribute to disease outbreaks. Floods foster fungal growth and provide new breeding sites for mosquitoes; while droughts concentrate micro-organisms and encourage aphids, locusts, whiteflies and – when interrupted by sudden rains – spur explosions of rodent populations (Epstein & Chikwenhere, 1994). Drought may favour outbreaks of west Nile virus (Epstein & Defilippo, 2001). Because of the strong influence of climatic factors, the prediction of weather patterns based on ENSO and other climatic modes such as the North Atlantic Oscillation, integrated with regional sea surface temperatures and local topography, may prove useful for anticipating conditions conducive or vulnerable to such "biological surprises" and epidemics (Bouma *et al.*, 1994a,b; Epstein *et al.*, 1995; Hales *et al.*, 1996; Checkley *et al.*, 2000).

Weather extremes, especially intense precipitation, have been especially punishing for developing nations, and the aftershocks ripple through economies. Hurricane Mitch – nourished by a warmed Caribbean – stalled over Central America in November 1998 for three days, dumping precipitation that killed over 11 000 people and left over $5 billion in damages. In the aftermath, Honduras reported 30 000 cases of cholera, 30 000 cases of malaria and 1000 cases of dengue fever (Epstein, 1999a,b). The following year Venezuela suffered a similar fate, followed by malaria and dengue fever. Then, in February 2000, torrential rains and a cyclone inundated large parts of Southern Africa. Floods in Mozambique killed hundreds, displaced hundreds of thousands and spread malaria, typhoid and cholera. The island nation of Madagascar, extremely vulnerable because of its shrinking forest coverage, suffered enormous damage.

A central challenge for climatologists is to better understand how the ENSO cycle interacts with other natural cycles, such as the North Atlantic Oscillation (that governs winds across Europe) and the Pacific Decadal Oscillation (a cooling of which could dampen global warming for decades). Furthermore, it will be crucial to understand how anthropogenically induced long-term changes (e.g. warm winters) interact with extreme events (e.g. summer droughts) to influence ecological systems and disease patterns. Of utmost concern is the potential impact of *sequential extremes* (e.g. droughts followed by intense precipitation) in destabilizing functional group relationships, such as those among predator and prey, essential for controlling the proliferation of pests and pathogens. Drought followed by anomalous winter rains was a key sequence in the explosion of mice populations in the U.S. south-west in 1993, which potentiated the surprise outbreak of hantavirus pulmonary syn-drome (Engelthaler *et al.*, 1999). Winter warming followed by prolonged drought appears to favour the emergence of urban mosquito-borne diseases such as St. Louis encephalitis and west Nile fever (Epstein, 2000).

7.5 Implications for future research

If recent global climate trends continue and it becomes more certain that this rep-resents the beginning of anthropogenic climate change, then epidemiologists must find better ways of studying the evidence for the public health impacts. This will include the formalization of pattern recognition, often across multiple impacted systems. Forms of pattern recognition will assist the assessment of climate-change impacts on heat-related deaths, the incidence and seasonality of allergic disorders, and the geographical range and seasonality of particularly climate-sensitive infec-tious diseases.

It is likely that some of the first detectable health impacts will be changes in the geographical range (latitude and altitude) of certain vector-borne infectious diseases and/or in the seasonality of these diseases. More local data are needed to assess changes in latitude and elevation, as per the methodology adopted by IPCC Working Group II in its 2001 assessment. In a warmer world, summer-time food-borne infections (e.g. salmonellosis) may show longer-lasting annual peaks. If such events become more, or less, severe, then it would be possible to detect changes in the magnitude of health impacts associated with such events. If extreme weather events (e.g. heatwaves, floods, droughts) become more frequent, then the detectability of altered health impacts will depend principally on the extent to which the frequency and magnitude of such events (or "exposures") have increased.

The time frame of the emergence of the health impacts of climate change will depend on two main factors:

- The "incubation" period, that is the delay between the environmental event and the onset of illness. The incubation period ranges from near zero (storm-induced injury for example), to weeks or months (vector-borne infections), to years and decades (UV-related skin cancers).
- Factors influencing "detectability". The time of first detectability of health impacts of climate change will depend on:
  (i) the sensitivity of response (i.e. how steep the rate of increase is);
  (ii) whether there is a threshold that, once surpassed, results in a "step function" change; and
  (iii) the density of nonclimatic cofactors as confounders (i.e. factors that diminish the signal-to-noise ratio).

We must also consider that we may underestimate both the rate of climate change, thus the exposure to which biological systems are being subjected to, *and* the sensitivity of biological systems to small changes in temperature and the accompanying extremes. In terms of the exposure, recent work estimated that the rate of warming accelerated to 3 °C per century in the late 1990s (McCarthy *et al.*, 2001). In general, since 1950, minimum temperatures are rising at twice the overall rate of warming (Easterling *et al.*, 1997). Recent work demonstrates that minimum temperature warming is occurring at even faster rates at high attitudes in the Himalayas (Thompson *et al.*, 2000), as well as at faster rates in Boreal and Arctic regions compared to temperate and tropical latitudes (Albritton *et al.*, 2001; Honghton *et al.*, 2001).

Furthermore, an increase in the intensity and frequency of extreme weather events may have the most severe impacts on public health via two pathways. The first is the association of clusters of rodent-, water- and mosquito-borne diseases in the aftermath of extreme events (Epstein, 1999a,b). The second is through the decoupling of biological controls over pests from early springs, excessive winter warming and weather volatility. Devising ways of monitoring, mapping and modelling the upsurges of nuisance organisms associated with long-term climate change and weather extremes poses new challenges for the burgeoning field of assessing the present and future biological consequences of global climate change. Full appreciation of the biological trends associated with climate change (McCarthy *et al.*, 2001) could provide the impetus for adopting more urgent precautionary measures.

7.6 Conclusions

There is increasing evidence of decadal-to-centennial warming at high elevations and at deep ocean depths, while the Earth's surface experiences increasingly volatile weather patterns. At high elevations, the overall trends regarding glaciers, plants, insects and temperatures show remarkable internal consistency, and there is consistency between model projections and the ongoing changes. Several implications may be drawn from the following conclusions:

- Montane regions – where isotherms may first noticeably shift, and where physical and biological responses may be most easily ascertained – can serve as sentinel areas to monitor climate change.
- Shifts in mountain isotherms may become evident before shifts in latitudinal isotherms are discernible.
- Mutually coherent physical and biological evidence, consistent with decadal-to-century climate change, is now present in mountain regions.
- The current and projected expansion of vector-borne diseases into the subtropics and to higher elevations calls for heightened vigilance by public health officials in

montane areas and for those populations living on the fringes of regions now
affected.

• Health early warning systems (HEWS) of climate conditions *conducive* to
outbreaks and disease clusters may become feasible – enabling early,
environmentally sound public health interventions (e.g. immunizations,
neighbourhood clean-ups, *Bacillus thuringiensis* applications, and others).

Control of insect populations is central to the health of forests, crops, livestock and
humans. Vulnerability to insect infestations is a function of social conditions and
abilities to reduce breeding sites and protect populations. Targeted pesticide use has
a role. But pesticide efficacy is being compromised by several factors: (i) increasing
insect resistance, (ii) direct human health concerns about food and ground water
contamination, and (iii) toxicity to friendly insects (e.g. dragon-flies and ladybugs),
birds, reptiles and fish (that consume mosquito larvae) – the diversity of predators
that biologically control the proliferation of biting and herbivorous insects.

Human activities are altering atmospheric chemistry and changing the Earth's
heat budget. Together, these chemical and physical changes, compounded by large-
scale land-use and land-cover changes, have begun to affect biological systems.
Policy-makers and the public must be increasingly concerned with the biological
consequences and the societal costs associated with climate change. The waste products of fossil fuel combustion have direct consequences for human and ecological
health through air pollution and acid precipitation. Now the aggregate of emissions
appears to be destabilizing Earth's heat budget, and changing the conditions under
which opportunistic pests and pathogens can thrive.

Acknowledgements

This chapter has drawn from Epstein *et al.* (1998) and Kovats *et al.* (2000, 2001).
P. Martens and A. J. McMichael provided helpful suggestions regarding research
criteria and methods.

References

Albritton, D. L., Allen, M. R., Baede, A. P. M. *et al.* (2001). *IPCC Working Group I
Summary for Policy Makers, Third Assessment Report: Climate Change 2001: The
Scientific Basis.* New York City, NY: Cambridge University Press.
Anderson, P. K. & Morales, F. J. (1993). The emergence of new plant diseases: the case
of insect-transmitted plant viruses. In *Disease in Evolution: Global Changes and
Emergence of Infectious Diseases*, ed. M. E. Wilson, R. Levins & A. Spielman,
pp. 181–94. New York: NY Academy of Sciences.
Barry, J. P., Baxter, C. H., Sagarin, R. D. Gilman, S. E. (1995). Climate-related, long-
term faunal changes in a California rocky intertidal community. *Science*, **267**,
672–75.

Billett, J. D. (1974). Direct and indirect influences of temperature on the transmission of parasites from insects to man. In *The Effects of Meteorological Factors Upon Parasites*, ed. A. E. R. Taylor & R. Muller, pp. 79–95. Oxford: Blackwell Scientific Publication.

Billings, D. W. (1995). What we need to know: some priorities for research biotic feedbacks in a changing biosphere. In *Biotic Feedbacks in the Global Climate System*, ed. G. M. Woodwell & F. T. Mackenzie, pp. 377–92. New York: Oxford University Press.

Bindoff, N. L. & Church, J. A. (1992). Warming of the water column in the southwest Pacific. *Nature*, **357**, 59–62.

Bouma, M. J., Sondorp, H. E. & van der Kaay, J. H. (1994a). Health and climate change. *Lancet*, **343**, 302.

Bouma, M. J., Sondorp, H. E. & van der Kaay, J. H. (1994b). Climate change and periodic epidemic malaria. *Lancet*, **343**, 1440.

Bradshaw, W. E. & Holzapfel, U. (2001). Genetic shift in photoperiodic response correlated with global warming. *Proceedings of the National Academy of Science of the USA*, **498**, 14509–11.

Burgos, J. J. (1990). Analogias agroclimatologicas utiles para la adaptacion al posible cambio climatico global de America del Sur. *Revista Geofisica*, **32**, 79–95.

Burgos, J. J., Curto de Casas, S. I., Carcavallo, R. U. & Galindez, G. I. (1994). Global climate change in the distribution of some pathogenic complexes. *Entomologia y Vectores*, **1**, 69–82.

Centers for Disease Control and Prevention CDC (1995). Local transmission of *Plasmodium vivax* malaria – Houston, Texas, 1994. *Morbidity and Mortality Weekly Review*, **44**, 295–303.

CDC (1996). Mosquito-transmitted malaria – Michigan, 1995. *Morbidity and Mortality Weekly Review*, **45**, 398–400.

CDC (1997). Probable locally acquired mosquito-transmitted *Plasmodium vivax* infection – Georgia, 1996. *Morbidity and Mortality Weekly Report*, **46**, 264–7.

Checkley, W., Epstein, L. D., Gilman, R. H., Figueroa, D., Cama, R. I. & Patz, J. A. (2000). Effects of El Niño and ambient temperature on hospital admissions for diarrhoeal diseases in Peruvian children. *Lancet*, **355**, 442–50.

Dahlstein, D. L. & Garcia, R. (eds.) (1989). *Eradication of Exotic Pests: Analysis with Case Histories*. New Haven, CT: Yale University Press.

Davis, M. B. (1989). Lags in vegetation response to greenhouse warming. *Climatic Change*, **15**, 75–82.

Davis, M. B. & Zabinski, C. (1992). Changes in geographical range resulting from greenhouse warming: effects on biodiversity in forests. In *Global Warming and Biological Diversity*, ed. R. L. Peters & T. E. Lovejoy, pp. 297–308. New Haven, CT: Yale University Press.

DeMeillon, B. (1934). Observations on *Anopheles funestus* and *Anopheles gambiae* in the Transvaal. *Publications of the South African Institute of Medical Research*, **6**, 195–248.

Diaz, H. F. & Graham, N. E. (1996). Recent changes in tropical freezing heights and the role of sea surface temperature. *Nature*, **383**, 152–5.

Diaz, H. F., Beniston, M. & Bradley, R. S. (1997). *Climatic Change at High Elevation Sites*, pp. 1–298. Dordrecht: Kluwer.

Dobson, A. & Carper, R. (1993). Biodiversity. *Lancet*, **342**, 1096–9.

Easterling, D. R., Horton, B., Jones, P. D., Peterson, T. C., Karl, T. R., Parker, D. E., Salinger, M. J., Razuvayev, V., Plummer, N., Jamason, P. & Folland, C. K. (1997). Maximum and minimum temperature trends for the globe. *Science*, **277**, 363–7.

Easterling, D. R., Meehl, G. A., Parmesan, C., Changnon, S. A., Karl, T. R. & Mearns, L. O. (2000). Climate extremes: observations, modeling, and impacts. *Science*, **289**, 2068–74.

Elias, S. A. (1994). *Quaternary Insects and Their Environments*. Washington DC: Smithsonian Institution Press.

Engelthaler, D. M. *et al.* (1999). Climatic and environmental patterns associated with hanta virus pulmonary syndrome, Four Corners region, United States. *Emerging Infectious Diseases*, **5**, 87–94.

Epstein, P. R. (ed.) (1999a). Extreme weather events: the health and economic consequences of the 1997/98 El Niño and La Niña. Harvard Medical School Boston, MA: Center for Health and the Global Environment. http://chge2.med.harvard. edu/enso/disease.html.

Epstein, P. R. (1999b). Climate and health. *Science*, **285**, 347–8.

Epstein, P. R. (2000). Is global warming harmful to health? *Scientific American*, August, 50–7.

Epstein, P. R. & Chikwenhere, G. P. (1994). Biodiversity questions (Ltr). *Science*, **265**, 1510–11.

Epstein, P. R. & Defilippo, C. (2001). West-Nile virus and drought. *Global Change and Human Health*, **2**(2), 105–7.

Epstein, P. R., Pena, O. C. & Racedo, J. B. (1995). Climate and disease in Colombia. *Lancet*, **346**, 1243.

Epstein, P. R., Diaz, H. F., Elias, S., Grabherr, G., Graham, N. E., Martens, W. J. M., Mosley-Thompson, E. & Susskind, J. (1998). Biological and physical signs of climate change: focus on mosquito-borne disease. *Bulletin of the American Meteorological Society*, **78**, 409–17.

Focks, D. A., Daniels, E., Haile, D. G. & Keesling, L. E. (1995). A simulation model of the epidemiology of urban dengue fever: literature analysis, model development, preliminary validation, and samples of simulation results. *American Journal of Tropical Medicine and Hygiene*, **53**, 489–506.

Gill, C. A. (1920). The relationship between malaria and rainfall. *Indian Journal of Medical Research*, **7**, 618–32.

Gill, C. A. (1921). The role of meteorology and malaria. *Indian Journal of Medical Research*, **8**, 633–93.

Grabherr, G., Gottfried, N. & Pauli, H. (1994). Climate effects on mountain plants. *Nature*, **369**, 447.

Graham, N. E. (1995). Simulation of recent global temperature trends. *Science*, **267**, 666–71.

Haeberli, W. (1995). Climate change impacts on glaciers and permafrost. In *Potential Ecological Impacts of Climate Change in the Alps and Fennoscandanavian Mountains*, ed. A. Guisan, J. I. Holton, R. Spichiger & L. Tessier, pp. 97–103. Geneva: Conservatoire et Jardin Botaniques de Genève.

Haines, A., McMichael, A. J. Epstein, P. R. (2000). Global climate change and health. *Canadian Medical Association Journal*, **163**, 729–34.

Hales, S., Weinstein, P. & Woodward, A. (1996). Dengue fever in the South Pacific: driven by El Niño Southern Oscillation? *Lancet*, **348**, 1664–5.

Harvell, C. D., Kim, K., Burkholder, J. M., Colwell, R. R., Epstein, P. R., Grimes, J., Hofmann, E. E., Lipp, E., Osterhaus, A. D. M. E., Overstreet, R., Porter, J. W., Smith, G. W. Vasta, G. (1999). Diseases in the ocean: emerging pathogens, climate links, and anthropogenic factors. *Science*, **285**, 1505–10.

Hastenrath, S. & Kruss, P. D. (1992). Greenhouse indicators in Kenya. *Nature*, **355** (6360), 503–4.

Houghton, J. T., Ding, Y., Griggs, D. J., Noguer, M., van der Linden, P. J., Dai, X., Maskell, K. & Johnson, C. A. (2001). *IPCC Working Group I Third Assessment Report: Climate Change 2001: The Scientific Basis.* New York City, NY: Cambridge University Press.

Intergovernmental Panel on Climate Change (IPCC) (1996). *Climate Change '95: The Science of Climate Change. Contribution of Working Group I to the Second Assessment Report of the IPCC*, ed. J. T. Houghton, L. G. Meiro Filho, B. A. Callander, N. Harris, A. Kattenberg & K. Maskell, p. 149, pp. 370–4. Cambridge, UK: Cambridge University Press.

Jacobson, G. L. Jr., Webb T. III & Grimm, E. C. (1987). Patterns and rates of vegetation change during the deglaciation of eastern North America. In *North America and Adjacent Oceans During the Last Deglaciation. The Geology of North America*, ed. W. F. Ruddiman & H. E. Wright Jr., pp. 277–88, vol. K-3. Boulder, CO: Geological Society of America.

Karl, T. R., Jones, P. D., Knight, R. W., Kukla, G., Plummer, N., Razuvayev, V., Gallo, K. P., Lindsay, J., Charlson, R. J. & Peterson, T. C. (1993). A new perspective on recent global warming: asymmetric trends of daily maximum and minimum temperature. *Bulletin of the American Meteorological Society*, **74**, 1007–23.

Karl, T. R., Knight, R. W., Easterling, D. R. & Quayle, R. G. (1995a). Trends in U.S. climate during the twentieth century. *Consequences*, **1**, 3–12.

Karl, T. R., Knight, R. W. & Plummer, N. (1995b). Trends in high-frequency climate variability in the twentieth century. *Nature*, **377**, 217–20.

Karl, T. R., Nicholls, N. & Gregory, J. (1997). The coming climate. *Scientific American*, May, 78–83.

Kaser, G. & Noggler, B. (1991). Observations on Speke Glacier, Ruwenzori Range, Uganda. *Journal of Glaciology*, **37**, 513–18.

Koopman, J. S., Prevots, D. R., Marin, M. A. V., Dantes, H. G., Aquino, M. L .Z., Longini, I. M. Jr. & Amor, J. S. (1991). Determinants and predictors of dengue infection in Mexico. *American Journal of Epidemiology*, **133**, 1168–78.

Kovats, R. S., Campbell-Lendrum, D., Reid, C. & Martens, P. (2000). *Climate and Vector-Borne Diseases: An Assessment of the Role of Climate in Changing Disease Patterns.* ICIS/ UNEP/ LSHTM, October 2000, Maastricht University.

Kovats, R. S., Campbell-Lendrum, D. H., McMichael, A. J., Woodward, A. & Cox, J. (2001). Early effects of climate change: do they include changes in vector-borne diseases? *Philosophical Transactions*, **356**, 1057–68.

Krabill, W., Frederick, E., Manizade, S., Martin, C., Sonntag, J., Swift, R., Thomas, R., Wright, W. & Yungel, J. (1999). Rapid thinning of parts of the southern Greenland Ice Sheet. *Science*, **283**, 1522–4.

Leaf, A. (1989). Potential health effects of global climate and environmental changes. *New England Journal of Medicine*, **321**, 1577–83.

Leeson, H. S. (1939). Longevity of *Anopheles maculipennis* race *atroparvus*, Van Theil, at controlled temperature and humidity after one blood meal. *Bulletin of Entomological Research*, **30**, 103–301.

Levitus, S., Antonov, J. I., Boyer, T. P. & Stephens, C. (2000). Warming of the world ocean. *Science*, **287**, 2225–9.

Lindgren, E. & Gustafson, R. (2001). Tick-borne encephalitis in Sweden and climate change. *Lancet*, **358**, 16–18.

Lindgren, E., Tälleklint, L. & Polfeld, T. (2000). Impact of climatic change on the northern latitude limit and population density of the disease-transmitting European tick *Ixodes ricans. Environmental Health Perspectives*, **108**, 119–23.

Lindsay, S. W. & Martens, P. (1998). Malaria in the African highlands: past, present and future. *Bulletin of the World Health Organization*, **76**(1), 33–45.

Loevinsohn, M. (1994). Climatic warming and increased malaria incidence in Rwanda. *Lancet*, **343**, 714–18.

MacArthur, R. H. (1972). *Geographical Ecology*. New York: Harper & Row.

Maldonado, Y. A., Nahlen, B. L., Roberto, R. R. *et al.* (1990). Transmission of *Plasmodium vivax* malaria in San Diego County, California, 1986. *American Journal of Tropical Medicine and Hygiene*, **42**, 3–9.

Mann, M. B., Bradley, R. S. Hughs, M. K. (1998). Global-scale temperature patterns and climate forcing over the past six centuries. *Nature*, **392**, 779–87.

Martens, P., Jetten, T. H. & Focks, D. (1997). Sensitivity of malaria, schistosomiasis and dengue to global warming. *Climatic Change*, **35**, 145–56.

Martin, P. H. & Lefebvre, M. G. (1995). Malaria and climate: sensitivity of malaria potential transmission to climate. *Ambio*, **24**, 200–9.

Maskell, K., Mintzer, I. M. & Callander, B. A. (1993). Basic science of climate change. *Lancet*, **342**, 1027–31.

Matola, Y. G., White, G. B. Magayuka, S. A. (1987). The changed pattern of malaria endemicity and transmission at Amani in the eastern Usambara mountains, north-eastern Tanzania. *Journal of Tropical Medicine and Hygiene*, **90**, 127–34.

Matsuoka, Y. & Kai, K. (1994). An estimation of climatic change effects on malaria. *Journal of Global Environment Engineering*, **1**, 1–15.

McCarthy, J. J., Canziani, O. F., Leary, N. A., Dokken, D. J. & White, K. S. (eds.) (2001). *Climate change 2001: Impacts, Adaptation, and Vulnerability.* IPCC Working Group III, Third Assessment Report Geneva: IPCC.

McMichael, A. J., Haines, A. & Slooff, R. (eds.) (1996). *Climate Change and Human Health.* [Task Force: McMichael, A. J., Ando, M., Carcavallo, R., Epstein, P. R., Haines, A., Jendritzky, G., Kalkstein, L. S., Odongo, R. A., Patz, J., Piver, W. T. & Slooff, R.] World Health Organization, World Meteorological Organization, United Nations Environmental Program, Geneva, Switzerland.

Morse, S. S. (1995). Factors in the emergence of infections diseases. *Emerging Infections Diseases*, **1**, 7–15.

Molineaux, L. (1988). In *Malaria, Principles and Practice of Malariology* (vol. 2), ed. W. H. Wernsdorfer & I. McGregor, pp. 913–98. New York: Churchill Livingstone.

Mosley-Thompson, E. (1997). Glaciological evidence of recent environmental changes. Presentation: AAG, Fort Worth, Texas, April 3.

Overpeck, J. T., Bartlein, P. J. & Webb, III T. (1991). Potential magnitude of future vegetation change in eastern North America: comparisons with the past. *Science*, **254**, 692–5.

Parkinson, C. L., Cavalieri, D. J., Gloersen, P., Zwally, H. J. & Comiso, J. C. (1999). Spatial distribution of trends and seasonality in the hemispheric sea ice covers. *Journal of Geophysical Research.* **104**, 20827–35.

Parmesan, C. (1996). Climate and species' range. *Nature*, **382**, 765.

Parmesan, C. (1999). The world as a patchwork. *Nature*, **399**, 747.

Parmesan, C., Ryholm, N., Stefanescu, C., Hill, J. K., Thomas, C. D., Descimon, H., Huntley, B., Kaila, L., Kullberg, J., Tammaru, T., Tennent, W. J., Thomas, J. A. & Warren, M. (1999). Poleward shifts in geographical ranges of butterfly species associated with regional warming. *Nature*, **399**, 579–83.

Parrilla, G., Lavin, A., Bryden, H., Garcia, M. & Millard, R. (1994). Rising temperatures in the sub-tropical North Atlantic Ocean over the past 35 years. *Nature*, **369**, 48–51.

Patz, J. A., Epstein, P. R., Burke, T. A. & Balbus, J. M. (1996). Global climate change and emerging infectious diseases. *Journal of the American Medical Association*, **275**, 217–23.

Pauli, H., Gottfried, M. & Grabherr, G. (1996). Effects of climate change on mountain ecosystems – upward shifting of alpine plants. *World Resource Review*, **8**, 382–90.

Peters, R. L. (1991). Consequences of global warming for biological diversity. In *Global Climate Change and Life on Earth*, ed. R. L. White. New York: Routledge, Chapman and Hall.

ProMED (1996). Malaria, autochthonous – Florida, USA. *ProMED Archives*, reported by Roger Spitzer, 2 Aug 1996.

ProMED (1997). Malaria – Indonesia (Irian Jaya). *ProMED Archives*, reported by Budi Subianto, 23 Dec 1997.

ProMED (1999). V99.n213, September, 1999. Malaria in New York City http://www.medscape.com/other/ProMED/public/archive.html

Rajagopalan, B., Kumar, K. K. & Cane, M. A. (1999). On the weakening relationship between the Indian monsoon and ENSO. *Science*, **284**, 2156–9.

Reeves, W. C., Hardy, J. L., Reisen, W. K. & Milby, M. M. (1994). Potential effect of global warming on mosquito-borne arboviruses. *Journal of Medical Entomology*, **31**, 323–32.

Regaldo, A. (1995). Listen up! The world's oceans may be starting to warm. *Science*, **268**, 1436–7.

Reisen, W. K., Meyer, R. P., Preser, S. B. & Hardy, J. L. (1993). Effect of temperature on the transmission of western equine encephalomyelitis and St. Louis encephalitis viruses by *Culex tarsalis* (Diptera: *Culicadae*). *Journal of Medical Entomology*, **30**, 51–160.

Retallack, G. J. (1997). Early forest soils and their role in Devonian global change. *Science*, **276**, 583–5.

Roemmich, D. & McGowan, J. (1995). Climatic warming and the decline of zooplankton in the California current. *Science*, **267**, 1324–6.

Rosenzweig, C. & Hillel, D. (1998). *Climate Change and the Global Harvest*, pp. 101–22. New York: Oxford University Press.

Rosenzweig, C., Iglesias, A., Yang, X. B., Epstein, P. R. & Chivian, E. (2000). *Climate Change and U.S. Agriculture: the Impacts of Warming and Extreme Weather Events on Productivity, Plant Diseases, and Pests*. Harvard Medical School, Boston, MA: Center for Health and the Global Environment.

Rosenzweig, C., Iglesias, A., Yang, X. B., Epstein, P. R. & Chivian, E. (2001). Climate change and extreme weather event. Implications for food production plant diseases, and pests. *Global Change and Human Health*, **2**(2), 90–104.

Rothrock, D. A., Yu, Y. & Marykut, G. A. (1999). Thinning of the Arctic Sea-ice cover. *Journal of Geophysical Research Letters*, **26**, 3469–72.

Rozendaal, J. (1996). *Assignment Report: Malaria. Pt. Moresby, Papua New Guinea*. Geneva: World Health Organization.

Santer, B. D., Taylor, K. E., Wigley, T. M. L., Johns, T. C., Jones, P. D., Karoly, D. J., Mitchell, J. F. B., Oort, A. H., Penner, J. E., Ramaswamy, V., Schwarzkopf, M. D., Stouffer, R. J. & Tett, S. (1996). A search for human influences on the thermal structure of the atmosphere. *Nature*, **382**, 39–46.

Shepherd, A., Wingham, D. J., Mansley, J. A. D. & Corr, H. F. J. (2001). Inland thinning of Pine Island Glacier, West Antarctica. *Science*, **291**, (5505), 862–4.

Shope, R. (1991). Global climate change and infectious disease. *Environmental Health Perspectives*, **96**, 171–4.

Some, E. S. (1994). Effects and control of highland malaria epidemic in Uasin Gishu District, Kenya. *East African Medical Journal*, **71**(1), 2–8.

Suarez, M. F. & Nelson, M. J. (1981). Registro de altitud del *Aedes aegypti* en Colombia. *Biomedica*, **1**, 225.

Susskind, J., Piraino, P., Rokke, L., Iredell, L. & Mehta, A. (1997). Characteristics of the TOVS Pathfinder Path A data set. *Bulletin of the American Meteorological Society*, **78**, 1449–72.

Sutherst, R. W. (1990). Impact of climate change on pests and diseases in Australasia. *Search*, **21**, 230–2.

Tälleklint, L. & Jaenson, T. G. T. (1998). Increasing geographical distribution and density of *Ixodes Ricans* (Acari: Ixodidae) in central and northern Sweden. *Journal of Medical Entomology*, **35**, 521–6.

Thompson, L. G., Mosley-Thompson, E., Davis, M., Lin, P. N., Yao, T., Dyurgerov, M. & Dai, J. (1993). "Recent warming": ice core evidence from tropical ice cores with emphasis on Central Asia. *Global and Planetary Change*, **7**, 145–56.

Thompson, L. G., Yao, T., Mosley-Thompson, E., Davis, M. E., Henderson, K. A. & Lin, P.-N. (2000). A high-resolution millennial record of the South Asian Monsoon from Himalayan ice cores. *Science*, **289**, 1916–19 .

Thwaites, T. (1994). Are the antipodes in hot water? *New Scientist*, 12 November, p. 21.

Travis, J. (1994). Taking a bottom-to-sky "slice" of the Arctic Ocean. *Science*, **266**, 1947–8.

World Health Organization (WHO) (1996). *The World Health Report 1996*: *Fighting Disease, Fostering Development*. Geneva: World Health Organization.

Yoon, C. K. (1994). Warming moves plants up peaks, threatening extinction. *The New York Times*, 21 June, p. C4.

Zucker, J. R. (1996). Changing patterns of autochthonous malaria transmission in the United States: a review of recent outbreaks. *Emerging Infectious Diseases*, **2**, 37.

8

Integrated Assessment modelling of human health impacts

PIM MARTENS, JAN ROTMANS & DALE S. ROTHMAN

8.1 Introduction

Although mathematical modelling is often used by epidemiologists – to gain insights into the observed dynamics of infectious disease epidemics, for example, or to estimate future time trends in diseases – the complex task of estimating future trends and outcomes in relation to global atmospheric change and human health requires the use of integrated, systems-based mathematical models (McMichael & Martens, 1995). Looking at the complexity of the interactions between global (environmental) changes and human health, we need an integrated approach to ensure that key interactions, feedbacks and effects are not inadvertently omitted from the analysis. The various pieces of this complex puzzle can no longer be examined in isolation. Integrated Assessment (IA) aims to fit the pieces of the puzzle together, thereby indicating priorities for policy.

There is increasing recognition and credibility for the rapidly evolving field of IA. Within the setting of the political arena, it is accepted that IA can be supportive in the long-term policy-planning process, while in the scientific community more and more scientists realize the complementary value of IA research.

At present modelling is the dominant method of performing IA, including looking at global-change impacts on human health. However, many studies that are either explicitly or implicitly integrated assessments are still qualitative in nature, without using any models. Furthermore, the complexity of the issues addressed and the value-laden character of assessment activities make it impossible to address this process using only one approach. It is clear that the issues under concern are generally more complex, subtle and ambiguous than can ever be expressed by formal mathematical equations in models. In particular the decision-making processes (which are sequential and involve multiple actors), human values, preferences and choices are not, and probably never will be, captured by formal models.

In this chapter, we will focus on one of the analytical methods related to global (climate) changes and human health: Integrated Assessment modelling. Integrated

Assessment models (IAMs) are computer simulation frameworks that try to describe quantitatively as much as possible the cause–effect relationships of a specific issue, and the interlinkages and interactions among different issues. Before going in detail into the use and limitations of IAMs in health-impact assessments, we will first briefly outline the available IA approaches and the process of performing an IA.

8.2 Integrated Assessment

One of the problems of IA is the many definitions and interpretations that circulate (Parson, 1996; Ravetz, 1997; Rotmans & Dowlatabadi, 1998). Irrespective of whatever definition is taken, IA can be described as (Rotmans, 1998):

a structured process of dealing with complex issues, using knowledge from various scientific disciplines and/or stakeholders, such that integrated insights are made available to decision-makers.

IA is an iterative, continuing process, where integrated insights from the scientific and stakeholder community are communicated to the decision-making community, and experiences and learning effects from decision-makers form one input for scientific and social assessment. Although the participation of stakeholders is not a necessary prerequisite for such an assessment, the conviction in the IA community is growing that the participation of stakeholders is a vital element in IA. The engagement of nonscientific knowledge, values and preferences into the IA process through social discourse improves the quality of IA by giving access to practical knowledge and experience, and to a wider range of perspectives and options.

IA attempts to shed light on complex issues by illuminating different aspects of the issue under concern: from causes to impacts, and from options to strategies. IA has been widely applied in the global-change research area. IA emerged as a new field in global-change research because the traditional disciplinary approach to global-change research has been unable to meet two significant challenges central to our understanding of global phenomena. The first challenge is the development of an adequate characterization of the complex interactions and feedback mechanisms among the various facets of global change. Such feedbacks and interactions are defined away or treated parametrically in traditional disciplinary research. The second challenge is that of providing support for public decision-making. IA offers an opportunity to develop a coherent framework for testing the effectiveness of various policy strategies, and exploring trade-offs among different policy options.

In general, two types of IA methods can be distinguished: analytical methods and participatory methods. The divergence of methods employed is due to the uneven state of scientific knowledge across different problem domains, from the differences in problem perception, and from the different perspectives of the scientists and stakeholders involved in the assessment process (see Fig. 8.1). While analytical

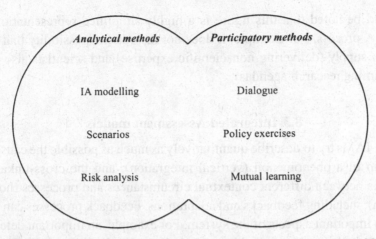

Fig. 8.1 Tool-kit for Integrated Assessment (Rotmans & van Asselt, 1999).

methods are often rooted in natural sciences or economics, participatory methods, also called interactive or communicative methods, stem from social sciences. The group of analytical methods is reasonably well defined and basically includes model analysis, scenario analysis and risk analysis. Their commonality is that they provide analytical frameworks for representing and structuring scientific knowledge in an integrated manner. The group of participatory methods, however, involves a plethora of methods, varying from expert panels, Delphi methods, to gaming, policy exercises and focus groups. They share the aim of involving nonscientists as stakeholders in the process, where the assessment effort is driven by stakeholder–scientist interactions. The aim of the various methods is to facilitate the IA process as sketched in Fig. 8.2.

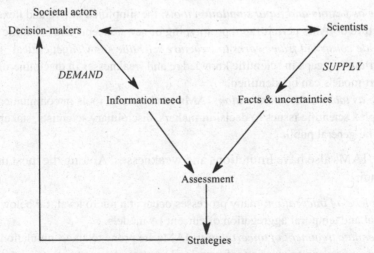

Fig. 8.2 Sketch of the IA process (Rotmans & van Asselt, 1999).

It should be noted that this figure is a highly simplified representation of the complex IA process. For example, it does not include the possibility that societal actors also supply (delivering nonscientific expertise) and scientists also demand (e.g. in framing research agendas).

8.3 Integrated Assessment models

In general, IAMs try to describe quantitatively as much as possible the cause–effect relationship of a phenomenon (vertical integration), and the cross-linkages and interactions between different contextual circumstances and processes (horizontal integration), including feedbacks and adaptations. Feedback processes can amplify or dampen important aspects of the system. For example, an important determinant of the number of people infected by malaria is the level of (temporary) immunity within the target population. In highly endemic regions with a high prevalence of immunity, the impact of a climate-related increase in the malaria transmission potential of the mosquito population will be low and will soon be counteracted by the further boost in immunity.

However, any attempt to fully represent a complex issue and its numerous inter-linkages with other issues in a quantitative model is doomed to failure. Nevertheless, even a simplified but integrated model can provide a useful guide to complex issues and complement highly detailed models that cover only some parts of complex phenomena.

Among the major strengths of IAMs are their ability to:

- *Explore interactions and feedbacks*: explicit inclusion of interactions and feedback mechanisms between subsystems can yield insights that disciplinary studies cannot offer.
- *Serve as flexible and rapid simulation tools*: the simplified nature and flexible structure of IAMs permit rapid prototyping of new concepts and scientific insights.
- *Provide consistent frameworks to structure scientific knowledge*: critical uncertainties, gaps in scientific knowledge and weaknesses in discipline-oriented expert models can be identified.
- *Serve as tools for communication*: IAMs can be useful tools in communicating complex scientific issues to decision-makers, disciplinary scientists, stakeholders and the general public.

Obviously, IAMs also have limitations and weaknesses. Among the most important ones are their:

- *High level of integration*: many processes occur at a micro level, far below the spatial and temporal aggregation of current IA models.
- *Inadequate treatment of uncertainties*: IAMs are prone to an accumulation of uncertainties, and to a variety of types and sources of uncertainty.

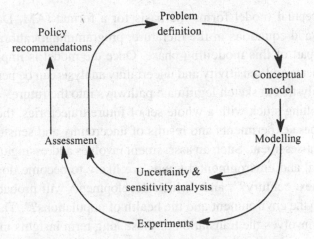

Fig. 8.3 IA cycle (Rotmans & van Asselt, 1999).

- *Absence of stochastic behaviour*: most IAMs describe processes in a continuous, deterministic manner, excluding extreme conditions that may significantly influence the long-term systems behaviour;
- *Limited calibration and validation*: the high level of aggregation implies an inherent lack of empirical variables and parameters, and current datasets are often too small and/or unreliable to apply.

These strengths and weakness of IAMs as well as the complementary value of other IA methods need to be considered when reading the present chapter.

In general, several steps can be distinguished when building an IAM (Fig. 8.3). Building an IAM starts with a problem definition that implies a certain problem perception. An example of a problem definition underlying IA projects in the field of climate change is, "Will a change in the global climate system in the future change the distribution of vector-borne diseases and, if so, how will that change manifest itself?"

The next step in this cycle is to design a conceptual model. One way is to adopt a systems-oriented approach inspired by systems theory which originated in the 1960s (e.g. Forrester, 1961, 1968; Bertalanffy, 1968; Goodman, 1974). Notwithstanding the relativity of the concept "system", the systems view always treats systems as integrated wholes of their subsidiary components and never as the mechanistic aggregate of parts in separate causal relations (Laszlo, 1972). Complex problems such as climate change, human health and sustainable development require an examination of the natural resilience and buffer capacity of natural reservoirs and the vulnerability of human populations in relation to anthropogenic disturbances. Such complex issues can be represented by a set of interconnected cause–effect chains. These inextricably interconnected cause–effect chains form a complex system, the properties of which are more than just the sum of its constituent subsystems.

Such a conceptual model forms the basis for a formal IAM. Data gathering, searching for valid equations in the literature, programming, calibration and validation are all part of this modelling phase. Once the model is implemented and tested, experiments and sensitivity and uncertainty analysis can be performed. This enables the analyst(s) to sketch legitimate pathways into the future.

Instead of getting stuck with a whole set of future trajectories, the challenge is to synthesize these experiments and results of uncertainty and sensitivity analysis into a coherent assessment. Such an assessment involves addressing questions such as "which social and environmental trends are likely to become dominant in the course of the next century?", and "which developments will probably constitute major threats to the environment and the health of populations?". The next step in the assessment involves the translation of these long-term insights into short-term policy recommendations.

The insights gathered by those assessment exercises reveal a description of the issue under concern that is likely to deviate from the problem perception in the initial phase of the scientific assessment. Ideally, the new understanding, in turn, evokes an iteration of the IA cycle.

BOX 8.1
Other modelling approaches

Next to IAMs, other modelling approaches have been used to assess the impact of global changes on human health (see Table 8.1). For example, in the estimation of future health impacts of climate change, biophysical impact and economic models are also being used. In general, IAMs make use of all of these approaches.

Table 8.1. *Models used to understand and assess potential health impacts of climate change*

Empirical-statistical models	Extrapolation of the climate–disease relationship in time and space, e.g. change of the distribution of vectors with change in climate
Process-based models	Models derived from accepted theory; can be applied universally, e.g. vector-borne disease risk forecasting with model based on vectorial capacity
Economic models	Model that estimates the likely effects of climate change on (measurable) economic quantities such as income
IAMs	Multidisciplinary linkage of process-based models, horizontally and vertically, e.g. impact of climate change on the transmission potential of the malaria mosquito and malaria prevalence

Biophysical impact models

Biophysical impact models are used to evaluate the biological and physical interactions between climate and health. There are two main types: (i) process-based models and (ii) empirical-statistical models.

Models are *empirical-statistical* if they are based on statistical relationships observed between climate and health (or health-related) outcome. This may range from applying simple indices of risk (for example, identifying the minimum temperature threshold for malaria transmission) to using complex multivariate models that combine the various important environmental factors that affect risk. Where these models are founded on good knowledge and where there are good grounds for extrapolation, they can be useful tools in human-health impact assessment. However, one limitation of this modelling approach is that there may be limited data points to calibrate the projections, making results difficult to validate. Generally, this approach relies little on underlying mechanisms, and often these models are not able to easily consider interactions or other variables not included in the available observations. However, empirical-statistical models are often simpler to use and less data-demanding than process-based models.

Process-based (or mechanistic or biological) models draw entirely or substantially on accepted theory (the product of previously accumulated, coherent, research findings). The models incorporate mathematical equations that represent processes that can be applied universally to similar systems in different environments. For example, an index that encapsulates many of the important processes in the transmission of vector-borne diseases is the vectorial capacity, which may apply to a range of vector-borne diseases and circumstances. In practice, however, these models are not always able to consider all of the important interactions. Outcomes in the form of projected changes in disease risk due to climatic changes are based on current interactions of physical and biological variables, and have not considered various kinds of adaptation or evolution in the many factors that determine transmission or host response. Although process-based models provide more insights in the underlying processes than do empirical-statistical models, they are also more data-demanding.

Economic models

Economic models (see also Section 8.4.2) are useful tools for estimating the likely effects of climate change on (measurable) economic quantities such as income, gross domestic product (GDP), employment and savings. However, most of the economic models fail to account for nonmarket and indirect effects of climate change, which are prevalent in considering health impacts. For example, many inputs to production are directly affected by climate change, such as labour loss due to ill-health, but are not contained in most (macro)economic models. Economic models are also widely used to consider the relative cost-effectiveness of mitigation and adaptation options for issues other than climate change, along with the associated economic, social and environmental impacts of these options.

These classes of models are not mutually exclusive. A process-based model may be used to model the essence of the process under study, combined with statistical approaches to extend the process to the population level. No single approach is superior for the creation of reliable models. Choosing the most appropriate modelling approach depends largely on the type of data available and the type of output needed for decision-making. An elaboration of existing models that analyse the climate–malaria link can be found in Box 8.2.

8.4 IA modelling of population and health

In modelling population dynamics, fertility modelling approaches are widely accepted but only recently have existing mathematical techniques been introduced in the broader health area (e.g. Weinstein *et al.*, 1987; Martens, 1998). Usually, regression techniques are used to explore the relationships between broad health determinants, such as literacy, income status, nutritional status, water supply and sanitation, education and medical services, and the health status measured as healthy life expectancy. However, these regression techniques can only give some suggestive evidence as to the causes of population and health changes. Statistical models to estimate future fertility and health levels are based on the extrapolation of past and current data. They operate on a short time horizon, and are static in terms of specifying the dynamics behind changing fertility and health patterns. Therefore, there is a need for integrated approaches that take into account the simultaneous occurrence of multiple risk factors and diseases, as well as cause–effect relationships. Such an integrated approach cannot be used in the clinical area on an individual basis, but is appropriate at the population level.

In Rotmans and de Vries (1997), an integrated systems approach to population and health is presented. A generic model was designed, which simulates the driving forces (socioeconomic and environmental factors), fertility behaviour, disease-specific mortality, burden of disease, life expectancy, the size and structure of the population, and a number of fertility and health policies. The major objective of the population and health model is to simulate changes in morbidity and mortality levels under varying social, economic and environmental conditions. Hilderink (2000) elaborated on this modelling framework and developed the PHOENIX model.

A second but related example is the MIASMA framework, which is designed to describe the major cause and effects relationships between atmospheric changes and human population health (Martens, 1999a). The model focuses on climate change (in terms of changes in temperature and precipitation) and ozone depletion (in terms of changes in UV-B radiation). Under varying climate and ozone regimes, changes in the dynamics and distribution of vector-borne diseases are simulated (malaria, schistosomiasis and dengue), changing patterns of skin cancer incidences and changing mortality levels as a result of thermal stress.

8.4.1 Three examples

8.4.1.1 Future food production and risk of hunger

Long-term changes in world climate would affect the amount of food available worldwide. IA modelling studies have recently addressed the potential global impacts of climate change on food supply. An integrated assessment of the impact of climate change on global food production (Rosenzweig & Parry, 1994; Parry *et al.*, 1999) used the definition of a population at "risk of hunger", based upon methods developed by the Food and Agriculture Organization (FAO) of the UN: defined as the population with an income insufficient to either produce or procure their food requirements.

Regionally based crop yield models are used to simulate the effects of climate change (a GCM-based scenario) and increased CO_2 (which has a fertilization effect) on the yield of the major cereal crops. An established world food trade model (the Basic Linked System) is then used to simulate the economic consequences of yield changes – these include changes in world food output and in world food prices. Assuming no change in climate, the model (Parry *et al.*, 1999) estimates that world cereal production grows from about 1800 million metric tonnes (mmt) in 1990 to about 4012 mmt in 2080, matching global food requirements throughout the period. Food prices are estimated to rise but the relative risk of hunger will decrease. These projections are consistent with those of FAO (based on expert judgement). They assume a 50 % liberalization of trade by 2020 and an annual increase in cereal yields of just under 1 %. Some consider these assumptions to be optimistic, but they are consensus best estimates.

Worldwide, the additional number of people at risk of hunger due to climate change is projected to be about 80 million by the 2080s. It is important to note that the direction and order of magnitude of the forecast changes are more important than absolute quantitative estimates; these are given here to indicate the relative importance of population growth, economic trends and climate change on future estimates of populations at risk of hunger. Further, global estimates mask important regional and local differences. Yields decrease in low latitude countries, in particular the arid and sub-humid tropics. Thus, effects on hunger will be exacerbated where countries are unable to afford adaptive measures, particularly the purchase of food from abroad.

Note also that these assessments are very long term and focus on average effects over space and time. At the local level (e.g. in very vulnerable areas) and over short periods (e.g. in spells of drought or flooding) the effects of climate change will be more adverse. There is also a need to include estimates of the impact of climate change upon plant pests and pathogens, taking into account local ecological circumstances.

8.4.1.2 Stratospheric ozone depletion, ultraviolet irradiation and skin cancer incidence

Stratospheric ozone depletion is, strictly speaking, a separate process from that of tropospheric accumulation of greenhouse gases and the resultant climate change. However, there are some physical interactions between the two processes. The modelling of future changes in skin cancer incidence in response to stratospheric ozone depletion illustrates well the use of relatively clear-cut, integrated process-based modelling.

Information on the general relationship between solar exposure and skin cancer is available from a large body of epidemiological research. It provides estimates of risk increments associated with different amounts of time, by life-stage, spent exposed to solar radiation. It has not been generally possible to measure (especially in retrospect) an individual's actual radiation exposure; rather, the exposure has been expressed in terms of such indices as person-time outdoors, frequency of severe exposure episodes, or category of occupation. Yet, in order to estimate how future changes in ultraviolet irradiance might alter the incidence of skin cancer, we require estimates of risk gradients associated with actual levels of UVR radiation exposure. It has been necessary to glean these from broad-brush population-level epidemiological studies that describe the relationship between average ambient local exposure levels and local skin cancer rates.

A model developed by Martens, Slaper and colleagues (Martens *et al.*, 1996; Slaper *et al.*, 1996) integrate the dynamic aspects of the source-to-risk causal chain: from production and emission of ozone-depleting substances, global stratospheric chlorine concentrations, local depletion of stratospheric ozone, the resulting increases in UV-B levels, and finally, the effects on skin cancer rates. Several delay mechanisms (Fig. 8.4) in the effect of ozone depletion on skin cancer rates are important, such as tumour development. In the case of ozone depletion, the separate scenarios modelled relate to various international agreements which restrict the production of ozone depleting substances. Thus, full compliance with the Copenhagen Amendments to the Montreal Protocol (the latest and most restrictive agreement), would lead to a peak in stratospheric chlorine concentration and ozone depletion around 2000, and to a peak in skin cancer by about 2050. The latter delay is mainly due to the fact that skin cancer incidence depends on the cumulative UV-B exposure.

Another factor described in the model is a "lifestyle factor" because skin cancer rates are very sensitive to sun exposure habits. In addition, as skin cancer occurs primarily among the elderly, the changing age-profile of the population is modelled.

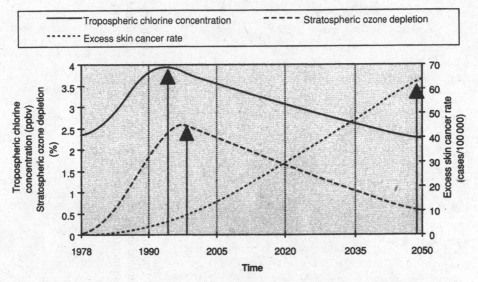

Fig. 8.4 Delay mechanisms in the cause–effect chain regarding the impact of stratospheric ozone depletion on skin cancer rates. Results are generated with the Copenhagen Amendments to the Montreal Protocol for skin cancer rates in Australia (Martens *et al.*, 1996). It should be noted that recent scientific developments foresee the problem continuing for one or more decades because of the interactions between global climate change and stratospheric ozone depletion.

An ageing population may experience a 50–60 % increase in overall incidence compared to a "younger" population with the same level of UV exposure (Fig. 8.5).

8.4.1.3 Climate change and vector-borne diseases

Vector-borne diseases are one of the most obvious examples of a category of health problems with complex, climate-related and ecologically based dynamics. The distribution of vector-borne diseases is limited by the climatic tolerance of their vectors and by biological restrictions that limit the survival of the infective agent in the vector population. In addition, certain human activities that help to prevent the

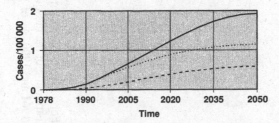

Fig. 8.5 Excess skin cancer rates (SCC) (for Australia as simulated with the Copenhagen Amendments to the Montreal protocol) with a "constant population" scenario (dotted line), an "ageing population" scenario (full line). The dashed line represents an ageing population with a 50 % decrease of UV exposure (Martens *et al.*, 1996).

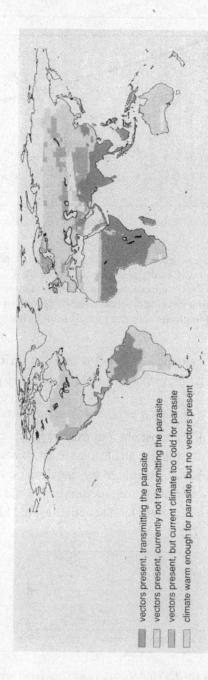

vectors present, transmitting the parasite
vectors present, currently not transmitting the parasite
vectors present, but current climate too cold for parasite
climate warm enough for parasite, but no vectors present

Fig. 8.6 For a colour version of this figure, see www.cambridge.org/9780521114028 Potential for the spread of malaria as shown by the current distribution of the mosquito vector and the climatic limits of the parasite, as estimated under conditions matching the average climate between 1961 and 1990. As of 1997, malaria is endemic (red) for only a fraction of its potential worldwide distribution (Martens, 1999b).

spread of pathogens and reduce vector populations restrict the distribution of many diseases in countries that can afford them. However, climate can play a dominant role in determining the distribution and abundance of insects, either directly or indirectly through its effects on host plants and animals. Therefore, it is anticipated that climate change will have a significant effect on the geographical range of many vector species (see Box 8.2 for studies related to changes in malaria risk).

Simulations with vector-borne diseases models (Martens, 1995a,b; Martens *et al.*, 1997, 1999) using climate change scenarios from several GCMs combined with the Transmission or Epidemic Potential index (which incorporates the basic dynamics of climatic influences on vector-borne disease transmission) show an increase of the populations at risk of malaria, dengue and schistosomiasis (see Fig. 8.6), three of the world's most prevalent vector-borne diseases. At present some 2400, 1800 and 600 million people are regarded as being at risk of contracting malaria, dengue and schistosomiasis, respectively. There would be an increase of the risk of local transmission of these three vector-borne diseases in developed countries, associated with imported cases of the disease. However, given the fact that effective control measures are economically feasible in these countries, it is not to be expected that human-induced climate changes would lead to a return of a state of endemicity in these areas.

Unfortunately, past events also teach us that local human-made environmental change, either directly or via changes in the local climate, can have an important effect on vector abundance and disease transmission, in particular deforestation, agriculture and water resource development (Lines, 1993; Lindsay & Martens, 1998). Untangling these relationships to find a climate change "signal" in observed changes in insect distributions is the subject of much current debate.

8.4.2 Economic analysis of health impacts in IAMs

The incorporation of economic analysis in the modelling of climate change and human health has, ironically, been limited but at the same time a source of great controversy. Perhaps this should not be surprising, as trying to place monetary values on changes in morbidity and mortality is an extremely difficult exercise and inherently raises moral and ethical issues. This is further complicated by the need to consider impacts in regions at very different stages of economic development and across generations.

In a review of the representation of impacts in 19 IAMs of climate change as of the year 1996 (little has happened since this time to change the general picture presented here), only eight include specific mention of health-related impacts as a damage category considered and, for two of these, the measurement of impact is only in physical units. Heat and cold stress, specifically mortality related to these, are the most common categories considered. Malaria and other vector-borne diseases are also considered in two of the models providing impact measures in monetary terms (Tol & Frankhauser, 1998). Other health-related impacts, including the impact of mortality and morbidity on economic production, are not included in any of the models.

Since it is the only aspect of human health that is commonly included in IAMs, and it has raised such controversy, it is important to mention a few points about how changes in mortality are translated into monetary values. Pearce *et al.* (1996) are careful to note that in economic and policy analysis, "what is valued is a change in the risk of death, not human life itself". When normalized to a change resulting in the risk of one more life being lost or saved, this is termed the Value Of a Statistical Life (VOSL). One of two perspectives is usually adopted for estimating the VOSL – prescriptive and descriptive. The former, driven primarily from a moral perspective, asks how much a life should be valued; the latter, driven from an economic perspective, asks how much a life has been or is valued based on the actual actions of individuals and societies. The descriptive approach has been the one adopted in almost all economic studies related to climate change. Two general approaches have been developed under the descriptive view – the human capital and Willingness-To-Pay/Willingness-To-Accept Compensation approaches. In the former, estimates of cumulative lost earnings or "contributions to society" are used to determine the VOSL. Obviously, this leads to higher values being placed on younger individuals with higher incomes. In the latter, hedonic methods, which examine wage and price differentials for jobs and products of varying risk, public and private expenditures on insurance and policy programmes, or stated preference methods that directly ask individuals how much they are willing to pay to avoid or accept to bear an increased risk of dying, are used to estimate the VOSL. These, too, will generally provide higher values for individuals with higher incomes.

One final point needs to be raised about how economic analysis has been incorporated in IAMs, specifically those that include monetary estimates of climate change impacts. Since estimates generally only exist for the temperature change associated with an equivalent doubling of atmospheric CO_2, the customary practice has been to use one of these estimates along with an assumption of zero damage at the pre-industrial level of CO_2, and to extrapolate a linear or curvilinear path through these two points (Tol & Frankhauser, 1998). Darmstadter and Toman (1993), among others, have discussed in more detail the importance of the assumption of the degree of nonlinearity assumed. Tol (1999) has explored how the rate, i.e. not just the level, of climate change may affect the degree of impacts. There also remains the question of whether it is appropriate to use a benchmark estimate based upon impacts of an equilibrium climate change on a present-day society and economy when the actual change will not be at equilibrium and it will affect a very different society.

BOX 8.2
Modelling the impact of climate change on malaria

Models incorporating a range of meteorological variables have been developed to describe a specific "bioclimate envelope" for malaria. Multivariate statistical techniques can be used to select predictive variables (whether meteorological or environmental, ground-based or remotely sensed). Models that match the presence of a particular species with a discrete range of temperature and precipitation parameters can be used to project the effect of climate change on vector redistribution.

Some mosquito species have been successfully mapped in Africa using meteorological data (e.g. Lindsay *et al.*, 1998). Satellite data are often used as surrogates for instrumental meteorological data; weather variables are usually measured at ground level but coverage can be relatively sparse or inappropriate, especially in developing countries (Hay *et al.*, 1996). In addition, more complex indices may be useful: the Normalized Difference Vegetation Index (NDVI) correlates with the photosynthetic activity of plants, rainfall and saturation deficit. Rogers has mapped the changes of three important disease vectors (ticks, tsetse flies and mosquitoes) in Southern Africa under three climate change scenarios (Hulme, 1996). The results indicate significant changes in areas suitable for each vector species, with a net increase for malaria mosquitoes (*Anopheles gambiae*). The final objective of such work is to map human disease risk but the relationship between vector-borne disease incidence and climate variables is complicated by many socioeconomic and environmental factors.

Another example of an empirical-statistical model is the CLIMEX model. This model, developed by Sutherst *et al.* (1995), maps the translocation of species between different areas as they respond to climate change. The assessment was based on an "ecoclimatic index" governed largely by the temperature and moisture requirements of the malaria mosquito. CLIMEX analyses conducted in Australia indicate that the indigenous vector of malaria would be able expand its range 330 km south under one typical scenario of climate change. However, these studies clearly cannot include all factors that affect species distributions. For example, local geographical barriers and interaction/competition between species are important factors that determine whether species colonize the full extent of suitable habitat (Davis *et al.*, 1998). Assessments may also include additional dynamic population (process-based) models (e.g. DYMEX (Sutherst, 1998)).

Martin and Lefebvre (1995), using a similar approach, developed a Malaria-Potential-Occurrence-Zone (MOZ) model. This model was combined with five GCMs (general circulation models) to estimate the changes in malaria risk based on moisture and minimum and maximum temperatures required for parasite development. This model corresponded fairly well with the distribution of malaria in the nineteenth century and the 1990s, after allowing for areas where malaria had been eradicated. An important conclusion of this modelling exercise was that all simulation runs showed an increase in seasonal (unstable) malaria transmission, under climate change, at the expense of perennial (stable) transmission.

Rogers and Randolph (2000), using a multivariate empirical-statistical model, found that, for the IS92a (business as usual) climate change scenario, there is no significant net change by 2050 in the estimated portion of world population living in malaria-transmission zones: malaria increased in some areas and decreased in others. The outcome variable in this model is based on present-day distribution limits of malaria. However, using current distribution limits in the estimate yields a biased estimation of the multivariate relationship between climatic variables and malaria occurrence since the lower temperature range, in temperate zones (especially Europe and southern U.S.), would have been treated as nonreceptive to malaria.

An integrated, process-based model to estimate climate-change impacts on malaria (that is part of the MIASMA modelling framework) has been developed by Martens and colleagues (Martens, 1995a, b, 1999a). This model differs from the others in that it takes a broad approach in linking GCM-based climate-change scenarios with a module that uses the formula for the basic reproduction rate (R_o) to calculate the "transmission or epidemic potential" of a malaria mosquito population. The use of R_o is defined as the number of new cases of a disease that will arise from one current case when introduced into a nonimmune host population during a single transmission cycle (Anderson & May, 1991). This goes back to classical epidemiological models of infectious disease. Model variables within R_o which are sensitive to temperature include: mosquito density, feeding frequency, survival and extrinsic incubation period. The extrinsic incubation period (i.e. the development of the parasite in the mosquito) is particularly important. The minimum temperature for parasite development is the limiting factor for malaria transmission in many areas. Tol and Dowlatabadi (2001) integrated the results of MIASMA within the FUND framework, developed by Tol, to estimate the trade-off between climate change and economic growth on malaria risk. The first results of this exercise show the importance of economic variables in estimating changes in future malaria risk.

All of the examples discussed above have their specific disadvantages and advantages. For example, the model developed by Rogers and Randolph (2000) incorporates information about the current social, economic and technological modulation of malaria transmission. It assumes that those contextual factors will apply in future in an unchanged fashion. This adds an important, though speculative, element of multivariate realism to the modelling – but the model thereby addresses a qualitatively different question from the biological model. The biological model of, for example, Martens and collegues (e.g. Martens, 1998, 1999a) assumes that there are known and generalizable biologically mediated relationships. Also, this modelling is only making a start to include the horizontal integration of social, economic and technical change. The statistical model is based on socioeconomically censored data. It derives its basic equation from the existing (constrained) distribution of

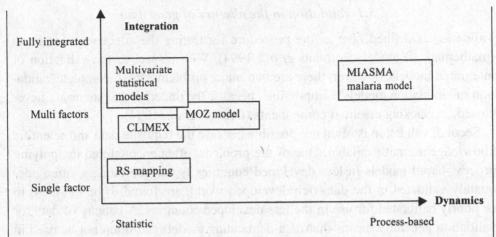

Fig. 8.7 Dimensions of integration and dynamics in climate-malaria modelling.

malaria in today's world and climatic conditions, and foregoes much information on the malaria/climate relationship within the temperate-zone climatic range. Yet this range is likely to be considerably important in relation to the marginal spread of malaria under future climate change.

In Fig. 8.7 the above models are depicted according to their dimension of integration and dynamics.

8.5 Critical methodological issues in IA modelling of human health

When dealing with complex, long-term issues surrounded with uncertainties, IA models seem to be an appropriate tool. Even a simplified but integrated model can provide a useful guide to complex issues and complement highly detailed models that cover only some parts of the complexity of disease epidemiology.

However, to improve the health risk assessment by means of integrated modelling, there are several key methodological issues that need our attention. First of all, there is a clear need to validate highly aggregated assessment models, like the MIASMA model, on a local or regional scale. The equations within a global model may well be inappropriate for particular local conditions. Second, the unavoidable uncertainties introduced within these models must be treated adequately. Third, quantitative and qualitative approaches need to be combined to obtain a better understanding of the complex human–environment interactions. Finally, although IA modelling deals with a variety of processes on different spatial and temporal scales, there is hardly any experience in linking different scale levels.

In the following sections, these methodological issues are discussed in more detail.

8.5.1 *Validation in the absence of good data*

Validation is defined here as the procedure for testing the adequacy of a given (mathematical) model (Rotmans *et al.*, 1994). With respect to the validation of integrated models, however, there are two major problems. First, complete validation of simulation models is impossible, because the underlying systems are never closed, i.e. lacking essential components (Oreskes *et al.*, 1994).

Second, validation is often not possible because the requisite data and scientific knowledge are not available. One of the problems often encountered in applying process-based models in less developed countries is that the models, often adequately validated in the data-rich developed world, are found to be ill suited to or poorly calibrated for use in the less developed countries. A paucity of data for validation generally means that data-demanding models can often not be used in such circumstances, and reliance has to be placed on less data-demanding models (Carter *et al.*, 1994). Unavailability of data necessitates reliance on simplified assumptions to generate an initial framework for analysis; this framework can be used to focus interdisciplinary communication on assessing health risks and identify priorities for future research. Although the use of such assumptions and simplifications potentially decreases the quantitative accuracy of the assessment, it should still allow for adequate prioritization and estimation of relative risk (Patz & Balbus, 1996).

Validation can be divided into different types (Rodin, 1990; Rykiel, 1996). Data (or pragmatic) validation requires concordance of the model's projections with observational datasets; this concordance can often be assessed by "testing" the model on historical datasets. However, as mentioned above, the relative inaccuracy and imprecision of ecological and epidemiological data places limits on the model's testability in this way. Conceptual validation requires that the hypotheses and the theoretical structures of the model reasonably describe the perceived real world. This implies that the model structure, relations, parameters and dynamic behaviour reflect the prevailing theoretical insights and the key facts relating to that part of reality that the model is supposed to represent.

However, although the underlying model relationships may reflect the prevailing theoretical insights well (e.g. the use of the basic reproduction rate or its derivatives to estimate malaria transmission dynamics), this does not necessarily mean that a conceptually valid model will make accurate projections. It remains difficult to validate the highly aggregated outcomes of the global models presented earlier. For example, there is great uncertainty regarding the "natural" limits of malaria distribution (Molineaux, 1988). This is due in part to the lack of historical records describing the presence of malaria, and also because, at the fringes, malaria transmission is highly unstable; that is, only infrequent transmission occurs.

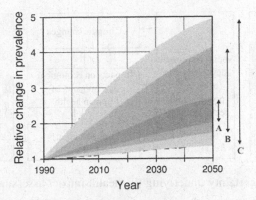

Fig. 8.8 The cumulation of uncertainties associated with climate-change projections, malaria mosquito transmission dynamics and malaria prevalence, based on variations of only some crucial parameters. This example, for *Plasmodium falciparum* in regions of low endemicity, shows how the uncertainty range widens as one moves to more remote parts of the cause–effect chain (Martens, 1998). Uncertainties in the estimates of changes in malaria prevalence are influenced by uncertainties in the outcomes of climate-change models (A), cumulated with uncertainties in the climate-related change of the transmission potential of a vector population (B), cumulated with uncertainties in the spread of the disease in the human population (C).

However, more confidence in global model outcomes can be obtained by validating the models on a local or regional scale, where data are at hand. For example, the malaria model developed by Martens *et al.* (Martens, 1995a, b, 1999a) has been (partially) validated at the country and regional level. For the African continent, and more specifically for Zimbabwe, the model was able to simulate both the seasonal pattern of transmission and the distribution of transmission intensity by altitude within the country (Lindsay & Martens, 1998). The model was also validated using a time series of malaria cases in Colombia. The model output reproduced the inter-annual variation in malaria cases and the historical peaks, the latter in part related to El Niño (Poveda *et al.*, 2000). Both exercises showed a qualitative agreement of the model and the observed risk of malaria, illustrating the role of temperature variability in modulating malaria transmission in these regions. Nevertheless, one needs to interpret the results with caution and realize that in other regions climate may not be as dominant in determining the seasonality of the transmission of malaria.

8.5.2 Cumulating uncertainties

Any exploration of future developments inevitably involves a considerable degree of uncertainty. Because of the cross-disciplinary character of IA modelling, it includes many different types and sources of uncertainty. Because IA models are end-to-end approaches, they also contain an accumulation of uncertainties (Fig. 8.8). Uncertainties may arise from incomplete knowledge of key physiological,

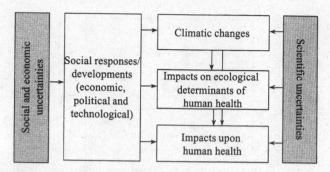

Fig. 8.9 Layers of uncertainty underlying the health-impact assessment of global climate change.

chemical and biological processes. Many uncertainties are of a socioeconomic nature – related to people's behaviour – and reflect inadequate knowledge with respect to the driving forces of human behaviour.

Various attempts have been made to classify the different types and sources of uncertainty. Morgan and Henrion (1990) distinguish uncertainty about empirical quantities and uncertainty about the functional form of models, which may have arisen from: subjective judgement; disagreement among experts; systematic errors; approximation; and inherent randomness. Funtowicz and Ravetz (1990) classified uncertainties in three categories: (i) technical uncertainties (concerning observations versus measurements); (ii) methodological uncertainties (concerning the right choice of analytical tools); and (iii) epistemological uncertainties (concerning the conception of a phenomenon).

Here, the various types and sources of uncertainty are aggregated into two categories (Rotmans *et al.*, 1994) (Fig. 8.9).

(i) Scientific uncertainties arise from the degree of unpredictability of global climate-change processes and their impact upon human health, and may be narrowed as a result of further scientific research or more detailed/appropriate modelling. These uncertainties include, for example, incomplete knowledge about the relationships between precipitation patterns and mosquito abundance.

(ii) Social and economic uncertainties arise from the inherent unpredictability of future geopolitical, socioeconomic, demographic and technological evolution. Examples of these future eventualities are vaccine development, trends in urbanization and levels of poverty, all of which affect the vulnerability and response of human populations to global changes.

Whatever classification is chosen, the various types and sources of uncertainties in IAMs need to be addressed in an adequate manner. One way of presenting uncertainties is by specifying a set of future scenarios, where the scenarios selected are expected to span a range of plausible, representative futures. The Intergovernmental Panel on Climate Change (IPCC), for example, adopted this approach for its greenhouse gas emissions scenarios. The difficulty with this approach is that it does not give an indication of the cumulative uncertainty as well as the origin and meaning of the uncertainty range.

Another approach is the probabilistic method, using a (subjective) probability distribution for empirical quantities. In this method a large numbers of inputs are specified as probability distribution functions, and a number of repeated model runs are done to determine the uncertainties surrounding the output(s), as in Fig. 8.8. The major difficulty with this method is that it requires specific knowledge about the nature of the distribution functions and the number of runs required. Furthermore, the usage of probability density functions is merely useful to address technical uncertainties. These techniques are not suitable for analysing methodological and epistemological uncertainties, which primarily arise from subjective judgements and fundamental disagreement among experts. Another problem is that classical uncertainty analysis methods only address uncertainties in model inputs and neglect the interactions among multiple, simultaneous uncertainties which are crucial in IA. Therefore, new methods are needed; for example, methods in which not only parameters but also relationships within the model are varied (according to the bias and preference of a particular perspective) (Janssen, 1998; van Asselt, 2000).

8.5.3 Blending qualitative and quantitative knowledge

In most frameworks for IA – including those related to human health – quantitative and qualitative knowledge are considered and treated as mutually exclusive. For instance, usually those aspects of a problem under concern that are not well known, or about which there is only vague and qualitative knowledge, are left out of the modelling process. This means, however, that we miss crucial links in the causal chains that form archetypal patterns of human–environment interactions. Quantitative rigour therefore prevents IA from being comprehensive, in the sense of studying all relevant aspects of a complex problem. It is therefore illusory to think that the full complexity of human–environment interactions could be integrated into a formal, quantitative modelling framework.

IA thus needs modelling frameworks that are combinations of quantitative and qualitative approaches, from the perspective that they complement each other. A promising way to blend quantitative and qualitative information is to incorporate vague and qualitative knowledge in IAMs, by using fuzzy logic (see e.g. Snow *et al.*, 1998). Contrary to Boolean logic, fuzzy logic makes use of a continuum of values, which represent elements of fuzzy knowledge.

8.5.4 A matter of scale

One of the most critical issues in IA modelling is that of aggregation versus disaggregation. The problem of modelling the impacts of global (environmental) change processes on human health is that it has to cope with a variety of processes that operate on different temporal and spatial levels, and differ in complexity.

First, IA modelling has to connect disciplinary processes that differ by nature: physical processes, monetary processes, social processes and policy processes. Because of the multitude of disciplinary processes to be combined, a simple as possible representation of disciplinary knowledge is preferable. There is, however, no unifying theory on how to do this. In addition, the processes to be linked are usually studied in isolation from each other. This isolation is needed as part and parcel of the classic model of scientific progress and discovery. However, when the constraints of isolation are removed, there is a variety of ways in which to connect the reduced pieces of disciplinary knowledge. This manifold of possible integration routes, for which there is no unifying theory, is one of the reasons why quality control is so difficult to achieve in IA modelling. For instance, in order to link the reduced pieces of disciplinary knowledge in a systemic way, one can use elements from classical systems analysis, the method of system dynamics, a sequential input-output analysis, a correlation-based approach, or a pressure-state-impact-response approach.

Second, IA modelling has to deal with different spatial scale levels. One of the ultimate challenges in IA modelling is to connect higher scale assessments with lower scale ones. So far, there has been hardly any experience of playing around with scale levels in IA modelling. Down-scaling or up-scaling the spatial level of a model has profound consequences. This is related to the question of to what extent the processes considered are generic, or spatially bound in character. In other words: does a relationship at one scale hold at larger or lower scale levels?

Third, IA modelling is faced with a multitude of temporal scales. Short-term needs and interests of stakeholders have to be considered. However, biogeochemical processes usually operate on a long time scale, whereas economic processes operate on short to medium time scales. Another challenging aspect of IA modelling is to interconnect long-term targets as specified as a result of analysing processes operating on longer term time scales, with short-term goals for concrete policy actions. Unfortunately, there is not yet a sound scientific method for doing this; thus far only heuristic methods have been used.

In Fig. 8.10, for example, some important factors determining malaria risk are depicted along "temporal" and "spatial" scale axes. Looking at the climate, human and mosquito system, it is apparent that they vary in their spatial and temporal scale: mosquito larval development takes place at the level of puddles and at time scales varying from days to weeks, climate change is a process influencing the global

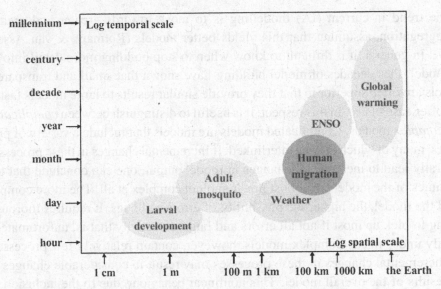

Fig. 8.10 Some processes on different temporal and spatial scales that affect malaria epidemiology. ENSO, El Niño Southern Oscillation.

climate system in time frames of years to centuries or more, whereas economic processes operate on short to medium time scales. Furthermore, short-term needs and interests of stakeholders have to be considered.

Although the assessment of malaria risk may be done on a variety of geographical scales – varying from a village to an entire country, region, or the world as a whole – so far there has been hardly any experience with altering scale levels. This can highlight important issues. For instance, an early version of the malaria model developed by Martens *et al.* (1995) uses a composite measure of different species of *Anopheles*. This globally aggregated model assumed that there are universal relationships that are sufficiently dominant to ensure a valid approximate overall forecast. Later versions of the model (Martens *et al.*, 1999) include species-specific relationships (as far as are available in literature) between climate and transmission dynamics. Even so, the equations within a global model may well be inappropriate for particular local conditions.

In this context, an interesting approach is proposed by Root and Schneider (1995): the so-called Strategic Cyclical Scaling (SCS) method. This method involves continuous cycling between large- and small-scale assessments. Such an iterative scaling procedure implies that a specific global model is disaggregated and adjusted for a specific region or country. The new insights are then used to improve the global version, after which implementation for another region or country follows. In malaria modelling some progress has been made (Lindsay & Martens, 1998). This SCS method can also be used for the conceptual validation of models.

The trend in current (IA) modelling is to move towards greater and greater disaggregation, assuming that this yields better models (Rotmans & van Asselt, 1999). In general, it is difficult to know when to stop building more detail into an IA model. Past decades of model building have shown that small and transparent models are often superior in that they provide similar results to large models faster, and offer ease of use. In this respect, it is useful to distinguish between *complicated* and *complex* models. Complicated models are models that include a variety of processes, many of which may be interlinked. If incremental changes in these processes generally lead to incremental changes in model output, one can conclude that the dynamics of the model are almost linear and not complex at all. The more complicated the model, the higher the possibility of errors and bugs. It requires thorough testing to pick up most if not all errors and bugs, an activity that is, unfortunately, heavily under-rated. Complex models, however, contain relatively few processes, but incremental changes in these processes may result in considerable changes in the results of the overall model. This nonlinear behaviour, due to the inclusion of feedbacks, adaptation, self-learning and chaotic behaviour, is often unpredictable.

Practically speaking, this means that disaggregation of IAMs has profound consequences for the dynamics of the model. Breaking down a global model into various regions requires that the regional dynamics be dealt with in an adequate manner. Current regional IAMs use grid cells or classes for representing geographical differences and heterogeneities in regional IAMs. They do not capture, however, the regional dynamics with regard to population growth and health, economic development, resource use and environmental degradation, let alone regional interactions through migration and trade.

8.6 The way ahead

During recent decades, IAMs have proven themselves as legitimate and powerful approaches to complex issues. Especially in the climate-change debate, IAMs have contributed by exploring the impacts of climate change and evaluating mitigation and abatement strategies. IAMs have also provided useful information on balancing the carbon budget, the role of sulphate aerosols and on various aspects of land use. IA modelling of human health is one of the latest branches of the IA tree.

However, there is still a long way to go before IAMs will be fully accepted by the scientific modelling community on the one hand, and by the decision-making community on the other. Therefore, an ultimate challenge for IA modelling is to build up scientific and political credibility. To improve this, IA scholars should enhance IA models, improve the communication with disciplinary scientific and policy-making communities, and enrich and augment communication techniques.

Enhancing IAMs is a continuous process and various strategies for further improving IAMs can be followed. First, realizing that an IA framework is only as good as its weakest part of the whole model chain, it is much more effective to improve the rather weak or poorly defined model parts, rather than refining and disaggregating the already adequate submodels. Second, empirical testing is a crucial element in enhancing the quality of IAMs. To this end, various model components should be tested against disciplinary expert models and field and laboratory studies to validate the scientific hypotheses, process formulation, boundary conditions and model structure.

Finally, most IAMs are unbalanced with regard to the representation of social and natural systems. While in many macroeconomic-oriented IAMs the socioeconomic aspects are adequately represented, but the biological–physical aspects only poorly so, in most biosphere-oriented IAMs the natural system is reasonably well represented, but the socioeconomic system only as a caricature. Also in the current IAMs related to human health, an adequate representation of social dynamics is missing.

One of the reasons is that the social scientific knowledge base is smaller, resulting in fewer disciplinary models being available to be represented in a reduced form in IAMs. In addition, the modelling of socioeconomic aspects suffers from a dearth of basic data, particularly outside industrialized countries. Therefore, the models developed to describe social dynamics have not been calibrated and validated adequately. There is, however, sufficient knowledge of vital socioeconomic processes to use in the IA modelling context. However, so far, this knowledge available has hardly been used in IA modelling of human health.

With respect to the IAMs described in this chapter, the first point to make is that aggregating data about the "natural" world (including human populations) necessarily involves simplifications, aggregation and averaging. As discussed, a further difficulty of an integrated modelling approach is that vulnerability to global changes will vary greatly between different segments of the world's population. Poorly resourced populations, such as those of Bangladesh and sub-Saharan Africa, will be more vulnerable to environmental and climate-change impacts than will rich nations. Globally aggregated models average over all populations, but specific projections will often be required for more localized populations. Furthermore, the complexity of influences of various factors upon human health defies a ready quantitative analysis of net effects.

As the *full* complexity of the interactions between global developments and human health cannot be satisfactorily reduced to modelling, what is the role of such modelling? Despite the difficulties and limitations of the modelling process, models first of all draw attention to the potential health impact of these global changes. Second, they may indicate the relative importance of the factors that influence these outcomes. This enhances public discussion, education and policy-making. However, even more important is the role that (integrated) modelling plays in the systematic linkage of multiple cause-and-effect relationships based on available knowledge and reasoned guesses. This should increase our understanding of the health impacts of global changes, and identify key gaps in the data and knowledge needed to improve the analysis of these effects.

Considerable scientific effort is therefore now going into the development of integrated models, and into imbuing top-down models with the capacity for down-scaled application to regions and countries, taking account of local physical, ecological and demographic particulars. Future health risk assessments of global environmental and climate change will ultimately need to integrate the global environmental change scenarios with local socio-economic and environmental factors into an integrated modelling framework (McMichael *et al.*, 2000). There has also been recent recognition of the need for integrated mathematical models to take even greater account of feedback processes within and between linked systems. Complex systems may often need to be modelled as complex adaptive systems, displaying spontaneous or socially based adaptive responses. The use of genetic algorithms (Janssen & Martens, 1997) and artificial neural networks (Sethi & Jain, 1991), incorporating a capacity for adaptive change and "learning" processes, holds much promise for this purpose.

References

Anderson, R. M. & May, R. M. (1991). *Infectious Diseases of Humans: Dynamics and Control.* New York: Oxford University Press.

Bertalanffy, L. V. (1968). *General Systems Theory: Foundations, Development, Applications.* New York: Braziller.

Carter, T. R. *et al.* (1994). *IPCC Technical Guidelines for Assessing Climate Change Impacts and Adaptations.* London: University College London.

Darmstadter, J. & Toman, M. A. (1993). Nonlinearities and surprises in climate change: an introduction and overview. In *Assessing Surprises and Nonlinearities in Greenhouse Warming. Proceedings of an Interdisciplinary Workshop*, ed. J. Darmstadter & M. A. Toman, pp. 1–10. Washington, DC: Resources for the Future.

Davis, A. J. *et al.* (1998). Making mistakes when predicting shifts in response to global warming. *Nature*, **391**, 783–6.

Forrester, J. W. (1961). *Industrial Dynamics.* Cambridge, MA: MIT Press.

Forrester, J. W. (1968). *Principles of Systems.* Cambridge, MA: Wright-Allen Press Inc.

Funtowicz, S. O. & Ravetz, J. R. (1990). *Uncertainty and quality in science for policy.* In *Philosophy and Methodology of the Social Sciences* ed. W. Leinfellner & G. Eberlein, p. 229. Dordrecht: Kluwer Academic Publishers.

Goodman, M. R. (1974). *Study Notes in System Dynamics.* Cambridge, MA: Wright-Allen Press Inc.

Hay, S. I. *et al.* (1996). Review: remotely sensed surrogates of meteorological data for the study of the distribution and abundance of arthropod vectors of disease. *Annals of Tropical Medicine and Parasitology*, **90**, 1–19.

Hilderink, H. (2000). *World population in transition, an integrated regional modelling framework. Population studies.* Amsterdam: Thela Thesis.

Hulme, M. (1996). *Climate Change and Southern Africa: an Exploration of Some Potential Impacts and Implications in the SADC Region.* Norwich: WWF International Climate Research Unit.

Janssen, M. (1998). *Modelling Global Change: the Art of Integrated Assessment Modelling.* Cheltenham: Edward Elgar Publishing Limited.

Janssen, M. A. & Martens, P. (1997). Modelling malaria as a complex adaptive system. *Artificial Life*, **3**, 213–36.

Laszlo, E. (1972). *The Systems View of the World: the Natural Philosophy of the New Development in Sciences.* New York, NY: Braziller.

Lindsay, S. W. & Martens, P. (1998). Malaria in the African highlands: past, present and future. *Bulletin of the World Health Organization*, **76**, 33–45.

Lindsay, S. W., Parson, L. & Thomas, C. J. (1998). Mapping the ranges and relative abundance of the two principal African malaria vectors, *Anopheles gambiae sensu stricto* and *An. Arabiensis*, using climate data. *Proceedings of the Royal Society of London*, **265**, 847–54.

Lines, S. (1993). *The effects of climatic and land-use changes on insect vectors of human disease.* In *Insects in a Changing Environment.* Harpenden: Academic Press.

Martens, P. (1995a). Climate change and malaria: exploring the risks. *Medicine and War*, **11**, 202–13.

Martens, P. (1995b). *Modelling the Effect of Global Warming on the Prevalence of Schistosomiasis.* Bilthoven, The Netherlands: RIVM (National Institute of Public Health and the Environment).

Martens, P. (1998). *Health and Climate Change: Modelling the Impacts of Global Warming and Ozone Depletion.* London, UK: Earthscan Publications Ltd.

Martens, P. (1999a). *MIASMA: Modelling Framework for the Health Impact Assessment of Man-Induced Atmospheric Changes – CD ROM.* Electronic Series on Integrated Assessment Modeling **2**.

Martens, P. (1999b). How will climate change affect human health? *American Scientist*, **87**, 534–41.

Martens, P. *et al.* (1995). Potential impacts of global climate change on malaria risk. *Environmental Health Perspectives*, **103**, 458–64.

Martens, P. *et al.* (1996). The impact of ozone depletion on skin cancer incidence: an assessment of the Netherlands and Australia. *Environmental Modelling and Assessment*, **1**, 229–40.

Martens, P., Jetten, T. H. & Focks, D. A. (1997). Sensitivity of malaria, schistosomiasis and dengue to global warming. *Climatic Change*, **35**, 145–56.

Martens, P. *et al.* (1999). Climate change and future populations at risk of malaria. *Global Environmental Change*, **S9**, S89–S107.

Martin, P. & Lefebvre, M. (1995). Malaria and climate: sensitivity of malaria potential transmission to climate. *Ambio*, **24**, 200–7.

McMichael, A. J. & Martens, P. (1995). The health impacts of global climate change: grappling with scenarios, predictive models and multiple uncertainties. *Ecosystem Health*, **1**, 23–33.

McMichael, A. J. *et al.* (2000). Climate change and human health: mapping and modelling potential impacts. In *Spatial Epidemiology: Methods and Applications*, ed. P. Elliott *et al.*, pp. 444–61. Oxford: Oxford University Press.

Molineaux, L. (1988). The epidemiology of human malaria as an explanation of its distribution, including some implications for its control. In *Malaria: Principles and Practice of Malariology*, ed. W. H. Wernsdorfer & S. I. McGregor, pp. 913–98. Edinburgh: Churchill Livingstone.

Morgan, G. M. & Henrion, M. (1990). *Uncertainty – a Guide to Dealing with Uncertainty in Quantitative Risk and Policy Analysis*, p. 332. New York, NY: Cambridge University Press.

Oreskes, N., Shrader-Frechette, K. & Belitz, K. (1994). Verification, validation, and confirmation of numerical models in the earth sciences. *Science*, **263**, 641–6.

Parry, M. L. *et al.* (1999). Climate change and world food security: a new assessment. *Global Environmental Change*, **9**(SI), S51–S67.

Parson, E. A. (1996). Three dilemmas in the integrated assessment of climate change. *Climatic Change*, **34**, 315–26.

Patz, J. A. & Balbus, J. M. (1996). Methods for assessing public health vulnerability to global climate change. *Climate Research*, **6**, 113–25.

Pearce, D. W., Cline, W. R. *et al.* (1996). The social costs of climate change: greenhouse damage and the benefits of control. In *Climate Change 1995: Economic and Social Dimensions of Climate Change*, ed. J. P. Bruce, H. Lee & E. F. Haites, pp. 179–224. New York: Cambridge University Press.

Poveda, G. *et al.* (2000). Climate and ENSO variability associated with vector-borne diseases in Colombia. In *El Nino and the Southern Oscillation*, H. F. Diaz & V. Markgraf, pp. 183–204. New York: Cambridge University Press.

Ravetz, J. R. (1997). *Integrated Environmental Assessment Forum: Developing Guidelines for Good Practise*. Darmstadt: Darmstadt University of Technology.

Rodin, E. Y. (ed.) (1990). *Mathematical Modelling: a Tool for Problem Solving in Engineering, Physical, Biological and Social Sciences*. Oxford: Pergamon Press.

Rogers, D. J. & Randolph S. E. (2000). The global spread of malaria in a future, warmer world. *Science*, **289**, 1763–6.

Rosenzweig, C. & Parry, M. L. (1994). Potential impact of climate change on world food supply. *Nature*, **367**, 133–8.

Root, T. & Schneider, S. (1995). Ecology and climate: research strategies and implications. *Science*, **269**, 334–41.

Rotmans, J. (1998). Methods for IA: the challenges and opportunities ahead. *Environmental Modelling and Assessment*, **3**, 155–79. [Special issue: *Challenges and Opportunities for Integrated Environmental Assessment*, ed. J. Rotmans & P. Vellinga]:

Rotmans, J. & Dowlatabadi, H. (1998). Integrated Assessment of Climate Change: Evaluation of Methods and Strategies. In *Human Choice and Climate Change: An International Social Science Assessment*, ed. S. Rayner & E. Malone. Washington: Battelle Press.

Rotmans, J. & de Vries, H. J. M. (1997). *Perspectives on Global Change: the TARGETS Approach*. Cambridge: Cambridge University Press.

Rotmans, J. & van Asselt, M. B. A. (1999). Integrated assessment modelling. In *Climate Change: an Integrated Perspective*, ed. P. Martens & J. Rotmans, pp. 239–75. Dordrecht: Kluwer Academic Publishers.

Rotmans, J. *et al.* (1994). *Global Change and Sustainable Development*. Bilthoven, The Netherlands: RIVM (National Institute of Public Health and the Environment).

Rykiel, E. J. (1996). Testing ecological models: the meaning of validation. *Ecological Modelling*, **90**, 229–44.

Sethi, I. & Jain, A. K. (eds.) (1991). *Artificial Neural Networks and Pattern Recognition: Old and New Connections*. Amsterdam: Elsevier.

Slaper, H. *et al.* (1996). Estimates of ozone depletion and skin cancer incidence to examine the Vienna Convention achievements. *Nature*, **384**, 256–8.

Snow, R. W. *et al.* (1998). Models to predict the intensity of *Plasmodium falciparum* transmission: applications to the burden of disease in kenya. *Transactions of the Royal Society of Tropical Medicine and Hygiene*, **92**, 601–6.

Sutherst, R. W. (1998). Implications of global change and climate variability for vector-borne diseases: generic approaches to impact assessments. *International Journal of Parasitology*, **28**, 935–45.

Sutherst, R. W., Maywald, G. F. & Skarratt, D. B. (1995). Predicting insect distributions in a changed climate. In *Insects in a Changing Environment, 17th Symposium of the Royal Entomological Society of London*, pp. 60–91, ed. R. Harrington & N. E. Stork.

Tol, R. S. J. (1999). *New Estimates of the Damage Costs of Climate Change. Part I: Dynamic Estimates.* Amsterdam: Free University, Institute for Environmental Studies.

Tol, R. S. J. & Dowlatabadi, H. (2001). Vector-borne diseases, development and climate change. *Integrated Assessment*, **2**(4), 173–181.

Tol, R. S. J. & Frankhauser, S. (1998). On the representation of impact in integrated assessment models of climate change. *Environmental Modeling and Assessment*, **3**, 63–74.

van Asselt, M. B. A. (2000). *Perspectives on Uncertainty and Risk: the PRIMA Approach to Decision Support.* Dordrecht: Kluwer Academic Publishers.

Weinstein, M. C. et al. (1987). Forecasting coronary heart disease incidence, mortality and cost: the coronary heart policy model. *American Journal of Public Health*, **77**, 1417–26.

9

Remote sensing, GIS and spatial statistics: powerful tools for landscape epidemiology

LOUISA R. BECK, URIEL KITRON & MATTHEW R. BOBO

Acronyms

AATSR	Advanced Along Track Scanning Radiometer
ADEOS	Advanced Earth Observation Satellite
ALOS	Advanced Land Observing Satellite
ARIES	Australian Resource Information & Environment Satellite
ASTER	Advanced Spaceborne Thermal Emission & Reflection Radiometer
AVHRR	Advanced Very High Resolution Radiometer
AVIRIS	Advanced Visible and Infrared Imaging Spectrometer
AVNIR	Advanced Visible & Near Infrared Radiometer
CBERS	China Brazil Earth Resources Satellite
ENVISAT	Environmental Satellite
EO-1	Earth Orbiter-1
ERS	ESA (European Space Agency) Remote Sensing
ETM+	Enhanced Thematic Mapper (Landsat)
GLI	Global Imager
IRS	Indian Remote Sensing Satellite
LISS	Linear Imaging Self Scanning System
MARA	Mapping Malarial Risk in Africa
MODIS	Moderate Resolution Imaging Spectro Radiometer
MSS	Multispectral Scanner
NOAA	National Oceanographic & Atmospheric Administration
PAN	Panchromatic
RS	Remote Sensing
SAR	Synthetic Aperture Radar
SPOT	Système Pour l'Observation de la Terre
TM	Thematic Mapper (Landsat)
WiFS	Wide Field Scanner
XS	Multispectral (SPOT)

9.1 Introduction

There has been much speculation about the potential impacts of climate change on the map of human health, particularly on the patterns of vector-borne diseases (e.g. Longstreth & Wiseman, 1989; WHO, 1990; Dobson & Carper, 1992; Shope, 1992; Kovats, 2000; Chapters 7 & 8). The impact of climate change on the transmission patterns of these diseases can be both direct (e.g. effect of changes in precipitation on populations of arthropod vectors) and indirect (e.g. human population dynamics and their effects on exposure risk, changes in vegetation, hydrology and other landscape features). These impacts need to be assessed at a variety of spatial and temporal scales from short-term, year-to-year local variations to long-term, climate-driven, global landscape changes.

In other chapters of this book, the use of data from weather satellites to monitor changes in temperature and precipitation in order to predict continental and global patterns of disease outbreaks has been covered extensively; in this chapter, we will also discuss the application of low-spatial resolution data for disease monitoring, but emphasize how higher resolution satellite data have been used in a landscape epidemiological approach to modelling patterns of disease-transmission risk at local to regional scales. A comprehensive model of disease risk due to climate change is improved by incorporating the temporal aspects of the climate models integrated with the spatial forecasting made possible by the use of Geographical Information Systems (GIS) technologies and spatial analyses (Kitron, 2000). In this chapter, we review the basics of remote sensing (RS), GIS and spatial statistics, how these tools have been used to develop models of vector-borne diseases, and discuss further uses of these tools. The discussion will end with a brief summary of new RS systems and how their data might be used for landscape epidemiology.

9.2 Background

The spatial distribution of many diseases is restricted geographically by landscape, climate and anthropogenic factors. For vector-borne diseases, the spatial distribution of vectors, reservoir hosts and even the pathogens themselves may be limited by habitat conditions. Changes in climate, land cover and land use may then result in changes in the risk of disease transmission. Landscape epidemiology and landscape ecology provide a conceptual approach to study environmental risk factors and disease transmission under these considerations. RS and GIS technologies provide unprecedented amounts of data and data-management capabilities. New analytical techniques, such as cellular automata models, Fourier series and geostatistics, have made landscape ecology into a major research avenue with applications in resource management, conservation biology and risk assessment (Gustafson, 1998; Hunsaker *et al.*, 1990).

The concept of landscape epidemiology originated in a book entitled *The Natural Nidality of Transmissible Diseases*, by Pavlovsky (1966). In his book, Pavlovsky related the spatial pattern of disease transmission to biotopes of geographical landscapes (Galuzo, 1975). Nidi are disease foci (typically vector-borne diseases) where several habitat features result in a niche where all the necessary components for disease transmission occur. Landscape epidemiology is a holistic approach that involves the interactions and associations between elements of the physical and cultural environments. By knowing the vegetation, hydrological and geological conditions necessary for the maintenance of specific pathogens in nature, one can use the landscape to identify the spatial and temporal distribution of disease risk. A related concept is landscape ecology, which is the study of the structure, function and change of spatial patterns of ecosystems (Forman & Godron, 1986; Lavers & Haines-Young, 1993; Pickett & Cadenasso, 1995). Landscape ecology considers the environment as a mosaic and deals with patches (e.g. sampling area size, shape), edges (limits of areas) and corridors (connecting routes between suitable habitats). Landscapes are the outcome of continuous interactions among their biotic and abiotic components and are spatially heterogeneous on a range of scales (Risser *et al.*, 1984; Turner & Gardner, 1991). Measures of heterogeneity are affected by two primary scaling factors (Gustafson, 1998): grain (the resolution of the data – minimum mapping unit, time interval) and extent (size of area mapped or studied).

Various environmental elements, including elevation, temperature, rainfall and humidity, directly influence the presence, development, activity and longevity of pathogens, vectors, zoonotic reservoirs of infection, and their interactions with humans (Meade *et al.*, 1988). These elements influence biotic variables (e.g. vegetation, alternative hosts, natural enemies of vectors and reservoir hosts) that indirectly determine transmission dynamics and disease risk. Vegetation type and distribution are particularly influenced by these environmental variables, and can be expressed as landscape elements that can be sensed remotely and whose relationships can be modelled spatially. Since the early 1970s, RS technology has proved to be a valuable tool for describing the Earth's landscape, including its vegetation, water, soils, geology and urban areas, at a range of spatial and temporal scales (Washino & Wood, 1993; Hay, 1997; Hay *et al.*, 1997; Curran *et al.*, 2000; Goetz *et al.*, 2000). RS systems are designed to measure and record reflected, absorbed or emitted electromagnetic energy from features on the Earth. Because every object on the Earth's surface has a different spectral response, RS data can be used to infer the physical properties of those objects (e.g. colour, moisture content, texture, temperature, vigour). In general, these data have been used to derive environmental parameters that are surrogate measures for monitoring, surveillance, or disease transmission risk mapping. Table 9.1 includes

Table 9.1. *Studies that have used remotely sensed data for human health applications (partial list)*

Disease	Vector	Sensor	Citation
Dracunculiasis	*Cyclops* spp.	TM[a]	Clarke *et al.*, 1990
	Cyclops spp.	TM	Ahearn & De Rooy, 1996
Eastern equine encephalitis	*Culiseta melanura*	TM	Freier, 1993
Filariasis	*Culex pipiens*	AVHRR[b]	Thompson *et al.*, 1996
	Cx. pipiens	TM	Hassan *et al.*, 1998a,b
Leishmaniasis	*Phlebotomus papatasi*	AVHRR	Cross *et al.*, 1996
Lyme disease	*Ixodes scapularis*	TM	Dister *et al.*, 1993, 1997
	I. scapularis	TM	Kitron & Kazmierczak, 1997
	I. ricinus	MSS	Daniel & Kolár, 1990
	I. ricinus	TM	Daniel *et al.*, 1998
Malaria	*Anopheles albimanus*	TM	Pope *et al.*, 1993
	An. albimanus	SPOT[c]	Rejmánková *et al.*, 1995
	An. albimanus	SPOT	Roberts *et al.*, 1996
	An. albimanus	TM	Rodríguez *et al.*, 1996
	An. spp.	AVHRR, Meteosat	Thomson *et al.*, 1996, 1997
	An. albimanus	TM	Beck *et al.*, 1994, 1995, 1997
	An. gambiae	AVHRR, Meteosat	Connor, 1999
		AVHRR, Meteosat	Hay *et al.*, 1998a,b
Onchocerciasis	*Simulium* spp.	TM	WHO, 1985
Rift Valley fever	*Aedes* & *Cx.* spp.	AVHRR	Linthicum *et al.*, 1987, 1990
	Cx. spp.	TM, SAR[d]	Pope *et al.*, 1992
	Cx. spp.	SPOT, AVHRR	Linthicum *et al.*, 1994
Schistosomiasis	*Biomphalaria* spp.	AVHRR	Malone *et al.*, 1994, 1997
Tick-borne encephalitis	*I. ricinus*	AVHRR	Randolph, 2000
Trypanosomiasis	*Glossina* spp.	AVHRR	Rogers, 1991
	Glossina spp.	TM	Kitron *et al.*, 1996
	Glossina spp.	AVHRR	Rogers & Randolph, 1993, 1994
	Glossina spp.	AVHRR	Rogers & Williams, 1993, 1994
	Glossina spp.	AVHRR	Robinson *et al.*, 1997a,b

[a]Landsat Thematic Mapper.
[b]Advanced Very High Resolution Radiometer.
[c]Système Pour l'Observation de la Terre.
[d]Synthetic Aperture Radar.

a partial list of vector-borne disease research involving the use of satellite RS data.

GIS technology has great potential for landscape epidemiology because it provides users the ability to store, integrate, query, display and analyse data from the molecular level to that of satellite resolution through their shared spatial attributes (Clarke *et al.*, 1996). Field observations on environmental conditions, including vegetation, water and topography, can be combined in a GIS to enable interpretation of RS data and to facilitate characterization of the landscape in terms of vector and pathogen prevalence. The associations between disease-risk variables (e.g. vector, pathogen and reservoir host abundance and distribution) and environmental variables can be quantified using the spatial analysis capabilities of the GIS (Kitron, 1998).

Spatial statistics provides a tool for quantifying the patterns and associations of hosts, vectors and environmental variables in determining the distribution, spread and transmission of infectious diseases. The environmental information may come from RS, GIS, or both. The last ten years has seen a renaissance in the application of statistical models for infectious diseases to empirical datasets (Anderson & May, 1991; Kitron, 1998; Robinson, 2000). Statistical analysis of remotely sensed landscape patterns, combined with spatial analysis, allows one to define landscape predictors of disease risk that can be applied in larger regions where field data are unavailable (Kitron, 2000). This makes RS/GIS and spatial analysis a powerful set of tools for disease surveillance, predicting potential disease outbreaks and targeting intervention programmes.

Landscape epidemiology has been applied to various vector-borne diseases including African trypanosomiasis (Ford, 1971; Kitron *et al.*, 1996; Rogers *et al.*, 1996), Chagas' disease (Schofield *et al.*, 1982), cutaneous leishmaniasis (Tolezano, 1994), rabies (Carey *et al.*, 1978), Crimean Congo haemorrhagic fever (Hoogstraal, 1979), bovine tuberculosis (White *et al.*, 1993), eastern equine encephalitis (Komar & Spielman, 1994) and other arboviruses (Reisen *et al.*, 1997). Satellite imagery, GIS and spatial analysis tools were applied to varying extents in these studies. Malaria and Lyme disease studies using RS/GIS and spatial analysis are reviewed in more detail later in this chapter.

9.3 Basics of RS, GIS and spatial analysis

9.3.1 Remote sensing

Remote sensing can be defined as the science and art of obtaining information about an object, area or phenomenon through the analysis of data acquired by a device not in direct contact with that feature (Lillesand & Kiefer, 1999). The science of RS

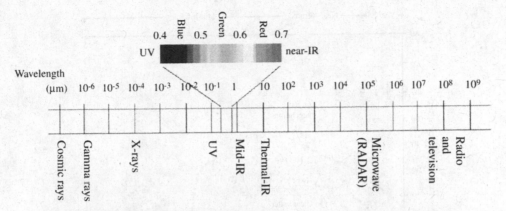

Fig. 9.1 The electromagnetic spectrum. For a colour version of this figure, see www.cambridge.org/9780521114028

rests on the fact that every object reflects or emits energy at specific and distinctive wavelengths of the electromagnetic spectrum (EMS) (Washino & Wood, 1993). The art lies in the ability to exploit this basic energy and matter relationship to identify, map and/or monitor features of interest.

Visible light (i.e. that visible to humans) constitutes only a narrow range of wavelengths within the EMS (Fig. 9.1). One distinct advantage of RS is that it provides the opportunity to utilize wavelengths outside the visible wavelengths, thus making it possible to study phenomena we cannot directly see with our eyes. For example, data from satellite sensors have been used to identify diurnal temperature differences of the surface to provide information on soil moisture and subsequently snail habitat for mapping schistosomiasis and tsetse habitat for mapping trypanosomiasis (see Table 9.1).

By measuring the energy that is reflected (or emitted) by objects on the Earth's surface over a range of wavelengths, a spectral response for that object can be built. Fig. 9.2 depicts the general spectral response functions of different land-cover types with the corresponding Landsat Thematic Mapper (TM) bands (highlighted in blue). It is the exploitation of these different spectral responses that allows researchers to analyse, map and monitor various processes and phenomena.

Although remotely sensed reflectance can be captured on photographic film, the major advantage of RS is that digital images may be generated simultaneously from hundreds of spectral regions of the EMS depending on the design of the sensor. A digital image is composed of pixels, the smallest, nondivisible element of an image. An important advantage of digital imaging is that it can easily be integrated with other data within a GIS. With the exception of a few satellites for reconnaissance purposes, most data from satellites are in a digital format.

232 *Powerful tools for landscape epidemiology*

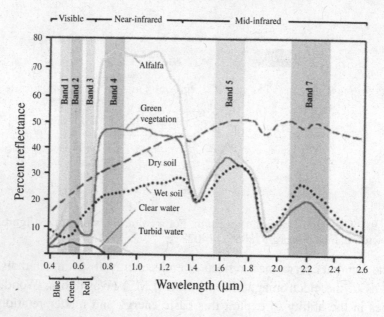

Fig. 9.2 Spectral response functions of different land-cover types. For a colour version of this figure, see www.cambridge.org/9780521114028

9.3.2 Resolution

RS data can be collected at multiple resolutions and times making it possible to analyse various phenomena from local to global scales over time (Quattrochi & Pelletier, 1990). When selecting the appropriate sensor for a given application, one needs to consider the sensor's resolution. There are four types of resolution that affect the quality and nature of the data a sensor collects: radiometric, spatial, spectral and temporal. It is not possible to increase all resolutions simultaneously. For example, as the spatial resolution becomes finer, spectral and temporal resolutions become coarser. With increased resolution of any type comes the need for increased storage space, more powerful data-processing tools (i.e. hardware and software) and often more highly trained individuals to perform or guide analysis (Jensen, 1986). For these reasons, it is important to determine from the outset the minimum resolution requirements needed to accomplish a given task.

Radiometric resolution refers to the sensitivity of the RS system to incoming radiation (i.e. how much change in radiation there must be before a difference is recorded in brightness value). This sensitivity to radiation levels will determine the total number of data values that can be recorded by a sensor (Jensen, 1986). Landsat TM data have 8-bit radiometric resolution; this means that there are 256 brightness values, or grey levels, in each band of data. By contrast, the older Landsat

   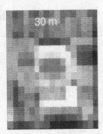

Fig. 9.3 Effect of decreasing spatial resolution on image quality (Source: ERSC, University of Wisconsin).

Multispectral Scanner (MSS) data were only in 6 bits, and the new Ikonos sensor records data in 11 bits. The greater the range of brightness values, the more precisely one can measure the radiance recorded by the sensor (Jensen, 2000).

Spatial resolution is a measurement of the minimum distance between two objects that will allow them to be differentiated on an image. A sensor that has a spatial resolution of 30 m results in the integration of all the energy reflected from the surface over a 900 square meter area (i.e. 30 m × 30 m). The finer the spatial resolution, the more precisely the shapes of objects are captured, the more readily objects can be identified and the more accurately the location, extent and area of objects can be determined. However, as the spatial resolution increases, the size of the image increases quadratically. For example, in Fig. 9.3, the 1 m image is 900 times larger than the 30 m image. Higher resolution sensors, such as Landsat TM and France's SPOT XS, have been extensively employed in studies operating at local to regional scales. Coarse spatial resolution sensors, such as the Advanced Very High Resolution Radiometer (AVHRR), are much more applicable to countrywide, continental and global scales.

Temporal resolution refers to the frequency with which a given ground location is imaged by an individual sensor. Spatial and temporal resolution are also correlated; i.e. as the pixel size increases, the revisit frequency increases as well (Table 9.2). One way the temporal resolution of high-spatial-resolution sensors can be increased is by having pointable sensors, such as those onboard the French SPOT satellites. This allows the same area to be imaged on successive overpasses, albeit with different viewing geometries.

Many ground objects have characteristic temporal patterns. This refers to how the reflectance for a given object in a given waveband, or band combination, changes through time. For vegetated surfaces, the temporal pattern is related to the physiological changes of the surface vegetation over time as it goes through its seasonal cycles. Fig. 9.4 illustrates the temporal effects on the spectral response of a hardwood forest for two Landsat TM scenes of Ohio. The maroon regions in the left-hand image are areas where deciduous trees have lost their leaves during the fall

Table 9.2. *Sensor resolution comparison chart*

Sensor	Pixel size (m)	Swath width (km)	Return cycle (days)	Number of bands
Meteosat	2500 & 5000	Hemisphere	30 min	3
AVHRR	1100	2300	2x	5
Landsat TM	28.5	185	16	7
Landsat-7 Pan	15	185	16	1
SPOT XS	20	60	26	3
SPOT Pan	10	60	26	1
IRS-1C LISS III	24 & 71	142	24	4
IRS-1C Pan	6	70	24	1
Radarsat	10–100	35–500	6	1
ERS-1	30	100	35	1

senescence. The right-hand shows the same area with full canopy during the summer. The frequency with which seasonal changes may be observed is dependent on the return cycle (Table 9.2) of the individual sensor.

Spectral resolution refers to the number and width of wavelength intervals in the EMS that can be recorded by a sensor. The higher the spectral resolution, the more completely and precisely the spectral pattern of each individual pixel will be characterized, and the more readily scene elements can be discriminated based on those patterns. Bands are typically selected to maximize contrast among objects of interest.

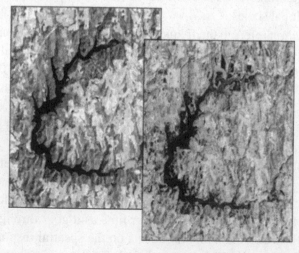

Fig. 9.4 Seasonal variation in forest cover of a hardwood forest in northeastern Ohio. For a colour version of this figure, see www.cambridge.org/9780521114028

Sensors can be divided into groups based oñ their spectral resolution:

- *Panchromatic* sensors record energy from a broad range of wavelengths between the ultraviolet and the near-infrared regions of the EMS in a single wide band. Typically a panchromatic image comes in the form of a black-and-white image. Panchromatic sensors are able to produce very high spatial resolution images. Landsat-7 has a 15 m panchromatic sensor, SPOT has a 10 m sensor, and the Indian IRS-1C has a 6 m sensor. Several new commercial RS satellites, such as Ikonos, are capable of producing 1 m resolution panchromatic images.
- *Multispectral* sensors record energy from multiple areas of the EMS into individual bands. Landsat TM, for example, is a 7-band multispectral sensor. Multispectral sensors are able to record data in the visible, near- and thermal-infrared regions of the EMS.
- *Hyperspectral* sensors may record in hundreds of very narrow, continuous bands of the EMS. Currently, there are no operational, satellite-based hyperspectral systems, although there is an experimental, 220-band hyperspectral instrument onboard NASA's EO-1 satellite. The Advanced Visible and Infrared Imagine Spectrometer (AVIRIS) is an example of an airborne system, which records in 224 continuous bands from the visible to mid-infrared portion of the EMS.

9.3.3 Geographical Information Systems

A GIS is a computer-based system that enables the capture, storage, retrieval, manipulation, analysis and presentation of geographically referenced data. Essentially, a GIS is a database management system for geographical data. A GIS provides a set of powerful tools for studying spatial patterns and their relationships. It allows for the organization of information about a given region as a series of maps where each map (called a layer or coverage) displays information about a specific characteristic of the region or from a specific time period. Once the layers have been registered to a common geographical reference system, the information contained in the individual layers can be compared and analysed in combination with other layers.

Spatial data may be represented in a GIS in either vector and/or raster formats. In the vector data format, layers are represented by three basic geometrical entities: points, lines and polygons (Star & Estes, 1990), which represent the location and boundaries of geographical features. The raster (or grid-cell) data format is similar to aerial and satellite imaging technologies, which represent geographical objects as pixels. Raster data are much better at representing continuous phenomena such as radiance derived from RS, whereas vectors are better for representing discrete entities such as roads and streams.

Geographical analysis is based on the association between the spatial elements and their attribute data. GIS provides the technology to perform sophisticated analysis that makes use of the links between the two. Queries to a GIS can thus be graphics- or data-driven. Graphics-driven queries involve spatial-based searches for objects and retrieval of the associated attribute data. Data-driven queries involve the use of data values to display the matching spatial features.

Operations to retrieve, summarize, analyse and display geographical data are embedded within a GIS. Examples of the types of spatial analysis include the following:

- *Choropleth mapping* involves grouping area features according to some attribute or set of attributes. A legend acts as a look-up table with each range of attribute values being associated with a particular colour or shading pattern. A typical choropleth map for health application would be the shading of administrative districts according to disease prevalence such as cancer.

- *Buffer generation* concerns the establishment of boundaries around points, lines, or areas at a user-defined distance in all directions. This distance should relate to some real-world requirement of the phenomena under investigation. The results of the buffering are circular, corridor, or polygon-shaped features, which represent areas at set distances from the original object. Beck *et al.* (1994) used 1-km buffers around villages, which related to the flight range of the mosquito under investigation, to estimate the percentage of land-cover types favourable as breeding habitat. This was done to distinguish villages at high risk for malaria transmission from those villages at relatively low risk.

- *Contouring and surface generation* relates to the interpolation of lines or surfaces from sample point data to fill in the gaps where point data do not exist. To produce these surfaces, the variable under consideration must be continuous, such as rainfall or soil pH. These new surfaces can then be treated as another layer in a GIS to support the spatial analysis to be performed. An example of surface generation in health application includes the calculation of rainfall maps from weather station data. However, wide local discrepancies may result in large errors. An alternative is the use of weather satellite data to produce such maps (Randolph, 2000).

- *Polygon overlay* entails combining two or more GIS layers in order to calculate areas that have a certain combination of attributes. Rainfall, temperature and land-cover data could be overlaid to produce a habitat suitability map for malaria vectors, for example. Alternatively, layers of the same attributes at different time periods can be overlaid to show temporal changes.

- *Terrain analysis* involves the creation of three-dimensional views, digital terrain models, or slope/aspect from elevation data. Slope and aspect can provide information about other variables of interest such as vegetation. For example, many tree species only grow on defined ranges of slopes or at well-defined elevations; also, certain mosquito species may only be found above or below threshold elevations. Digital elevation data and derived variables can be used to generate a three-dimensional flow model of pollution runoff to examine potential human exposure and health risks.

- *Network analysis* traces a network of connected line features in order to simulate flows of traffic, water, or some other phenomenon. This type of analysis is often used to calculate the shortest or quickest paths between two points in a network. For example, emergency response units use network analysis tools to calculate an optimum route for responding to calls.
- *Area and length calculations* can be employed to highlight only those features with attributes satisfying certain criteria. This type of analysis is also useful for determining the allocation of resources; for example, calculation of the area of oil spills so that the proper resources can be allocated for cleanup.

9.3.4 Spatial analysis: tools

A wide variety of tools have been used by epidemiologists and disease ecologists to quantify the spatial patterns of disease and vector distribution, as well as to assess and predict the risk of disease transmission. Many methodologies have been adapted from other disciplines, such as geology, geography and engineering (e.g. geostatistics, scanning for clusters, Fourier series and wavelet transformations), but commonly used ecological and epidemiological tools (e.g. discriminant analysis, logistic regression, dynamic and simulation models) have also been used. We will only present the most commonly applied tools that have been used in studies of vector-borne disease in association with RS data and GIS, and provide some examples of their applications.

Spatial statistics have been developed for the analysis of points, lines and polygons (areas). Point data include disease case distribution, location of larval mosquito sites and human residences. Point data (e.g. centroids) can be used as surrogates for areal data for which analytical techniques are less readily available. Line data include rivers, roads and vector flight routes, while polygon data include administrative units, forests and water bodies. The most commonly used geostatistics in epidemiology are cluster analysis, autocorrelation, second-order neighbourhood analysis, trend surface analysis and various forms of kriging based on variograms (Goovaerts, 1998; Kitron, 1998; Moore & Carpenter, 1999; Robinson, 2000). All spatial statistics are based on the idea that attributes of neighbouring locations are likely to be more similar than what is predicted under assumptions of complete spatial randomness. An extensive review of geostatistics, including the use of variograms and various types of kriging, is provided by Goovaerts (1998). Robinson (2000) reviewed the use of spatial statistics in epidemiology and public health, with descriptions and examples of disease distribution mapping, spatial patterns, spatial relationships and decision support.

Nonspatial statistics, such as discriminant analysis, logistic regression and tree-based classifications, have also been used in analysis of RS data and other risk factors for vector-borne diseases. Factors, such as the number of negative and positive sites, size of training area and threshold of acceptance, play a major role when interpreting the results of statistical analysis in terms of underlying biological processes (Kitron, 2000; Randolph, 2000). Fractal methods have been used to measure the stability of patches of tick habitat (Meltzer, 1991), and spectral density analysis (Hay *et al.*, 2000b) can be used to study periodicity in epidemiological and meteorological data.

Fourier series have been applied by Rogers and his colleagues in studies of tsetse and African trypanosomiasis (Rogers *et al.*, 1996), and later by Hay *et al.* (1998a) for malaria. Meteorological and vegetation data were described by Fourier-processed surrogates derived from multitemporal NOAA AVHRR and Meteosat images, and analysed using discriminant analysis resulting in a prediction of the distributions of vectors and disease in Africa. Wavelet transforms are an alternative approach for the processing of multitemporal data with the general advantage that they can be locally tailored and that periodicity does not have to be true for the entire series.

A comparison of many of these methods was conducted by Cullinan and Thomas (1992), who concluded that a combination of methods is needed to study ecological pattern and scale because of the variable sensitivity of each method and the need to consider multiple temporal and spatial scales. They classified these methods according to the questions that they could address, including estimation of patch size, examination of spatial variation and correlation, estimation of scale of pattern or period, detection of multiple scales and detection of ecological change.

While these statistical tools allow for better understanding of available data and can be used to derive some insight into the operating biological mechanisms, dynamic processes need to be incorporated into population models. Models that have been used in epidemiology and landscape ecology include dynamic host–parasite models (Anderson & May, 1991); the Kermack–McKendrick model and cellular automata models (Holmes, 1997), in which nonlinear Markov chains are applied to a matrix of cells (patches) to describe population dynamics; simulation models based on percolation networks; diffusion processes; cell interactions in grid-mosaics; neutral models; metapopulation models; and hierarchical models of heterogeneity (Gardner *et al.*, 1987; Wiens, 1989; Kotlar & Wiens, 1990; Levin, 1992). (See Hunsaker *et al.* (1993) for a review of spatial modelling of ecological systems and processes, and Robinson (2000) for an overview of the use of spatial statistics and GIS in epidemiology.) Successful integration of GIS and spatial models is needed for the application of such models to epidemiological research and applications.

The challenge in all of these approaches is twofold: how to consider multiple scales or select the most appropriate one(s), and how to relate the scale of measurement to the scale at which ecological or epidemiological processes are taking place. A specific challenge is to translate processes that occur on a global scale, such as climate change, to what takes place on the microhabitat level where vectors of disease operate (Levin, 1992; Kitron, 2000).

9.4 Applications to health

As can be seen in Table 9.1, satellite RS data have been used to investigate a variety of diseases; this list can be roughly divided into two distinct groups – those that use low-spatial resolution data (e.g. AVHRR or Meteosat) to map broad-scale patterns and those that use high-spatial resolution data (e.g. Landsat or SPOT) to conduct more local- to regional-scale applications.

9.4.1 Low resolution

Low-spatial resolution data, particularly the NOAA AVHRR 1, 4, and 8 km datasets, have been used to explore the temporal patterns of disease outbreaks. In many cases, this type of data is available at low or no cost through the Internet. A large body of work conducted by Rogers and others (e.g. Rogers *et al.*, 1996) has explored the relationship between vegetation (NDVI, a vegetation index), temperature (channel 4-brightness) and rainfall (Cold Cloud Duration index) derived from AVHRR and Meteosat data and the distribution of tsetse flies throughout Africa. These studies were based on a good understanding of fly biology and the environmental determinants of its distribution. Advantage was taken of the multitemporal availability of these low-spatial resolution data, and sophisticated transformation of the data using a time-series described by a Fourier series representation.

In a recent study, Linthicum *et al.* (1999) presented a model that combined a time-series of AVHRR-derived sea surface temperature (SST) anomalies in the Pacific and Indian Oceans with NDVI data to explain the temporal patterns of Rift Valley fever outbreaks in East Africa. Lobitz *et al.* (2000) used AVHRR SST data and TOPEX/Poseidon sea surface height anomalies (SSA) for the Bay of Bengal to understand the patterns of cholera outbreaks in Bangladesh, while Pascual *et al.* (2000) used RS-derived climate and atmospheric information to investigate the linkages between El Niño Southern Oscillation (ENSO) and cholera dynamics in south Asia. They found that the cholera patterns are linked to changes in atmospheric circulation and to regional temperature anomalies.

In Africa, the MARA (Mapping Malaria Risk in Africa) initiative on the continent level (Snow *et al.*, 1996, 1999) is developing an atlas of malaria for Africa using GIS to integrate spatial malaria data with environmental information, and is producing maps of the type and severity of malaria transmission. On the country level, studies in the Gambia (Thomson *et al.*, 1999) using AVHRR data have shown

an association between the amount of seasonal vegetation greenness (measured using NDVI) and malaria infection in children. In Kenya and the Gambia, Hay *et al.*
(1998a,b) used Fourier-processed NOAA-AVHRR and Meteosat data (NDVI, land
surface temperature, middle-infrared radiance and Cold Cloud Duration) to predict
seasonal patterns of malaria. See Hay *et al.* (2000a) for a overview of the use of RS
and GIS for monitoring malaria in Africa. Climate-based models, such as CLIMEX
(Sutherst & Maywald, 1985) and BIOCLIM (Nix, 1986), were used to explain tick
distributions in Australia and Africa before meteorological satellite imagery became widely available (Kitron & Mannelli, 1994); both models use interpolation
of data from weather stations. For a discussion of these models, see Norval *et al.*
(1992) and Randolph (2000).

For Lyme disease in the U.S., maps have been developed on the country level for
tick distribution (Dennis *et al.*, 1998) and for risk of transmission (Fish & Howard,
1999). In Wisconsin, Kitron and Kazmierczak (1997) found an association between
human cases and AVHRR-NDVI on the county level. Nicholson and Mather (1996)
collected nymphal ticks from sites in Rhode Island and considered the distribution
of human cases with a resolution of 10 km^2. Environmental data consisting of land
use, vegetation, hydrography and roads were associated with the epidemiological
and entomological data. Surface pattern analysis using simple kriging was applied
to model the risk of Lyme disease.

In Europe, Daniel and Kolár (1990) used Landsat MSS imagery to associate
habitats characterized by different vegetation with tick infestation levels in the
Czech Republic. Randolph and colleagues used logistic regression to create a predictive map of tick-borne encephalitis risk based on digital elevation models and
Fourier-processed AVHRR data (Randolph, 2000).

9.4.2 High resolution

While the low-spatial resolution data are more efficient and cost-effective for continental to global risk modelling, they cannot account for finer scale differences in
risk factors and transmission patterns. For example, long-term climate forecasts of
higher-than-normal precipitation over East Africa might well indicate an increased
risk of Rift Valley fever, but these regional forecasts cannot be used to identify
specific areas or foci (e.g. dambos) where risk is highest. In a recent paper, Rogers
and Randolph (2000) posited that climatic variables used in malaria risk models
are probably more appropriate for modelling risk at the edges of malaria's current
range, although the ecological and epidemiological studies needed to support this
hypothesis have not been done (Dye & Reiter, 2000). These studies could make
use of higher resolution satellite data, as there is a large body of ecological research conducted using such data. However, the higher resolution data can be very

expensive (both to purchase and process), and often only one image for a single day is used, resulting in a static picture that may not offer insight into the dynamic process underlying observed patterns.

The examples described below illustrate how higher resolution data have been used for modelling disease-transmission risk, as well as how spatial analysis can be used to develop RS-based risk models.

9.4.3 Lyme disease

Daniel *et al.* (1998) used Landsat TM imagery to detect micro-foci with vegetation suitable for high tick survival. Merler *et al.* (1996) used a tree-based classification model to show that geology and altitude were better predictors of tick distribution than a biotic variable related to vegetation type and cover density. In the U.S., Glass *et al.* (1994, 1995) used a GIS to predict tick abundance and assess environmental risk factors for Lyme disease using human case data and tick collections from deer in Maryland. Environmental data were extracted from digital maps and a Landsat TM image, and multiple linear and logistic regression models were used to generate the risk models and predictive maps. Kitron *et al.* (1991, 1992) used digitized environmental coverages and second-order neighbourhood analysis to demonstrate the clustering of tick-infested deer and association with landscape/landcover features.

A model based on a landcover map derived from Landsat TM data was developed to predict Lyme disease transmission risk in Westchester County, New York (Dister *et al.*, 1993). Using a GIS, the proportions of each cover type were extracted by municipality. Canine seroprevalence (CSR) data, which consisted of the percentage by municipality of dogs infected with the Lyme disease spirochete, were then compared with the landcover proportions. When all deciduous forest classes were grouped into a single class, the deciduous forest group was positively correlated with the CSR data. Another class representing residential areas with a large percentage of ornamental vegetation was also compared with the canine data. When used alone, this class was not significantly correlated with CSR; however, when only the proportion of pixels in this class adjacent to a forest pixel was used, there was a strong correlation with CSR. This analysis showed that the spatial arrangement of landcovers is a critical consideration when analysing the links between the environment and disease-transmission risk.

9.4.4 Malaria

Landsat TM data were also used in a spatial approach to model malaria transmission risk in Chiapas, Mexico (Beck *et al.*, 1994). In that study, a landcover map of the Tapachula area was generated from multitemporal TM data. The habitat requirements of the main malaria vector, *Anopheles albimanus*, include flooded grassland (larval habitat), bloodmeal sources (cattle in pastures) and resting sites (trees or houses). The study focused on analysing landcover classes surrounding the villages with the intention of identifying villages with high vector–human contact risk. Buffering, a GIS function described earlier, was used to quantify the proportions of landcover classes surrounding villages; the buffer size reflected the 1-km flight range of the mosquito. Using stepwise discriminant analysis and regression approaches, it was found that the proportions of two landcover classes could predict villages with high abundances of adult mosquitoes, with an overall accuracy of 90 %. The statistical models generated in the Tapachula were subsequently tested in another area of Chiapas (Beck *et al.*, 1997). In that blind test, the regression model was able to predict seven of the ten highest abundance villages.

9.4.5 Hantavirus

Landsat TM data and spatial statistics were used to develop a model of hantavirus risk in the American Southwest (McGwire, unpublished). In the study, field data on relative densities and infection rates in deer mouse populations were spatially analysed using semivariograms. The analysis indicated that rodent populations and infection rates clustered at two different scales. On a broad scale, these variables were determined in part by the distribution of suitable habitat; on a finer scale, rodent population clustering was apparent over smaller distances in the areas of suitable habitat. By plotting the squared differences in values over distance, semivariograms provided information on the spatial autocorrelation within the two datasets, thus indicating the range over which the clusters of infection extended. The initial results of the analysis indicated that spatial clustering of infection rates in the study area typically extended over a distance of less than a kilometre.

Glass *et al.* (2000) associated cases and control sites with precipitation data from weather stations, elevation data and Landsat TM data collected the year before the 1992–93 outbreak in the southwestern U.S. A logistic regression model showed an association of cases with elevation and local environmental conditions, but not directly with rainfall. The model was applied to satellite data from 1995 to predict risk for 1996, and the predicted high-risk areas were practically eliminated. Boone *et al.* (2000) associated infections in deer mice populations with environmental data from satellite imagery and digitized maps in an area in the western U.S. Canonical linear discriminant analysis was used to determine risk factors for infection.

9.4.6 Tsetse flies

Kitron *et al.* (1996) used a single Landsat TM image to associate the distribution of tsetse flies in western Kenya with environmental features. Spatial autocorrelation and spatial filtering were applied to determine the spatial association of the data. A high proportion of the variance in fly distribution could be explained through association with spectral bands, particularly those bands that detect soil moisture, a known determinant of successful fly survival.

9.4.7 Combining high- and low-resolution data

At a recent workshop sponsored by the Liverpool School of Tropical Medicine and Hygiene (Thomson *et al.*, 2000), a stratified approach to mapping malaria risk in Mali was demonstrated by Beck *et al.* (1999). In that approach, a time series of coarse-resolution data acquired from the Internet (i.e. AVHRR 8 km NDVI data) were used to stratify Mali into four strata based on greenness. The stratum whose temporal green-up patterns indicated potential for mosquito breeding and larval habitat persistence was then further stratified using 1 km AVHRR NDVI data (also from the Internet) into zones based on maximum greenness. Finally, more costly, high-resolution Landsat TM data were used in the greenest 1-km zones to identify villages at highest risk based on landscape features such as vegetation and soil moisture. By overlaying a GIS layer of villages, roads and health centres created by the World Health Organization's (WHO) HealthMap (http://www.who.int/emc/healthmap/healthmap.html) onto the classified TM data, GIS functions could then be used to identify villages that are both at high risk and distant from health centres. This multiscaled approach combines cheaper, more frequent datasets with more costly, selective acquisitions of higher-resolution data, enabling researchers and public health agencies to focus on local areas of risk while keeping costs down.

9.5 Future systems, future applications

The concept of global climate change encompasses more than merely an alteration in temperature; it also includes spatial and temporal covariations in precipitation and humidity (Dye & Reiter, 2000), and more frequent occurrence of extreme weather events. The impact of these variations, which can occur at a variety of temporal and spatial scales, could have a direct impact on disease transmission through their environmental consequences for pathogen, vector and host survival, as well as indirectly through human demographic and behavioural responses. New and future sensor systems will allow scientists to investigate the relationships between climate change and environmental risk factors at multiple spatial, temporal and spectral scales (Beck *et al.*, 2000; Wood *et al.*, 2000). Some of these satellite systems, and the opportunities offered by them, are mentioned below; for more information on these and other systems and how they might contribute to understanding disease risk, visit the sensor evaluation web pages located at http://geo.arc.nasa.gov/sge/health/sensor.

9.5.1 Spatial resolution

Higher spatial resolution will provide better opportunities for mapping urban features previously only possible with high-resolution aerial photography. These opportunities include housing quality (e.g. Chagas' disease, leishmaniasis) and urban mosquito habitats (e.g. dengue fever, filariasis, LaCrosse encephalitis). Current and future sensors with high spatial resolution include ALOS AVNIR-2; Orbview-3, -4; QuickBird-2; SPOT-4, -5; ARIES; IRS-1C,D PAN; and Ikonos. Sensors with both high spatial resolution and multispectral capabilities could help scientists to identify ecotonal habitats associated with disease risk, such as Lyme disease, Chagas' disease or Yellow fever. The sensors with these capabilities include ALOS AVNIR-2; ARIES; CBERS; Ikonos; IRS-1C, D LISS III; Landsat ETM+; OrbView-3, -4; QuickBird-2; SPOT-4, -5; and Terra ASTER. The Ikonos instrument, with 1 m PAN and 4 m multispectral resolutions, could also be used for mapping snail habitats, such as ponds and canals, thereby making these sensors useful for schistosomiasis studies.

9.5.2 Spectral resolution

There are or will be many new sensors that have higher spectral resolution, enabling scientists to acquire more information about parameters such as soil moisture, soil type, better vegetation discrimination and ocean colour, to name a few. Although soil moisture content is now detectable using Landsat, the new thermal, shortwave infrared and radar sensors will be able to provide this information at a variety of scales not achievable using Landsat. Soil moisture could become a key component in transmission risk models for Lyme disease (tick survival), helminthiases (worm habitat), malaria (vector-breeding habitat) and schistosomiasis (snail habitat). Sensors capable of detecting soil moisture include ADEOS II GLI; ARIES; CBERS; ENVISAT AATSR; MODIS; ERS-1, -2 instruments; IRS-1C, D LISS III; WiFS; AVHRR; RADARSAT-1, -2; SPOT-4, -5 Vegetation; and Terra ASTER, MODIS. The recently launched OrbView-2 SeaWiFS sensor was specifically designed for sensing ocean colour, a parameter that could be very useful for modelling cholera outbreaks or mapping red tides.

9.6 Conclusions

Remotely sensed environmental and climatological data are continuously becoming available in larger quantities and finer detail. Simultaneously, our abilities to manage, analyse and integrate these data with epidemiological, entomological, malacological (related to snails) and demographic data are becoming more sophisticated. Several questions arise that will need to be addressed by researchers, health agencies, funding sources and decision-makers. These include the following.

(i) How do we decide which changes to follow, and which ones are the most important as far as actual and potential health impacts? Depending on our objectives and level of understanding, we will need to consider both the biological importance and our detection ability of various changes, including climate and anthropogenic impacts. Some of these changes include deforestation (and conversely, reforestation), vegetation changes (e.g. maple replacing oak), vegetation fragmentation, urbanization (and attendant impacts such as roads, pollution) and hydrology (e.g. through urbanization, dams, agriculture, or precipitation changes). Overlaying these are short-term impacts of weather and long-term impacts of climate change. Given a number of simple models associating one change (e.g. temperature) with predictions of drastic impacts on the potential distribution of diseases such as malaria, it is clear that a good understanding of the biological and epidemiological mechanisms underlying disease transmission is necessary to derive biologically meaningful models and create plausible and useful risk maps (Kitron, 2000; Rogers & Randolph, 2000).

(ii) Is a vertical approach based on a unique disease more productive than a horizontal approach based on overall health status? A vertical approach has an emphasis on a specific disease (usually on a national or coarser scale), as exemplified by the global malaria eradication and national malaria control programmes. In a horizontal programme, overall health status, or at least several diseases and their underlying determinants, are considered (often on a community or district scale). Budgetary constraints often result in conflicts between proponents of these two approaches. The advantage of the landscape epidemiology approach discussed here is that the underlying environmental and climate data that are gathered can be applied to research and programmes targeting individual diseases as well as to more general public health programmes.

(iii) How do we decide which resolutions are most important, and how do we select the most important data from the immense amount associated with new sensors? As we move across spatial and spectral resolutions, different information becomes available, and the sources and degree of heterogeneity change (Kitron, 2000). For some diseases (e.g. dengue, schistosomiasis), epidemiological data at the household level, and consequently fine-resolution data such as Ikonos imagery, need to be considered for understanding transmission patterns and recommending treatment regimes. For other diseases, such as cholera, transmission patterns and prevention measures occur on a much coarser scale, and fine-resolution imagery is superfluous; in fact, they may even mask important determinants of risk, and consequently the effectiveness of intervention strategies. Similarly, with regard to spectral resolution, for some vectors such as ticks in wooded areas or snails in aquatic habitats, identification to the level of tree species or unique water qualities may be relevant, while for other vectors the presence of any source of shade or water may be of most importance.

Ultimately, as we strive for improved understanding of the processes underlying patterns of disease transmission, human risk and the effectiveness of control and prevention measures, we need to apply our biological understanding and the new analytical tools to select the most appropriate data and consider them sequentially or simultaneously across spatial, temporal and spectral resolutions.

References

Ahearn, S. C. & De Rooy, C. (1996). Monitoring the effects of dracunculiasis remediation on agricultural productivity using satellite data. *International Journal of Remote Sensing*, **17**, 917–29.

Anderson, A. M. & May, R. M. (1991). *Population Biology of Infectious Diseases of Humans*. Oxford: Oxford University Press.

Beck, L. R., Rodríguez, M. H., Dister, S. W., Rodríguez, A. D., Rejmánková, E., Ulloa, A., Meza, R. A., Roberts, D. R., Paris, J. F., Spanner, M. A., Washino, R. K., Hacker, C. & Legters, L. J. (1994). Remote sensing as a landscape epidemiologic tool to identify villages at high risk for malaria transmission. *American Journal of Tropical Medicine and Hygiene*, **51**, 271–80.

Beck, L. R., Wood, B. L. & Dister, S. W. (1995). Remote sensing and GIS: new tools for mapping human health. *Geo. Info. Systems*, **5**, 3237.

Beck, L. R., Rodríguez, M. H., Dister, S. W., Rodríguez, A. D., Washino, R. K., Roberts, D. R. & Spanner, M. A. (1997). An assessment of a remote sensing based model for predicting malaria transmission risk in villages of Chiapas, Mexico. *American Journal of Tropical Medicine and Hygiene*, **56**, 99–106.

Beck, L. R., Lobitz, B. M. & Wood, B. L. (1999). A stratified approach to mapping malaria risk in Mali using multi-scaled remotely sensed data. Center for Health Applications of Aerospace Related Technologies, NASA Ames Research Center, Moffett Field, CA. Unpublished.

Beck, L. R., Lobitz, B. M. & Wood, B. L. (2000). Remote sensing and human health: new sensors and new opportunities. *Emerging Infectious Diseases*, **6**, 217–27.

Boone, J. D., McGwire, K. C., Otteson, E. W., DeBaca, R. S., Kuhn, E. A., Villard, P., Brussard, P. F. & St. Jeor, S. C. (2000). Remote sensing and geographic information systems: charting Sin Nombre virus infections in deer mice. *Emerging Infectious Diseases*, **6**, 248–57.

Carey A. B., Giles, R. H. & McLean, R. G. (1978). The landscape epidemiology of rabies in Virginia. *American Journal of Tropical Medicine and Hygiene*, **27**, 573–80.

Clarke, K. C., Osleeb, J. R., Sherry, J. M., Meert, J. P. & Larsson, R. W. (1990). The use of remote sensing and geographic information systems in UNICEF's dracunculiasis (Guinea worm) eradication effort. *Preventive Veterinary Medicine*, **11**, 229–35.

Clarke, K. C., McLafferty, S. L. & Tempalski, B. J. (1996). On epidemiology and geographic information systems: a review and discussion of future directions. *Emerging Infectious Diseases*, **2**, 85–92.

Connor, S. J. (1999). Malaria in Africa: the view from space. *Biologist*, **46**, 22–5.

Cross, E. R., Newcomb, W. W. & Tucker, C. J. (1996). Use of weather data and remote sensing to predict the geographic and seasonal distribution of *Phlebotomus paptasi* in Southwest Asia. *American Journal of Tropical Medicine and Hygiene*, **54**, 530–6.

Cullinan V. I. & Thomas, J. M. (1992). A comparison of quantitative methods for examining landscape pattern and scale. *Landscape Ecology*, **7**, 211–27.

Curran, P. J., Atkinson, P. M., Foody, G. M. & Milton, E. J. (2000). Linking remote sensing, land cover and disease. *Advances in Parasitology*, **47**, 37–72.

Daniel, M. & Kolár, J. (1990). Using satellite data to forecast the occurrence of the common tick *Ixodes ricinus* (L.). *Journal of Hygiene, Epidemiology, Microbiology and Immunology*, **34**, 243–52.

Daniel, M., Kolár, J., Zeman, P., Pavelka, K. & Sádlo, J. (1998). Predictive map of *Ixodes ricinus* high-incidence habitats and a tick-borne encephalitis risk assessment using satellite data. *Experimental and Applied Acarology*, **22**, 417–33.

Dennis, T. D., Nekomoto, T. S., Victor, J. C., Paul, W. S. & Piesman, J. (1998). Reported distribution of *Ixodes scapularis* and *Ixodes pacificus* in the United States. *Journal of Medical Entomology*, **35**, 629–38.

Dister, S. W., Beck, L. R. & Wood, B. L. (1993). The use of GIS and remote sensing technologies in a landscape approach to the study of Lyme disease transmission risk. In *Proceedings of the GIS '93 Seventh Annual Symposium: Geographic Information Systems in Forestry, Environmental and Natural Resource Management*, 15–18 February 1993, Vancouver, B.C., Canada.

Dister, S. W., Fish, D., Bros, S., Frank, D. H. & Wood, B. L. (1997). Landscape characterization of domestic risk for Lyme disease using satellite imagery. *American Journal of Tropical Medicine and Hygiene*, **57**, 687–92.

Dobson, A. & Carper, R. (1992). Global warming and potential changes in host-parasite and disease-vector relationships. In *Global Warming and Biodiversity*, ed. R. L. Peters & T. E. Lovejoy. New Haven, CT: Yale University Press.

Dye, C. & Reiter, P. (2000). Temperatures without fevers? *Science*, **289**, 1697–8.

Fish, D. & Howard, C. A. (1999). Methods used for creating a national Lyme disease risk map. *Morbidity and Mortality Weekly Report*, **48**, 21–4.

Ford, J. (1971). *The Role of Trypanosomiasis in African Ecology*. Oxford: Clarendon.

Forman, R. T. & Godron, M. (1986). *Landscape Ecology*. New York: Wiley.

Freier, J. W. (1993). Eastern equine encephalomyelitis. *The Lancet*, **342**, 1281–3.

Galuzo, I. G. (1975). Landscape epidemiology (epizootiology). In *Advances in Veterinary Science and Comparative Medicine*, Vol. 2, ed. C. A. Brandly & C. E. Cornelius, pp. 73–95. New York: Academic Press.

Gardner R. H., Milne, B. T., Turner, M. G. & O'Neill, R. O. (1987). Neutral models for the analysis of broad-scale landscape pattern. *Landscape Ecology*, **1**, 19–28.

Glass, G. E., Amerasinghe, F. P., Morgan, J. M. I. & Scott, T. W. (1994). Predicting *Ixodes scapularis* abundance on white-tailed deer using geographic information systems. *American Journal of Tropical Medicine and Hygiene*, **51**, 538–44.

Glass, G. E., Schwartz, B. S., Morgan, J. M., Johnson, D. T., Noy, M. P. M. & Israel, E. (1995). Environmental risk factors for Lyme disease identified with geographic information systems. *American Journal of Public Health*, **85**, 944–8.

Glass, G. E., Cheek, J. E., Patz, J. A., Shields, T. M., Doyle, T. J., Thoroughman, D. A., Hunt, D. K., Enscore, R. E., Gage, K. L., Irland, C., Peters, C. J. & Bryan, R. (2000). Using remotely sensed data to identify areas at risk for hantavirus pulmonary syndrome. *Emerging Infectious Diseases*, **6**, 238–47.

Goetz, S. J., Prince, S. D. & Small, J. (2000). Advances in satellite remote sensing of environmental variables for epidemiological applications. *Advances in Parasitology*, **47**, 289–304.

Goovaerts, P. (1998). Geostatistics in soil science: state-of-the-art and perspectives. *Geoderma*, **89**, 1–45.

Gustafson, E. J. (1998). Quantifying landscape spatial pattern: what is the state of the art? *Ecosystems*, **1**, 143–56.

Hassan, A. N., Beck, L. R. & Dister, S. (1998a). Prediction of villages at risk for filariasis transmission in the Nile Delta using remote sensing and geographic information system technologies. *Journal of the Egyptian Society of Parasitology*, **28**, 75–87.

Hassan, A. N., Dister, S. & Beck, L. R. (1998b). Spatial analysis of lymphatic filariasis distribution in the Nile Delta in relation to some environmental variables using geographic information system technology. *Journal of the Egyptian Society of Parasitology*, **28**, 119–31.

Hay, S. I. (1997). Remote sensing and disease control: past, present and future. *Transactions of the Royal Society of Tropical Medicine and Hygiene*, **91**, 105–6.

Hay, S. I., Packer, M. J. & Rogers, D. J. (1997). The impact of remote sensing on the study and control of invertebrate intermediate hosts and vectors for disease. *International Journal of Remote Sensing*, **18**, 2899–930.

Hay, S. I., Snow, R. W. & Rogers, D. J. (1998a). Prediction of malaria seasons in Kenya using multi-temporal meteorological satellite sensor data. *Transactions of the Royal Society of Tropical Medicine and Hygiene*, **92**, 12–20.

Hay, S. I., Snow, R. W. & Rogers, D. J. (1998b). From predicting mosquito habitat to malaria seasons using remotely sensed data: practice, problems and perspectives. *Parasitology Today*, **14**, 306–13.

Hay, S. I., Omumbo, J. A., Craig, M. H. & Snow, R. W. (2000a). Earth observation, geographic information systems, and *Plasmodium falciparum* malaria in sub-Saharan Africa. *Advances in Parasitology*, **47**, 173–215.

Hay, S. I., Myers, M. F., Burke, D. S., Vaughn, D. W., Endy, T., Ananda, N., Shanks, D., Snow, R. W. & Rogers, D. J. (2000b). The etiology of inter-epidemic periods of mosquito-borne disease. *Proceedings of the National Academy of Sciences of the USA*, **97**, 9335–9.

Holmes, E. E. (1997). Basic epidemiological concepts in a spatial context. In *Spatial Ecology: The Role of Space in Population Dynamics and Interspecific Interactions*, ed. D. Tilman & P. Kareiva, pp. 111–36. Princeton, NJ: Princeton University Press.

Hoogstraal, H. (1979). The epidemiology of tick-borne Crimean-Congo hemorrhagic fever in Asia, Europe, and Africa. *Journal of Medical Entomology*, **15**, 307–417.

Hunsaker, C. T., Graham, R. L., Suter, G. W., O'Neill, R. V., Barnthouse, L. W. & Gardnes, R. H. (1990). Assessing ecological risk on a regional scale. *Environmental Management*, **14**, 325–32.

Hunsaker, C. T., Nisbet, R. A., Lam, D. C. L., Browder, J. A., Baker, W. L., Turner, M. G. & Botkin, D. B. (1993). Spatial models of ecological systems and processes: the role of GIS. In *Environmental Modeling with GIS*, ed. M. F. Goodchild, B. O. Parks & L. T. Steyaert, pp. 248–64. Oxford: Oxford University Press.

Jensen, J. R. (1986). *Introductory Digital Image Processing: A Remote Sensing Perspective*. Englewood Cliffs, NJ: Prentice-Hall.

Jensen, J. R. (2000). *Remote Sensing of the Environment: an Earth Resource Perspective*. Englewood Cliffs, NJ: Prentice Hall.

Kitron, U. (1998). Landscape ecology and epidemiology of vector-borne diseases: tools for spatial analysis. *Journal of Medical Entomology*, **35**, 435-45.

Kitron, U. (2000). Risk maps: transmission and burden of vector-borne diseases. *Parasitology Today*, **16**, 324–5.

Kitron, U. & Kazmierczak, J. J. (1997). Spatial analysis of the distribution of Lyme disease in Wisconsin. *American Journal of Epidemiology*, **145**, 558–66.

Kitron, U. & Mannelli, A. (1994). Modeling the ecological dynamics of tick-borne zoonoses. In *Ecological Dynamics of Tick-borne Zoonoses*, ed. D. E. Sonenshine & T. N. Mathes, pp. 198–239. New York City, NY: Oxford University Press.

Kitron, U., Bouseman, J. K. & Jones, C. J. (1991). Use of the ARC/INFO GIS to study the distribution of Lyme disease ticks in Illinois. *Preventive Veterinary Medicine*, **11**, 243–8.

Kitron, U., Jones, C. J., Bouseman, J. K., Nelson, J. A. & Baumgartner, D. L. (1992). Spatial analysis of the distribution of *Ixodes dammini* (Acari: Ixodidae) on white tailed deer in Ogle County, Illinois. *Journal of Medical Entomology*, **29**, 259–66.

Kitron, U., Otieno, L. H., Hungerford, L. L., Odulaja, A., Brigham, W. U., Okello, O. O., Joselyn, M., Mohamed-Ahmed, M. M. & Cook, E. (1996). Spatial analysis of the distribution of tsetse flies in the Lambwe Valley, Kenya, using Landsat TM satellite imagery and GIS. *Journal of Animal Ecology*, **65**, 371–80.

Komar N. & Spielman, A. (1994). Emergence of eastern encephalitis in Massachusetts. *Annals of the New York Academy of Science*, **740**, 157–68.

Kotlar, N. B. & Wiens, J. A. (1990). Multiple scales of patchiness and patch structure: a hierarchical framework for the study of heterogeneity. *Oikos*, **59**, 253–60.

Kovats, R. S. (2000). El Niño and human health. *Bulletin of the World Health Organization*, **78**, 1127–35.

Lavers, C. P. & Haines-Young, R. (1993). Equilibrium landscapes and their aftermath: spatial heterogeneity and the role of new technology. In *Landscape Ecology and Geographic Information Systems*, ed. R. Haines-Young, D. R. Green & S. Cousins, pp. 57–73. London: Taylor & Francis.

Levin, S. A. (1992). The problem of pattern and scale in ecology. *Ecology*, **73**, 1943–67.

Lillesand, T. M. & Kiefer, R. W. (1999). *Remote Sensing and Image Interpretation*. New York: John Wiley and Sons.

Linthicum, K. J., Bailey, C. L., Davies, F. G. & Tucker, C. J. (1987). Detection of Rift Valley fever viral activity in Kenya by satellite remote sensing imagery. *Science*, **235**, 1656–9.

Linthicum, K. J., Bailey, C. L., Tucker, C. J., Mitchell, K. D., Logan, T. M., Davies, F. G., Kamau, C. W., Thande, P. C. & Wagateh, J. N. (1990). Applications of polar-orbiting, meteorological satellite data to detect flooding in Rift Valley fever virus vector mosquito habitats in Kenya. *Medical Veterinary Entomology*, **4**, 433–8.

Linthicum, K. J., Bailey, C. L., Tucker, C. J., Gordon, S. W., Logan, T. M., Peters, C. J. & Digoutte, J. P. (1994). Man-made ecological alterations of Senegal River basin on Rift Valley fever transmission. *Sistema Terra*, Year III, 45–7.

Linthicum, K. J., Anyamba, A., Tucker, C. J., Kelley, P. W., Myers, M. F. & Peters, C. J. (1999). Climate and satellite indicators to forecast Rift Valley fever epidemics in Kenya. *Science*, **285**, 397–400.

Lobitz, B., Beck, L., Huq, A., Wood, B., Fuchs, G., Faruque, A. S. G. & Colwell, R. (2000). Climate and infectious disease: use of remote sensing for detection of *Vibrio cholerae* by indirect measurement. *Proceedings of the National Academy of Science of the USA*, **97**, 1438–43.

Longstreth, J. D. & Wiseman, J. (1989). The potential impact of climate change on patterns of infectious disease in the United States. In *The Potential Effects of Global Climate Change on the United States: Appendix G, Health*, ed. J. B. Smith & D. A. Tirpak. Washington D.C.: U.S. Environmental Protection Agency.

Malone, J. B., Huh, O. K., Fehler, D. P., Wilson, P. A., Wilensky, D. E., Holmes, R. A. & Elmagdoub, A. (1994). Temperature data from satellite images and the distribution of schistosomiasis in Egypt. *American Journal of Tropical Medicine and Hygiene*, **50**, 714–22.

Malone, J. B., Abdel-Rahman, M. S., El Bahy, M. M., Huh, O. K., Shafik, M. & Bavia, M. (1997). Geographic information systems and the distribution of *Schistosoma mansoni* in the Nile Delta. *Parasitology Today*, **13**, 112–19.

McGwire, K. Desert Research Institute, Reno, Nevada, unpublished report.

Meade, M. S., Florin, J. W. & Gesler, W. M. (1988). *Medical Geography*. New York: The Guilford Press.

Merler, S., Furlanello, C., Chemini, C. & Nicolini, G. (1996). Classification tree methods for analysis of mesoscale distribution of *Ixodes ricinus* (Acari: Ixodidae) in Trentino, Italian Alps. *Journal of Medical Entomology*, **33**, 888–93.

Meltzer, M. I. (1991). The potential use of fractals in epidemiology. *Preventive Veterinary Medicine*, **11**, 255–60.

Moore, D. A. & Carpenter, T. E. (1999). Spatial analytical methods and geographic information systems: use in health research and epidemiology. *Epidemiologic Reviews*, **21**, 143–61.

Nicholson, M. C. & Mather, T. N. (1996). Methods for evaluating Lyme disease risks using geographic information systems and geospatial analysis. *Journal of Medical Entomology*, **33**, 711–20.

Nix, H. (1986). A biogeographic analysis of Australian elapid snakes. In *Atlas of Elapid Snakes of Australia*, ed. R. Longmore. *Australia Flora and Fauna series*, no. 7, pp. 4–15. Canberra: Australian Government Publishing Service.

Norval, R. A. I., Perry, B. D. & Young, A. S. (1992). *The Epidemiology of Theileriosis in Africa*. London: Academic Press.

Pascual, M., Rodó, X., Ellner, S. P., Colwell, R. & Bouma, M. J. (2000). Cholera dynamics and El Niño Southern Oscillation. *Science*, **289**, 1766–9.

Pavlovsky, E. N. (1966). *The Natural Nidality of Transmissible Disease*, ed. N. D. Levine. Urbana: University of Illinois Press.

Pickett, S. T. A. & Cadenasso, M. L. (1995). Landscape ecology: spatial heterogeneity in ecological systems. *Science*, **269**, 331–4.

Pope, K. O., Sheffner, E. J., Linthicum, K. J., Bailey, C. L., Logan, T. M., Kasischke, E. S., Birney, K., Njogu, A. R. & Roberts, C. R. (1992). Identification of central Kenyan Rift Valley fever virus vector habitats with Landsat TM and evaluation of their flooding status with Airborne Imaging Radar. *Remote Sensing of Environment*, **40**, 185–96.

Pope, K. O., Rejmánková, E., Savage, H. M., Arredondo-Jimenez, J. I., Rodríguez, M. H. & Roberts, D. R. (1993). Remote sensing of tropical wetlands for malaria control in Chiapas, Mexico. *Ecological Applications*, **4**, 81–90.

Quattrochi, D. A. & Pelletier, R. E. (1990). Remote sensing for analysis of landscapes: an introduction. In *Ecological Studies*, Vol. 82, *Quantitative Methods in Landscape Ecology*, ed. M. G. Turner & R. H. Gardner, pp. 51–76. New York: Springer-Verlag.

Randolph, S. E. (2000). Ticks and tick-borne disease systems in space and from space. In *Advances in Parasitology*, Vol. 47, ed. S. Hay, S. Randolph & D. Rogers, pp. 217–43. London: Academic Press.

Reisen, W. K., Lothrop, H. D., Presser, S. B., Hardy, J. L. & Gordon, E. W. (1997). Landscape ecology of arboviruses in southeastern California: Temporal and spatial patterns of enzootic activity in Imperial Valley, 1991–1994. *Journal of Medical Entomology*, **34**, 179–88.

Rejmánková, E., Roberts, D. R., Pawley, A., Manguin, S. & Polanco, J. (1995). Predictions of adult *Anopheles albimanus* densities in villages based on distance to remotely sensed larval habitats. *American Journal of Tropical Medicine and Hygiene*, **53**, 482–8.

Risser, P. G., Karr, J. R. & Forman, R. T. T. (1984). *Landscape Ecology: Directions and Approaches*. Illinois Natural History Survey Special Publication No. 2.

Roberts, D. R., Paris, J. F., Manguin, S., Harbach, R. E., Woodruff. R., Rejmánková, E., Polanco, J., Wullschleger, B. & Legters, L. J. (1996). Predictions of malaria vector distribution in Belize based on multispectral satellite data. *American Journal of Tropical Medicine and Hygiene*, **54**, 304–8.

Robinson, T. P. (2000). Spatial statistics and geographical information systems in epidemiology and public health. *Advances in Parasitology*, **47**, 82–120.

Robinson, T. P., Rogers, D. J. & Williams, B. (1997a). Univariate analysis of tsetse habitat in the common fly belt of Southern Africa using climate and remotely sensed vegetation data. *Medical Veterinary Entomology*, **11**, 223–34.

Robinson, T. P., Rogers, D. J. & Williams, B. (1997b). Mapping tsetse habitat suitability in the common fly belt of Southern Africa using multivariate analysis of climate and remotely sensed data. *Medical Veterinary Entomology*, **11**, 235–45.

Rodríguez, A. D., Rodríguez, M. H., Hernández, J. E., Dister, S. W., Beck, L. R., Rejmánková, E. & Roberts, D. R. (1996). Landscape surrounding human settlements and malaria mosquito abundance in southern Chiapas, Mexico. *Entomological Society of America*, **33**, 39–48.

Rogers, D. J. (1991). Satellite imagery, tsetse and trypanosomiasis. *Preventive Veterinary Medicine*, **11**, 201–20.

Rogers, D. J. & Randolph, S. E. (1993). Distribution of tsetse and ticks in Africa, past, present and future. *Parasitology Today*, **9**, 266–71.

Rogers, D. J. & Randolph, S. E. (1994). Satellite imagery, tsetse flies and sleeping sickness in Africa. *Sistema Terra*, Year III 40–3.

Rogers, D. J. & Randolph, S. E. (2000). The global spread of malaria in a warmer world. *Science*, **289**, 1763–9.

Rogers, D. J. & Williams, B. G. (1993). Monitoring trypanosomiasis in space and time. *Parasitology [Suppl.]*, **106**, 77–92.

Rogers, D. J. & Williams, B. G. (1994). Tsetse distribution in Africa: seeing the wood and the trees. Large scale ecology and conservation. In *Proceedings of the 35th Symposium of the British Ecology Society for Conservation Biology*, ed. P. J. Edwards, R. May & N. R. Webb, pp. 247–71. Southampton, UK: University of Southampton.

Rogers, D. J., Hay, S. I. & Packer, M. J. (1996). Predicting the distribution of tsetse flies in West Africa using temporal Fourier processed meteorological satellite data. *Annals of Tropical Medicine and Parasitology*, **90**, 225–41.

Schofield, C. J., Apt, W. & Miles, M. A. (1982). The ecology of Chagas' disease in Chile. *Ecology of Disease*, **1**, 117–29.

Shope, R. E. (1992). Impacts of global climate change on human health: spread of infectious disease. In *Global Climate Change: Implications, Challenges and Mitigation Measures*, ed. S. K. Majumdar, L. S. Kalkstein, B. Yarnal, E. W. Miller & L. M. Rosenfeld Chapter 25. Easton, PA: The Pennsylvania Academy of Science.

Snow, R. W., Marsh, K. & le Sueur, D. (1996). The need for a map of transmission intensity to guide malaria control in Africa. *Parasitology Today*, **12**, 455–7.

Snow, R. W., Craig, M. H., Deichmann, U. & le Sueur, D. (1999). A preliminary continental risk map for malaria mortality among African children. *Parasitology Today*, **15**, 99–104.

Star, J. & Estes, J. (1990). *Geographic Information Systems: An Introduction*. Englewood Cliffs, NJ: Prentice-Hall.

Sutherst, R. W. & Maywald, G. F. (1985). A computerised system for matching climates in ecology. *Agriculture, Ecosystems and Environment*, **13**, 281–99.

Thompson, D. F., Malone, J. B., Harb, M., Faris, R., Huh, O. K., Buck, A. A. & Cline, B. L. (1996). Brancroftian filariasis distribution and diurnal temperature differences in the southern Nile Delta. *Emerging Infectious Diseases*, **2**, 234–5.

Thomson, M. C., Connor, S. J., Milligan, P. J. M. & Flasse, S. P. (1996). The ecology of malaria – as seen from Earth-observation satellites. *Annals of Tropical Medicine and Parasitology*, **90**, 243–64.

Thomson, M. C., Connor, S. J., Milligan, P. J. M. & Flasse, S. P. (1997). Mapping malaria risk in Africa: what can satellite data contribute? *Parasitology Today*, **13**, 313–18.

Thomson, M. C., Connor, S. J., D'Alessandro, U., Rowlingson, B., Diggle, P., Cresswell, M. & Greenwood, B. (1999). Predicting malaria infection in Gambian children from satellite data and bed net use surveys: the importance of spatial correlation in the interpretation of results. *American Journal of Tropical Medicine and Hygiene*, **61**, 2–8.

Thomson, M. C., Connor, S. J., O'Neill, K. & Meert, J-P. (2000). Environmental information for prediction of epidemics. *Parasitology Today*, **16**, 137–8.

Tolezano, J. E. (1994). Ecoepidemiological aspects of American cutaneous leishmaniasis in the state of Sao Paulo, Brazil. *Mem Instit Oswaldo Cruz*, **89**, 427–34.

Turner, M. G. & Gardner, R. H. (1991). Quantitative methods in landscape ecology: an introduction. In *Ecological Studies, Vol. 82, Quantitative Methods in Landscape Ecology*, ed. M. G. Turner and R. H. Gardner, pp. 3–14. New York: Springer-Verlag.

University of Wisconsin-Madison, Environmental Remote Sensing Center, www.ersc. wisc.edu/ersc.

Washino, R. K. & Wood, B. L. (1993). Application of remote sensing to vector arthropod surveillance and control. *American Journal of Tropical Medicine and Hygiene*, **50**, 134–44.

Weins, J. A. (1989). Spatial scaling in ecology. *Functional Ecology*, **3**, 385–97.

White P. C., Brown, J. A. & Harris, S. (1993). Badgers (*Meles meles*), cattle and bovine tuberculosis (*Mycobacterium bovis*): an hypothesis to explain the influence of habitat on the risk of disease transmission in southwest England. *Proceedings of the Royal Society of London Series B Biological Sciences*, **253**, 277–84.

Wood, B. L., Beck, L. R., Lobitz, B. M. & Bobo, M. R. (2000). Education, outreach, and the future of remote sensing in human health. *Advances in Parasitology*, **47**, 331–44.

World Health Organization (WHO). (1985). *Ten Years of Onchocerciasis Control in West Africa: Review of the Work of the Onchocerciasis Control Program in the Volta River Basin Area from 1974–1984*. R 386-786-1086-388, Geneva, Switzerland.

World Health Organization (WHO). (1990). *Potential Health Effects of Climatic Change*. Report of a WHO Task Group. WHO/PEP/90/10, Geneva.

10

Monitoring the health impacts of global climate change

DIARMID H. CAMPBELL-LENDRUM, PAUL WILKINSON,
KATRIN KUHN, R. SARI KOVATS, ANDY HAINES,
BETTINA MENNE & TERRY W. PARR

10.1 Introduction

As the processes of global environmental change proceed, the importance of monitoring health outcomes of climate change increases (e.g. Haines *et al.*, 1993; Haines, 1999). Accurate, reliable and comparable data are necessary for detecting and quantifying the early impacts of these changes on health, and as an essential first step towards planned adaptation to minimize adverse health impacts in a future, environmentally changed, world (McMichael *et al.*, 1996; Balbus, 1998).

These issues are well illustrated by recent developments in relation to global climate change. This chapter developed from a report of a working group convened by the World Health Organization, European Centre for Environment and Health (WHO-ECEH), which prepared a background document for the Third Ministerial Conference on Environment and Health, held in London in June 1999. The document pointed to the need for the monitoring of potential impacts of climate change on human health, and highlighted the potential role of long-term integrated monitoring sites in investigating links between anthropogenic climate change, natural ecosystems and human health.

This chapter also draws on work of the NoLIMITS (Networking of Long term Integrated Monitoring Sites), preparatory action of the European Union ENRICH (European Network for Research in Global Change) programme. NoLIMITS aims to link current environmental monitoring sites throughout Europe, to make available policy-relevant scientific information to address environmental changes and their consequences at local to global scales, to provide a focus for collaborative interdisciplinary research between sites, networks and users, and to explore the possibility of introducing new measurements at existing monitoring sites to meet specific scientific and policy needs. The objectives of this work are to:

- Identify environmental/ecological/health indicators relevant for population health and suitable for long-term monitoring.

- Review determinants (biotic and abiotic factors) related to each relevant health effect.
- Identify key environmental processes involved (and associated models) and data required for detecting climatically related trends or extremes.
- Identify the characteristics of monitoring programmes required to provide the environmental and health data specified above.
- Make recommendations on the characteristics of data and information required in existing monitoring sites across Europe.
- Identify research needs related to integrated monitoring.

10.2 Background

A range of national, regional and international organizations routinely collect relevant data, most obviously those monitoring climate conditions and (usually separately) health status. Climatic conditions affect health indirectly as well as directly, however, so that relevant information may also be obtained from other monitoring systems; for example, information on ecosystems which harbour climate-sensitive infectious diseases, or the socioeconomic status of populations, which in turn affects vulnerability to climate change (Haines & McMichael, 1997).

While these systems constitute a potentially powerful resource, they were implemented for purposes other than studying climate-change effects on health. There is therefore a need to review the data quality, coverage, access and linkage of current systems, and identify changes that would be necessary for them to address the questions: "Is climate change already affecting health?", "How much impact is it having?", "What effect is it likely to have in the future?", and "What should we do to minimize the impacts on health?".

Such a review must take realistic account of the challenges that confront monitoring systems. These include the intrinsic uncertainties of measuring multiple health consequences of a long-term climatic process, modified by a wide range of socioeconomic and other effects. They also include severe logistical difficulties. For example, in the health sector, multiple national and international organizations carry out monitoring following different procedures, so that only the most basic measures of public health status (e.g. mortality) can be compared uniformly around the world. Despite a number of national and international initiatives to restore and improve surveillance and control of communicable diseases in the last decade (WHO, 2001), monitoring of disease incidence and associated morbidity still varies widely depending on the locality, the country and the disease. Even allowing for international standardization, many developing countries currently lack the resources to improve current monitoring systems substantially so as to give full geographical and population coverage.

This chapter describes the data and information requirements for investigating the effect of global climate change on the major classes of climate-sensitive health impacts, including direct effects of climate on health, and indirect effects for example through natural ecosystems. Current sources of these measurements are reviewed, and recommendations are made on how current shortfalls in data and understanding may best be addressed. This chapter focuses on Europe, which has a relatively well developed monitoring infrastructure, but principles established here could potentially be applied to other continents. Clearly, however, the relative and absolute effect of different impact categories on human health will vary according to geographical location and population vulnerability.

10.3 Aims and challenges of monitoring health impacts of climate change

Monitoring is "the performance and analysis of routine measurements, aimed at detecting changes in the environment or health status of populations" (Last, 1995).

We suggest that climate change/health monitoring should be directed towards the following five aims:

- Early detection of the health impacts of climate change.
- Improved quantitative analysis of the relationships between climate and health.
- Improved analysis of vulnerability to climate change.
- Prediction of future health impacts of climate change, and validation of predictions.
- Assessment of the effectiveness of adaptation strategies.

The data and analyses required to address each of these specific aims are discussed in the following sections. All, however, share some common challenges:

- The vast scale of potential impacts.
- The many biological and physical processes on which human health depends.
- The uncertainty of climate predictions.
- The long-term nature of the changes involved.

10.3.1 Early detection of health impacts of climate change

In principle, the gradual process of climate change (particularly global warming) should be reflected in corresponding and directly measurable changes in the baseline incidence of climate-sensitive diseases, in their seasonal pattern (e.g. longer transmission seasons for certain diseases), or in their geographical distribution (e.g. shifts towards the poles or to higher altitudes). But for most, perhaps all, health outcomes it is difficult to obtain direct evidence of this kind (see also Chapter 7).

There are two main reasons. First, most diseases show substantial year-to-year fluctuations, which means that modest shifts in incidence rate are detectable only with a long time series. Typically such a time series needs to be several decades long, but there are few health end-points for which we have reliable data series of this length, and prospective data collection will be slow to yield results. Second, even where we do have long data series, it is often difficult to exclude changes in disease recording or shifts in (unmeasured) nonclimate risk factors as alternative explanations of temporal shift in disease.

The feasibility of detecting effects of climate change on health depends not only on the year-to-year variability in the health end-point, but also on our knowledge of its determinants in space and time. Climate effects on communicable diseases, for example, are generally indirect, and often mediated through complex natural ecosystems. We often have only limited information on the relative importance of other environmental and nonenvironmental influences. For example, it is problematic to ascribe the increasing trend in cases of Lyme disease (transmitted from wildlife reservoirs to humans by ticks) recorded in the U.K. over the last decade to increasing summer temperatures (Fig. 10.1), without thorough investigation of other possible contributing factors. These include changes in availability of habitats or hosts for ticks, or of the effectiveness of the notification system for this recently characterized disease.

Variations in nonclimatic factors may often have much stronger influences than climate. For example, despite the increase in temperatures over the last century, malaria, once widespread throughout Europe, has disappeared from all but a few areas in the south of the continent (Jetten & Takken, 1994). The effects of changes in land-use, improved health-care and other factors have predominated over those of climate change, and unless we improve our understanding of how *all* these factors

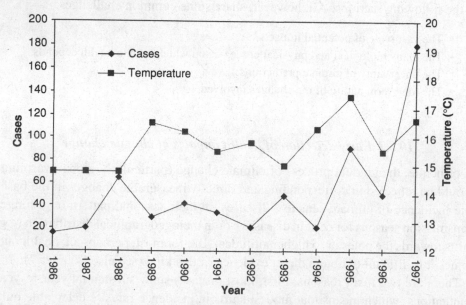

Fig. 10.1 Lyme disease reported in the U.K., and Central England summer temperature, 1986–97. Source: DETR, 1999.

contribute to malaria burden, we will be unable to draw robust inferences about climate-related health impacts.

If we restrict our aim to detection of any alteration in health due to climate change (rather than quantitative impact assessment), then some of the above pitfalls may be avoided by careful selection of "indicators" that are least likely to be affected by nonclimatic factors (DeGroot *et al.*, 1995; DETR, 1999). This approach has been widely used in the study of the effects of long-term climate change on natural ecosystems. There is now convincing evidence that over the last 20–30 years polewards shifts in the distribution of particular insect and bird species have occurred (Parmesan, 1996; Thomas & Lennon, 1999). Clear altitudinal shifts of plant communities (Grabherr *et al.*, 1994), and changes in the timing of life cycles (phenology) of insects, amphibians and birds (Beebee, 1995; Crick *et al.*, 1997; Sparks & Yates, 1997) have also been observed.

It may be possible to identify similar health indicators of climate change. These should fulfil the same rigorous criteria that are applied to studies of natural ecosystems, principally:

- Trends should be observed over long time periods, i.e. decades, and those confounding factors which cannot be excluded should be carefully measured (for the reasons described above).
- To detect geographical or phenological changes, multiple species or populations should be studied so as to maximize sample sizes.
- To detect polewards or altitudinal shifts in disease or disease agents (vectors or pathogens), populations should be monitored either over their full range (Parmesan, 1996), or at least the extremes of the range (Parmesan *et al.*, 1999).

These criteria are particularly hard to fulfil for health indicators. Most diseases are subject to control efforts, which constitute an important nonclimatic determinant of the geographical and temporal distribution of disease. The distribution and abundance of vectors or pathogens in areas which are not subject to control interventions are likely to prove more sensitive and robust indicators. Repeated sampling to monitor the full longitudinal and/or altitudinal range of specific disease agents, and their seasonal patterns, are probably the most cost-effective and robust methods of directly detecting early health-relevant effects of climate change. To our knowledge, no suitably standardized and comprehensive monitoring systems are currently in place.

10.3.2 Indirect health impact assessment

Long-term climate change effects are slow and difficult to measure with confidence, due to the confounding effects of changes in other causative factors. It is simpler

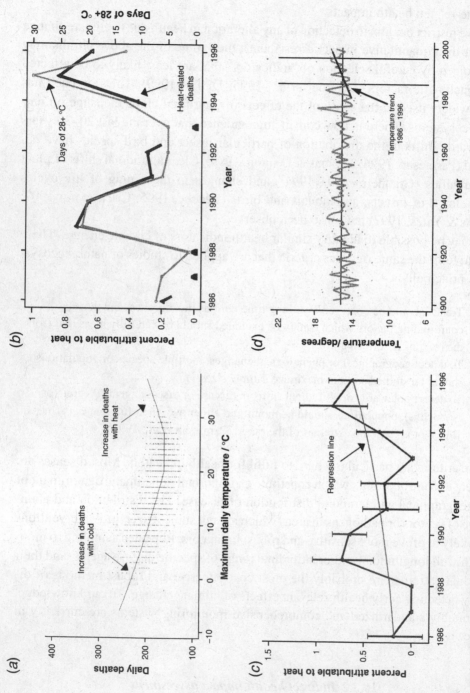

Fig. 10.2 (a) Daily deaths in London by maximum daily temperature. (b) Deaths attributable to heat based on time series analysis. (c) Trend in heat-attributable deaths, 1986–96. (d) Trend in central England mean June & July temperatures, 1900–99.

and quicker to measure the health effects of climatic variation either over short time periods (seasonal or interannual), or with geography. Such analyses of the effects of climate variability may be used to indirectly assess the impacts of long-term climate change.

For example, the relationship between mortality and temperature (described in Section 10.6) is well characterized. Time series analysis of available data over just a few years has enabled us to quantify the increase in mortality at low and high temperatures. In Britain, the temperature–mortality relationship is U-shaped, with mortality increasing by approximately 2 % per year for each degree Celsius below about 20 °C maximum daily temperature, and by a similar amount for every degree above 30 °C (Fig. 10.2(*a*)). Based on this relationship it is possible to ascertain the percentage of deaths attributable to hot days (over 28 °C) in any year (Fig. 10.2(*b*)).

Reliable records of daily temperature exist for thousands of sites across the globe, usually dating back at least several decades. Given such long data series and the ease and accuracy with which temperature can be measured, it is comparatively simple to gain evidence of even quite small shifts in climatic conditions as they occur. Combined with our knowledge of the temperature–mortality relationship, such evidence allows us to derive quantitative estimates of trends in heat-attributable mortality over time (Fig. 10.2(*c*)), and potentially to estimate the effects of climate change in the future.

Such analyses are a powerful tool for separating out and quantifying the effect of climate on health, but require careful interpretation. First, as for direct monitoring of climate-sensitive impacts, indirectly estimated trends are highly sensitive to the period chosen for analysis. Fig. 10.2(*d*) suggests that the upward trend in heat-attributable deaths from 1986–96 is not necessarily significant if other start and end-points are chosen. For example, investigation of individual heat-wave periods in London between 1976 and 1996 did not suggest any temporal trend in the effect size on deaths (S. Hajat, personal communication). Clear trends may perhaps only become apparent in much longer time series, or with an increasing rate of climate change. Second, temperature–mortality relationships are unlikely to remain static, but will be affected by physiological and behavioural adaptation to changing climates, and changes in the rates of diseases affecting vulnerability to heat and cold. Long-term health monitoring will remain necessary to update the relationships in our indirect analytical models, and provide a "reality check" of what actually happens to overall mortality and morbidity over time.

10.3.3 Improved understanding of vulnerability

As described above, monitoring of the effects of nonclimatic factors is necessary if we are to distinguish them from climate effects. However, these data may also serve another important purpose: identification of those populations which are likely to suffer the most severe health impacts of climate change (see Chapter 4).

Climate change may be expected to cause varying effects on different geographical populations, or on subpopulations within a single location. For example, a climate-driven increase in the distribution of mosquito-borne disease may have greater effects in poorer or politically unstable countries, which have limited resources for disease control. Within these vulnerable countries, the impact may be greatest on those individuals living in substandard housing (and therefore exposed to more mosquito bites), or whose immune systems are compromised by inadequate nutrition or HIV infection. Collection of sufficient data to test for interactions between climate and nonclimate risk factors should be an integral part of the quantitative health-impact assessments described above. Such analysis would serve the ultimate aims of identifying the main stakeholders in climate change and the effective direction of health interventions.

10.3.4 Generation and validation of predictive mathematical models

Mathematical models are increasingly used to summarize our quantitative understanding of the relationships between climate and health (see Chapters 6 and 10). Such models are usually formulated by mathematically summarizing either the climate effects on the biological processes that determine disease (process-driven or biological models), or the observed statistical associations between climate and disease in space or time, without necessarily elucidating the underlying biological mechanisms (statistical-empirical modelling). These models may be coupled to predictive models of climate change in order to estimate future changes in health states.

Monitoring data can make important contributions to mathematical modelling. First, monitoring of climate, health and other contributing factors is essential to define the associations upon which statistical-empirical models are based, and may also allow estimation of climate effects on biological processes that cannot be reliably defined from laboratory data. The amount of data required to define the mathematical relationship between climate and a particular health impact or biological process accurately will vary: it will tend to be greater for those impacts that are subject to multiple nonclimate factors, and may need to be collected from multiple regions if the relationships between climate and health vary significantly with geography.

Second, monitoring data can allow validation of model predictions, which may or may not accurately describe climate–health relationships (Davis *et al.*, 1998). Predictions of health states may be generated either from retrospective data on climate variations in time, or with geography, and these can then be compared with health monitoring data. Close associations between model predictions and monitoring data in the past would give confidence that the model may accurately predict future changes. However, relationships between climate and health are not static, and long-term monitoring of climate and health into the future is essential to tell us what actually happens. An important contribution of monitoring is therefore to supply the data necessary to continually refine models in an iterative fashion, so as to take account of the inevitable changes in the relationships between climate and health.

The interaction between modelling and monitoring is not all one-way: modelling can also help direct monitoring to the most appropriate locations by suggesting where shifts in disease are particularly likely to occur. Thus there is much to be said for developing close working relationships between researchers who are developing mathematical models of climate-change impacts and those involved in the design and management of monitoring systems.

10.3.5 Assessment of the effectiveness of adaptation strategies

In addition to detection, assessment and prediction, monitoring may also aid in assessing and refining adaptation strategies. The processes described above lead to quantitative expectations of the changes that may accompany climate change if no action is taken. If adaptation strategies are implemented, comparisons of these predictions with monitoring data can provide a test of their effectiveness.

As for the quantitative impact assessment described in Section 10.3.2, direct assessment through long-term monitoring is (on its own) likely to prove slow, and may not generate firm conclusions, as temporal changes in other contributing factors may be confounded with adaptation efforts. The most reliable measures of their effectiveness will probably derive from comparisons of the effect of short-term or geographical variation in climate on locations that have implemented adaptation strategies compared with those that have not. Long-term monitoring is necessary to assess gradual changes in the ways in which climate, adaptation strategies and other factors interact to influence health.

10.4 Which health impacts should be monitored?

Climate influences human health through a wide variety of often ill-defined mechanisms, so that comprehensive climate–health monitoring is unfeasible. Given that only limited resources are available to implement new monitoring systems or revise existing systems, priority health issues must be identified.

The principal criteria for such selection should be evidence of climate sensitivity, demonstrated either through observed health effects of temporal or geographical climate variation, or through evidence of climate (e.g. temperature, rainfall, humidity, etc.) effects on components of the disease-transmission process in the field or laboratory. For infectious diseases, detailed knowledge of transmission cycles is essential for identifying the major threats. Climate-change effects are likely to be most profound for diseases caused by organisms that replicate outside of human hosts (where they will be subject to ambient conditions), and will be less important and/or more difficult to detect for those where human-to-human transmission is common (Stanwell-Smith, 1998).

Monitoring should also be preferentially targeted towards significant threats to public health. These may be diseases with a high current prevalence and/or severity (WHO, 2000), resulting in a large loss of disability-adjusted life years (DALYs), or those considered likely to become prevalent under conditions of climate change.

Finally, logistic constraints should be taken into account. Consideration should be given to whether a proposed system is likely to generate the data necessary to permit a clear analysis of climate sensitivity, given potential problems of data quantity, quality and reliability.

Such considerations have been used to identify the most important health issues for both research and monitoring on a global scale. These include exposure to thermal extremes and extreme weather events, effects on the distribution and activity of vectors and infective parasites, waterborne and foodborne diseases, altered food (especially crop) productivity, sea level rise and flooding with population displacement, air pollution including pollens and spores, and social, economic and demographic dislocations due to effects on economy, infrastructure and resource supply (McMichael *et al.*, 1996). However, priority health issues vary with geography, and associated differences in climate and development. Within the European region for example, the following classes of health impacts have been identified as high priorities for monitoring in relation to climate change (WHO, 1999a):

(i) Direct effects of exposure to low and high temperatures.
(ii) Health impacts of extreme weather events (floods, high winds, etc.).
(iii) Increased frequency of food- and water-borne disease.
(iv) Geographical changes and altered transmission frequency of vector-borne disease, principally leishmaniasis, Lyme disease, tick-borne encephalitis and malaria.

The following have been identified as lower-priority threats:

(i) Other vector-borne diseases, including dengue and Toscana virus.
(ii) Aero-allergens, particularly pollen, as climate-sensitive causes of respiratory allergies.
(iii) Rodent-borne diseases, including hantavirus and leptospirosis.

These lists are neither exhaustive nor final, and monitoring systems should also be sensitive to changes in the importance of current problems and the appearance of new health concerns. Systems such as the e-mail mailing list of the Programme for Monitoring Emerging Diseases (PROMED) provide a valuable resource in identifying new climate-sensitive health concerns.

A framework for monitoring the health impacts of climate change is shown in Table 10.1, which summarises the types of impact which could be monitored, the sites where monitoring should be conducted and the main health outcomes that are appropriate to monitor. Criteria for the implementation of monitoring systems for infectious diseases have been outlined in a WHO working paper (summarized in Table 10.2). Similar principles might be applied to other health impacts (e.g. direct health effects of heat and cold).

Table 10.1. *A framework for monitoring potential health impacts*

What	Where	How
heat stress	urban centres in developed and developing countries	daily mortality data
changes in seasonal patterns (e.g. asthma and allergies)	"sentinel" populations at different levels	morbidity data from primary care, hospital admissions, emergency room attendances
natural disasters	all regions	persons affected and mortality data
effects on health of rise in sea level	low-lying regions	local population surveillance
freshwater supply	critical regions, especially in the interior of continents	measures of runoff, irrigation patterns, pollutant concentrations
food supply	critical regions	remote sensing, measures of crop yield, food access, nutrition (from local surveys); agricultural pest and disease surveillance
emerging diseases	areas of population movement or ecological change	identification of outbreaks of "new" syndromes or diseases, population-based time series, laboratory characterization
vector-borne diseases	margins of distribution (latitude and altitude)	primary care data, local field surveys, communicable disease centres, remote sensing
marine ecosystems	oceans	remote sensing, sampling of biotoxins, phytoplankton, essential nutrients. Epidemiology of cholera, other vibrios, shellfish and fish poisoning

From Haines *et al.* (1993).

Table 10.2. *Requirements and criteria for establishing surveillance of communicable diseases that may be influenced by climate change*

Surveillance should be:

- continuous
- frequent
- active as well as passive, i. e. eliciting additional information about variables required to interpret changes
- based on clear definitions and diagnostic criteria
- based on diseases with established information on aetiology and host factors
- based on diseases with environmental sources where climate is likely to affect incidence
- based on a minimum core dataset
- sustainable
- able to provide early warning of changes in incidence
- suitable for inclusion in a monitoring and surveillance network
- regularly reviewed, flexible and responsive to change, with appropriate training for users
- based on an holistic public health approach which includes collection of data on potential confounding factors, such as local environmental change and the implementation of disease-control measures

(Stanwell-Smith, 1998).

10.5 What kinds of measurements are relevant?

Several types of data are necessary for detecting climate effects on health, and quantitative analysis through either statistical or process-based/biological models. Many relevant variables (or surrogates of them) are already recorded by monitoring systems, and may require only access and cross-referencing with other data sources. For others, new monitoring systems or radical changes to existing systems may be necessary in order to carry out meaningful analyses. Relevant measurements fall into the following broad classes.

10.5.1 Climate

Temperature, rainfall, relative humidity and wind speed are known to have an important influence on a wide range of health processes, and all are predicted (with a greater or lesser degree of certainty) to be affected by future climate change. All are recorded through a comprehensive global network of ground meteorological stations (World Meteorological Office (WMO), 2000), and these measurements can be interpolated to give estimates for the entire globe (New *et al.*, 1999). Alternatively, satellite-based sensors record proxy measurements of temperature, rainfall and humidity with true global coverage (Hay *et al.*, 1996). Although these require analysis to generate estimates of climatic variables, new remote-sensing systems provide constantly increasing information content (NASA, 2000). The Global Oceanic Observing System (GOOS) monitors the climate over the oceans. Data from these sources can be geographically and temporally cross-referenced to health and environmental monitoring data in a Geographical Information System (GIS), providing powerful tools for broad-scale analysis of associations between climate and disease.

Their application is limited, however, by the relatively coarse nature of the data recorded. Measurements are either discontinuous point data or averages from a wide area, and may not describe either local variation in climate (e.g. temperature in city centres versus nearby rural areas), or microclimatic conditions in specific important environments (e.g. resting sites of adult mosquitoes). Climate measurements at a local level and in important microclimates (where these can be identified) should be recorded in intensively studied sites, to test whether they provide closer correlation with health outcomes. As it is unfeasible to make such measurements at all sites, their relationship with data from monitoring systems with global coverage should be modelled in order to allow "scaling up" across large areas.

10.5.2 Health outcomes

Analysis of climate effects on health depends ultimately on reliable recording of defined health states, at suitably high temporal and spatial resolution (Stanwell-Smith, 1998). Such monitoring should be sensitive and specific enough to detect and quantify changes in the intensity and temporal and geographical distribution of climate-sensitive health impacts. However, the utility of routinely collected data varies markedly, depending on the health issue and region.

Thermal extremes such as heatwaves in cities, for example, act as stressors on susceptible populations, increasing the frequency of mortality from cardiac and respiratory conditions. The health impact is calculated by assessing the number of excess deaths relative to long-term moving averages in mortality. Date of death is routinely recorded on death certificates in most developed regions of the world. This information is collated by regional and national health authorities, thus providing the basic data for an accurate quantitative assessment of the excess mortality caused by climate variation and, by inference, long-term climate change.

Even in developed regions, however, current monitoring systems are much less well equipped to measure climate effects on cause-specific (rather than total) mortality and morbidity. The data requirements are hardest to meet where climate affects health through complex ecological processes, such as infectious diseases transmitted through food, water or vectors. The major problems are:

- Temporal and geographical variation in diagnosis, case definition and completeness of case ascertainment, and reporting efficiency.

- Lack of accurate recording and quantification of the effects of nonclimatic causal factors and of adaptation strategies.
- Insufficient temporal or geographical resolution in health data and nonclimatic risk and protective factors.

One of the important aims for future monitoring should be to address these limitations. In some cases this may be achieved through revision of existing health datasets and linkage to climate records. For many climate-sensitive diseases, however, the coverage or quality of available data precludes this approach. Current monitoring practices may need to be revised.

10.5.3 Intermediate stages and environmental data

High quality health data are necessary, but not sufficient, to fully understand more indirect and complex effects of climate change on human health. Infectious diseases transmitted by water, food or vectors are often highly sensitive to climate, but human infections are only the end-product of a complex chain of environmental processes (Epstein, 1999). Ideally, climate and disease monitoring should be linked with parallel monitoring of intermediate stages in the transmission cycle (e.g. parasites, vectors and reservoir hosts), and of the wider ecosystem (e.g. the distribution of habitats suitable for vectors or reservoirs). The collection of accurate data on these variables would help to improve the explanatory value of models both in purely statistical terms, and in terms of identifying the key biological processes underlying changes in human disease. For example, integrated monitoring of the variables describe above would help to determine how much of the variation in the incidence of tick-borne encephalitis (TBE) in Europe is due to climate effects on tick population dynamics, compared to changes in the abundance of reservoir hosts, or in the distribution of habitats where ticks and reservoirs come into contact.

In some situations, it may be informative to define the relationship between climate and environmental intermediates, even in the absence of the disease itself. For example, competent vectors for both dengue and malaria are present in many areas of Western Europe, but there is no active parasite transmission. However, mosquito population density, biting rate and adult longevity (all known to be affected by temperature) are important determinants of the capacity of these vector populations to transmit the relevant pathogens (Garrett-Jones, 1964). It is therefore relevant to public health to assess the link between climate and vector populations, so as to assess changes in the probability of establishment of local transmission when parasites are imported to currently nonendemic areas. The specific parameters that should be recorded vary depending on the transmission cycle of specific diseases.

10.5.4 Socioeconomic factors affecting vulnerability

The vast majority of climate effects on health will be mediated by socioeconomic factors, both before and after exposure. Socioeconomic conditions should therefore be monitored for two reasons: first, to dissociate their effects from those of climate in quantitative analyses; and second, to identify particular socioeconomic factors that interact with climate (i.e. increase vulnerability to climate variation and climate change).

Relevant socioeconomic factors should be identified on the basis of previously observed effects on the health impact of interest. Some are likely to have a very wide range of effects. Poverty and living conditions will influence exposure to almost all health threats, from direct thermal effects and natural disasters to infectious diseases, and are also likely to increase the harmful effects of such exposures. Other socioeconomic factors may affect a smaller range of health issues; for example, occupation may influence exposure to some pathogens in the environment. Finally, the availability of infrastructure to mount dedicated disease-control programmes (e.g. insecticide spraying against malaria vectors or chlorination of water supplies) may affect only one or a narrow range of health impacts.

10.5.5 Other confounding variables and effect modifiers

Socioeconomic variables are not the only factors that obscure or modify climate effects on health. Some factors may simply act to obscure climate–disease relationships or to generate spurious correlations in analyses (for example, biases in health-recording systems over time or space). Others may themselves be of general relevance as modifiers of climate effects (e.g. effects of topography on the health impacts of heavy rainfall). Factors in both categories may be either general or disease specific, such as the variation in food-handling practices for food-borne infections, or variation in land-use for vector-borne diseases. As for socioeconomic factors, effect modifiers and confounders may only be identified on the basis of previously observed effects on each health outcome.

Two solutions are possible to address these effects: monitoring may be restricted to areas where modifiers or confounders are unlikely to exert a strong effect (for example, monitoring of climate effects on mosquito populations may take place in areas where there are no control activities). In this case, climate–health models can only be generalized to other areas under the assumption that the derived relationships hold true in other circumstances; for example, that temperature will affect mosquito population dynamics in the same way in habitats that are different from the original study site. Alternatively, where these factors are themselves of general relevance, it may be more useful to monitor their effect on the health outcome of interest so that they can be included in analytical models. This second approach has the advantage that the importance of climate effects may be assessed against other determinants (which may prove to be more important), and that interactions between climate and other variables may be directly measured. The amount of data necessary to carry out such an analysis, however, will increase exponentially with the number of factors that are being assessed. It should also be noted that it is often impossible either to identify or to eliminate all possible confounding effects on health.

In this section we describe the public health impact, aetiology, available evidence for climate sensitivity, and the main effects that climate change is likely to have on each of the main categories of health impacts. The data requirements for relating climate change to health are then described for both final health outcomes, and intermediate stages (environmental hazards to human health, including infected vectors and reservoirs).

10.6 Direct effects of thermal extremes on health

Direct effects of exposure to heat and cold are among the best characterized of all climate impacts on health although the relative contribution of temperature and viral infections to winter mortality are still under investigation. Such effects are evident both as seasonal fluctuations in health and more specifically as associations between daily temperature and mortality.

Methods of epidemiological analysis are well established, at least for investigating short-term (daily or weekly) associations. As a minimum, quantification of temperature-related mortality requires daily counts of deaths, ideally grouped by underlying cause, and temperature measured at similar temporal resolution. Data on such time-varying factors as respiratory infections are also necessary so that adjustments may be made for them as potential confounding factors because they are associated with increases in mortality, particularly amongst the elderly.

In developed regions such as Europe, these data are readily available from routine sources: mortality statistics from national or regional registries, and meteorological data from the wide network of quality-controlled monitoring stations across most regions. Temperature measurements of sufficient accuracy for epidemiological time series analysis are available from many sites over long periods of time, and fact of death (if not the cause) has been reliably recorded for many decades. Influenza and other seasonal infections are not as well recorded, but nonetheless data on them are available for recent decades from a number of infectious disease-reporting systems.

However, the scientific need is not just to quantify the temperature–mortality relationship, but also to understand how it depends on individual susceptibility and is modified by external factors. In particular, there are important questions of:

- Vulnerability (defined by such factors as age, sex, socioeconomic status and pre-existing disease).

- Physiological acclimatization and behavioural adaptation to changes in mean and extreme ambient temperatures.
- The impact of protective environments, such as controlled indoor temperatures achieved by use of air-conditioning, improved thermal insulation and efficient home heating.
- Co-existing effects of air pollution.

Changes in population vulnerability and adaptive processes are predicted to have a substantial influence on the burden of mortality and morbidity attributable to the direct effects of thermal extremes, as they may with all climate-related health impacts. It is therefore important that research also focuses on these modifying factors so that we may obtain more realistic estimates of health impact and can better understand how to protect against the adverse health effects of climate change.

The primary challenge in such research is the need to link temporal data on temperature and mortality with those on the personal, social and environmental characteristics of the population at risk. Because we are dealing here with effect modification, the required sample size is very large – larger than that of most existing individual cohorts. One of the main thrusts of epidemiological research must therefore be to establish the required data linkages, i.e. to define the characteristics of individuals or groups of individuals within much larger populations that may be subject to methods of time series analysis.

The ability to make the needed linkages between health on the one hand and social and environmental characteristics of the population on the other varies from country to country. In Europe, Scandinavian countries and the U.K. are better placed than most because of the accurate georeferencing of their routine health data, combined with the availability of data on population characteristics at the small-area level. But in all countries there is a need to search for, or to establish, surveys that can provide specific data on factors such as housing quality and individual behaviour that can be linked to health statistics. It is only from such linked data that researchers will be able to gain much-needed evidence on the complexity of factors that determine the direct health impacts of thermal extremes.

10.7 Natural disasters caused by extreme weather events: storms and floods

Extremes of wind and rainfall occurring over the synoptic time scale (three to six days) are amongst the most frequent causes of natural disasters, defined as

the result of a vast ecological breakdown in the relation between humans and the environment, a serious and sudden event (or slow, as in drought) on such a scale that the stricken community needs extraordinary efforts to cope with it, often with outside help or international aid

(Gunn, 1990)

Extreme weather events have a significant health impact. A total of 2042 flood and 2350 storm/hurricane disasters have been recorded throughout the world this century, with 8 650 681 deaths and 1 910 408 injuries attributed to these events (EM-DAT database, University of Louvain, Belgium). A strong association has already been established between El Niño climate variations and total health impact due to natural disasters (Bouma *et al.*, 1997). Global warming, through increasing the intensity of the hydrological cycle, may bring about increases in the occurrence of extreme precipitation (Hennessy *et al.*, 1997; McGuffie *et al.*, 1999), and consequently riverine and coastal flooding (Arnell, 1996). Evidence of long-term changes in windspeed and precipitation is weaker than for temperature, however, and predictions of future changes in these variables are consequently more uncertain (McGregor, 1999).

Much of the previously observed variation in human impact has been due to changes in total population, and in their vulnerability to particular natural disasters, such as geographical distribution relative to areas with high flood risk (Malilay, 1997a). Future variation in these factors is likely to be at least as important as changes in the climatic determinants of disasters, and the relative importance and interactions of these factors are still only poorly understood (Malilay, 1997a,b). Research is therefore required not only to identify and predict long-term changes in weather patterns, but to understand the relative importance and interactions of natural (e.g. meteorological, topographical) and anthropogenic (e.g. population distribution, housing quality) effects on the eventual health impacts of extreme weather events (Noji, 1997; McGregor, 1999).

10.7.1 Currently available information

Rainfall and windspeed data are routinely recorded at numerous ground meteorological stations throughout the world, accessible through WMO and national meteorological services at monthly resolution or higher. Although this relies on point measurements, these are sufficiently numerous and well spaced to serve as a representative sentinel system to detect changes in pattern and intensity. Point measurements can also be interpolated to generate continuous global coverage.

The principle international resource for data on health impacts of natural disasters is the EM-DAT Disaster Events Database (Sapir & Misson, 1992), developed by the Centre for Epidemiology on Disasters at the University of Louvain, Belgium. The database assembles, reviews and verifies information from governments and international agencies on all reported disasters (at least 10 deaths and/or 200 people affected and/or an appeal for outside assistance) and their primary cause (e.g. flooding, drought, storm). Estimates of number of deaths, persons injured, persons affected and persons made homeless are recorded, following well-defined criteria.

Although an important resource, the database is not perfect, as it relies on reporting from other agencies and an arbitrary disaster threshold, and there is currently no direct linkage to weather conditions or environmental data such as physical measures of flooding. Neither does the database record information on the specific causes of mortality and morbidity, particularly due to the impacts of disasters on mental health, which may occur long after the natural disaster, and are consequently extremely difficult to define and attribute. Detailed investigations of individual disasters utilizing data from a variety of sources (e.g. routine data on increases in suicide rate) are necessary to characterize better the full range of health impacts of natural disasters.

10.8 Food- and water-borne illness

10.8.1 Food-borne illness

Recent estimates from the U.K. (Wheeler *et al.*, 1999) and The Netherlands (Hoogenboom-Verdegaal *et al.*, 1994) suggest that approximately 20 % of the population develop infectious intestinal disease every year, with a substantial increase observed over the last 20 years. The wider picture is probably more serious than these results suggest: the former study suggested that only one in six cases presented to a general practitioner, while in the developing world the incidence of infection is much higher, and the consequences of each infection more severe (WHO, 2000). Infections may be fatal if they occur in vulnerable subpopulations, and/or are due to particularly pathogenic organisms such as *Vibrio cholerae*. Nonfatal infections have a large economic impact through treatment costs and loss of working time (Roberts *et al.*, 1989; Todd, 1989).

Food-borne illnesses are due mainly to bacterial or viral infection. The subset of bacteria which replicate in the environment, either in animal feed (affecting rates of infection in meat or poultry) or on food products, are likely to be most sensitive to climate change. Replication rates of most bacteria are positively correlated with temperature, although some, such as *Listeria monocytogenes*, are tolerant of cold temperatures and outcompete other bacteria on refrigerated foodstuffs (Lacey, 1993).

Convincing evidence for climate sensitivity is provided by analysis of all-cause food-poisoning notifications in England and Wales (Bentham, 1997), and diarrhoea cases in Lima, Peru (Checkley *et al.*, 2000). Both studies show a strong correlation between incidence and the temperature up to one month earlier, even after removing the effects of seasonal and long-term trends. There are several possible (nonexclusive) explanations for this variation. Variation in transmission rates may be due partly to the well-characterized effect on replication for some of the most frequent bacterial infections, with *Salmonella* replication rates rising between 7 and 37 °C (Baird-Parker, 1994). Such relationships do not explain all seasonal variation, however; *Campylobacter* infections show a strong peak in early summer in much of Europe, despite replicating only inside hosts.

Despite the significant association between variations in climate and food-poisoning, there is little evidence that long-term trends in food-poisoning notifications over the past 30 years may be explained simply by climate change. Changes in food production, handling and consumption practices and/or disease notification probably explain much of the long-term trends in the past (Baird-Parker, 1994) and may do in the future.

10.8.1.1 Currently available information

Comprehensive climate data are available from ground meteorological stations, and used for correlations against the incidence of food-borne diseases (as described above). They are probably of limited use, however, in identifying the specific mechanisms by which climate affects food-poisoning, as much of the food production–processing–consumption chain occurs in climatically controlled conditions.

Application of the Hazard Analysis Critical Control Point (HACCP) procedure may help to clarify which procedures in the food production–processing–consumption chain are most critical in influencing the risk of infection with particular pathogens. This may allow more efficient direction of monitoring to detect and quantify temperature or other climate effects on pathogens in food. Difficulties in measuring the temperatures to which foods are actually exposed before being consumed, and in quantifying the relationships between the dose of pathogen in food and human illness (Baird-Parker, 1994) will, however, continue to complicate full analysis of temperature effects on all components of the food chain.

The WHO programme for the surveillance of food-borne diseases collates data from national recording centres in 38 countries in Europe (WHO-ECEH, 1999), with most datasets continuous since the 1970s or 1980s. Coverage is more comprehensive in the EU than elsewhere in the region. Data are drawn from statutory notifications (case reporting), reporting of investigated outbreaks, laboratory reports and special surveys. Information includes: (i) number of ill persons; (ii) causative agent; and (iii) locations of contamination, acquisition and consumption; and (iv) factors contributing to the outbreak. National surveys have higher resolution information often with temporal resolution of one day, and geographical resolution to the second or third administrative level. In some countries, notification data for specific pathogens are continuously updated in a GIS format, allowing analysis of climate associations in time and space (Figs. 10.3 and 10.4; Swiss Federal Office of Public Health, 1999). Data on outbreaks of *Salmonella* and *E. coli* O157 throughout the European region are collated by the ENTER-NET network, which also works to standardize characterization methods.

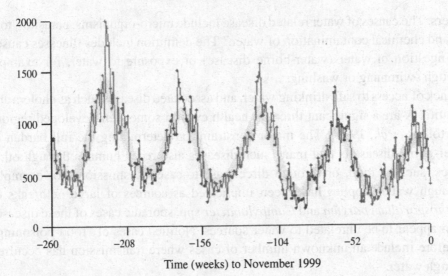

Fig. 10.3 Weekly time-series of salmonellosis cases reported in Switzerland: December 1994 to November 1999. Points show weekly observations (blue and red in alternate years); yellow line shows the moving average. For a colour version of this figure, see www.cambridge.org/9780521114028

10.8.2 Water-related illness

Water-related disease is defined by WHO as "any significant or widespread adverse affects on human health, such as death, disability, illness or disorders, caused directly or indirectly by the condition, or changes in the quantity or quality, of any

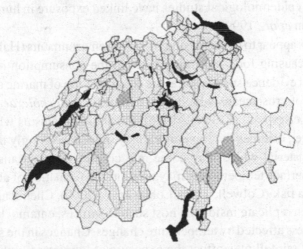

Fig. 10.4 Geographical distribution of salmonellosis incidence reported in each administrative region of Switzerland in 1999. Darker shading indicates higher incidence of disease.

waters. The causes of water related disease include micro-organisms, parasites, toxins and chemical contamination of water." The definition includes illnesses caused by ingestion of water (water-borne disease) or exposure to water, for example through swimming or washing.

Lack of access to safe drinking water, and associated diseases such as cholera and hepatitis A are a significant threat to health even in some well-developed regions (Bertollini *et al.*, 1996). The major constraint on determining the full burden of water-borne disease is that many such diseases also reach humans through other routes, such as food, or through direct case-to-case transmission. For example, although water supplies have been implicated as sources of large outbreaks of *Cryptosporidium parvum* and *Campylobacter* spp. sporadic cases of these diseases often appear to be unrelated to water sources. Notified cases of "food-poisoning" therefore include an unknown number of cases where transmission has occurred through water.

Few studies relating the incidence of water-borne infection to climate parameters have been carried out. Climate associations may potentially be sought with disease outbreaks where epidemiological investigation strongly implicates water-borne transmission, or with the minority of pathogens that are thought to be transmitted almost exclusively through water. These include *Legionella pneumophila*, *Vibrio cholerae* and diseases caused by cyanobacteria (blue-green algae) which multiply under favourable temperature, sunlight and nutrient conditions to form algal blooms. A significant proportion of freshwater cyanobacteria produce toxins which may be harmful to human health through causing skin irritation on contact, or gastrointestinal illness through ingestion. Cases are relatively rare, however, and no quantitative epidemiological studies have linked exposure in humans to climatic factors (Bartram *et al.*, 1999).

Algal blooms appear to be increasing in marine environments (Hallegraeff, 1993), in some cases causing food-poisoning through the consumption of contaminated seafood. Recent evidence suggests that the concentration of marine algae influences the abundance of crustaceans which may harbour *Vibrio cholerae* in Bangladesh, and potentially elsewhere (Colwell, 1996). For those organisms which persist and replicate in water, changes in temperature may influence strongly the total amount of hazardous material which may come into contact with humans. For example, increasing sea surface temperatures may promote the formation of algal blooms and increase cholera risk (Colwell, 1996; Lobitz *et al.*, 2000). Chemical pollutants and pathogens which replicate inside the host such as viruses, entamoeba and protozoa are likely to be less affected by temperature changes. Changes in the seasonal pattern and intensity of rainfall may affect the exposure of humans to both biological and chemical hazards, for example by increasing the frequency of contamination of water sources through flooding, but will probably have little impact on the total amount of pathogenic material in the environment (J. Bartram, personal communication).

Any analyses of climate-change effects on water-borne diseases must incorporate the effects of changes in human living conditions and behaviour, travel between endemic and nonendemic areas, and treatment programmes to combat water-borne disease.

10.8.2.1 Currently available information

Global climate data over land surfaces are available as described above, and temperature is often recorded in sampling of water bodies for pathogens or chemical toxins. Sea surface temperatures are monitored by remote sensing and floating monitoring stations.

National public health laboratories carry out surveillance of surface waters, ground waters and enclosed waters to make quantitative measurements of specific pathogens and toxins, including algal blooms in surface waters. Large-scale surveillance of marine algal blooms is carried out using remote sensing (Kahru & Brown, 1997).

Food-poisoning notifications to local and national recording centres and the WHO surveillance programme are described above. These records include many water-borne pathogens, but the mode of transmission is often unidentified. Epidemiological investigations of outbreaks may identify a water source with a greater or lesser degree of certainty. Some of the diseases with closest links to water (Legionnaire diseases and cholera) are notifiable to national surveillance systems in much of Europe. Data on cases of Legionnaire's disease throughout much of the WHO-EURO region are collated by the European Working Group for *Legionella* Infections (EWGLI), which also standardizes guidelines for case characterization and pathogen typing.

10.9 Vector-borne disease

For most vector-borne illnesses, current monitoring data can provide only very broad quantification of the relationship between climate and human disease (Kovats *et al.*, 2000). This is because disease reporting is often geographically and temporally inconsistent, and nonclimatic factors are important in determining vector abundance and capacity to transmit infection. An effective monitoring system should therefore gather data not just on climate and disease, but on factors such as land use, host abundance and intervention measures.

For any system requiring new data collection, limited availability of resources will result in a trade-off between intensity of monitoring and number of sites. Most information would probably be gained by a combination of monitoring strategies. This would include acquiring high-frequency (i.e. weekly) data on vector abundance, survival and (for some vectors) infection rate in a few endemic sites. These should be sited in areas where few other factors are likely to alter over time, and climate effects are therefore likely to be most detectable. There would also be merit in having a low frequency of measurements (i.e. monthly) at a much larger number of collection sites positioned on transects that cut across the margins of current distributions. Data from these systems could be combined with existing data in a GIS database, to facilitate linkage of data on climate, vector abundance and environmental conditions, giving additional insights into the mechanisms and relative magnitude of climate effects on health.

The following sections describe the current state of knowledge and data availability for research on the major vector borne-diseases in the European region.

10.9.1 Lyme borreliosis

Lyme disease was first identified in Europe in the late 1970s, and subsequently found to be prevalent across much of the region. The disease is caused by spirochaete bacteria of the *Borrelia burgdorferi* complex, and consists of a range of clinical symptoms including arthritis and cardiac and neurological disorders. *Borrelia burgdorferi* is transmitted by ixodid ticks (principally *Ixodes ricinus* in Western and Central Europe, and *I. persuculatus* to the east), which feed once every life stage (larva, nymph and adult). *Borrelia burgdorferi* transmission is maintained as a zoonosis, with small rodents and birds becoming infected by the bites of infective nymphs, and amplifying and transmitting infections to larvae. Adults are relatively unimportant in maintaining the transmission cycle, as they feed on different hosts to the other life stages (principally deer, sheep and cattle), and rarely feed on humans. Humans are at risk of infection when entering habitats with high densities of infective nymphs–heterogeneous deciduous woodlands with a recreational function and diverse fauna, including deer, have been identified as high-risk habitats (Gray *et al.*, 1998). In general, active transmission occurs throughout the distribution of *I. ricinus*.

The strong climate effects on transmission are shown by the very marked seasonality of cases, almost all of which occur during the spring-summer-autumn months when the ticks are active, and when people are most likely to enter tick habitats. The distribution of *I. ricinus* has recently expanded northwards and into higher altitudes, apparently as a result of global warming (Lindgren *et al.*, 2000). No published studies have yet statistically modelled climate effects on the geographical distribution or temporal variation of either Lyme disease or *I. ricinus*. Sufficient data are probably now available, however, to generate preliminary models of this type. The relatively long interfeed interval of ticks suggests that temperature effects on the rate of development of bacteria in the vector are unlikely to be important, although some evidence suggests that *B. burgdorferi* infections in North American ticks are suppressed by temperatures in excess of 37 °C (Shih *et al.*, 1995).

Climate warming may favour longer transmission periods and increased geographical range and abundance of ticks. The combined effect of temperature and rainfall on summertime soil moisture will strongly influence tick populations; drier summers may increase tick mortality and decrease questing activity. Climate influences on human behaviour may affect exposure to habitats harbouring infected ticks. The major confounding effects are likely to be caused by variation in land use affecting tick abundance and human exposure, and the effectiveness of case detection. These must be carefully monitored if their effects are to be distinguishable from those of climate change.

10.9.1.1 Currently available information

Climate datasets with global coverage are described above. Local temperature and humidity recordings accompany time series data on tick abundance in a limited number of published studies.

Published reports of time series of tick abundance (and in some cases infection rate) in human-frequented habitats are available for a limited number of sites, which could potentially be retrospectively linked to climate data from meteorological stations. Studies have identified the criteria for high-risk habitats for ticks and for Lyme disease transmission and these criteria have been used for risk stratification of the European region. In Sweden, questionnaire surveys have been used to define and retrospectively detect changes in the distribution of *I. ricinus*, and active surveillance has been used to verify the current latitudinal limits. These data have been linked to climate records, indicating an apparently climate-driven northward shift in the distributional limit.

Lyme disease is not subject to compulsory notification in most countries, although many have networks in place to collate and analyse voluntary notifications by general practitioners. Confirmed cases of Lyme disease are reported to the WHO Europe co-ordinating centre, and fewer aggregated data are available through national health statistics offices or local research networks.

10.9.2 Tick-borne encephalitis

In Europe, the potentially fatal viral disease tick-borne encephalitis (TBE) is transmitted by the same tick species that transmit Lyme disease. TBE is also maintained as a zoonosis in mammals, and humans are again incidental terminal hosts, usually infected through the bite of an infected tick. In contrast to Lyme disease, however, TBE occurs only in small subsets of the distribution of the tick vectors, in recognizable foci in south-central Scandinavia (especially along the Baltic sea coastline) Northern, Central and Eastern Europe, and extensively in Russia. Recent evidence suggests that this focality is due to the stringent conditions required for the maintenance of the zoonotic transmission cycle. In contrast to the relatively prolonged infections characteristic of *Borrelia burgdorferi*, mammals infected with TBE virus remain infective for only a few days. Most infections are apparently transmitted not by larvae feeding from a previously infected host, but by the passage of TBE virus directly from infected nymphs, through the host bloodstream, to uninfected larvae feeding at the same time, without establishment of significant host viraemia (Randolph *et al.*, 1996). Transmission is therefore only sustained in areas with strong seasonal variations in climate, which favour the synchronous feeding of nymphs and larvae (Randolph *et al.*, 1999).

TBE shows different seasonal incidence peak patterns in different regions. Such patterns may be influenced by changes in climate, especially by changes in the length of the period during spring and autumn seasons when ticks are active.

Long time series studies (four decades) from endemic regions in Sweden have shown that TBE incidence is strongly correlated with milder winter seasons (which favours increased vector survival) and extended spring and autumn seasons. (Lindgren, 1998; Lindgren & Gustafson, unpublished data). TBE incidence in succeeding years may be predicted in endemic areas by statistical models using numbers of days per season when temperatures exceeded the bioclimatic threshold for disease transmission (Lindgren, personal communication) expected in spring and autumn seasons.

The climate effects on tick populations and human behaviour described for Lyme borreliosis will apply equally to TBE transmission. In addition, changes in the amplitude of seasonal temperature variation are likely to affect the synchronous feeding of larvae and nymphs, a key factor in maintaining the zoonotic cycle. Increased seasonality may favour synchronicity and the maintenance of the zoonotic transmission cycle. The specific climatic requirements for the maintenance of the zoonotic cycle demand more detailed data collection than that required to characterize Lyme disease transmission, while TBE vaccination programmes and variation in frequency of transmission routes are possible confounding factors for the interpretation of human case data.

10.9.2.1 Currently available information

The availability of information and data on climate and intermediate stages for TBE is essentially the same as for Lyme disease. The stratification of the European region in terms of suitability for ticks is also equally applicable to TBE as to Lyme disease.

Available health data consist of numbers of TBE cases defined by clinical case definition or serological test, reported to the WHO Europe co-ordinating centre, national health statistics office or local research networks. The same organizations also identify outbreaks of TBE transmission, often differentiating between modes of transmission. Preliminary statistical models exist for climate effects on the distribution of TBE transmission.

10.9.3 Leishmaniasis

Leishmaniasis is a complex of zoonotic or anthroponotic diseases caused by proto-zoan parasites of the genus *Leishmania*, transmitted by the bite of phlebotomine sandflies. In Europe, sandflies and leishmaniasis are found mainly in arid and semi-arid areas of Mediterranean shrubland (Rioux *et al.*, 1967; Lucientes-Curdi *et al.*, 1991; Maroli *et al.*, 1994). The most common parasite is *Leishmania infantum*, maintained as a zoonosis amongst foxes, domestic dogs and occasionally black rats, with humans intermittently infected as incidental terminal hosts. Sandflies feed exclusively at night and are often found close to dwellings, so that infection often occurs in or around houses. A low proportion of infections (principally in young children or immune-compromised adults) leads to clinical symptoms, including potentially fatal visceral pathology or self-healing cutaneous lesions. There is some evidence that DDT spraying campaigns against malaria in some countries caused a decline of leishmaniasis cases but, as these campaigns ceased, leishmaniasis resurged. In the past ten years, leishmaniasis has emerged as an opportunistic disease in HIV-infected individuals throughout the region (WHO/UNAIDS, 1998).

The optimal temperature for development both of sandflies and *Leishmania* par-asites is approximately 25 °C (Killick-Kendrick & Killick-Kendrick, 1987; Rioux *et al.*, 1985). The biting activity of sandflies in Europe is strongly seasonal, restricted to the summer months in most areas. The geographical distribution of transmis-sion foci also suggests climate effects–disease foci are usually located below 45 °N latitude, and below 400–600 metres above sea level. Preliminary statistical mode-lling studies have confirmed that the geographical distributions of sandfly vectors and leishmaniasis transmission in Italy are significantly influenced by temperature (Kuhn, 1999).

Climate effects on parasite development and on the distribution, abundance and seasonal activity of the vector are likely to expand distributions into more north-ern latitudes and higher altitudes. Sandflies have a medium to short lifespan rela-tive to the extrinsic development period of *Leishmania*, thus the balance between temperature-driven acceleration in parasite development and possible decreased vector survival is likely to significantly influence sandfly infection rates. The per-sistence in Europe of endemic leishmaniasis with relatively high transmission rates despite the many socioeconomic changes which caused malaria to disappear indi-cates that sandfly-borne diseases are perhaps less sensitive to human modifications of the landscape and changes in lifestyle. There is thus a reasonable chance that the distribution of leishmaniasis in Europe will change with climate, especially along the northern edge of the disease range where there is currently "phlebotomism with-out leishmaniasis". Variation in land-use and vector-control programmes are poten-tial confounding variables for the interpretation of changes in sandfly abundance, while climate effects on domestic dog reservoir infection rates may be obscured by travel with their owners between endemic and nonendemic areas, and drug therapy for infected dogs. Migration and especially HIV coinfections are possible confounders for interpretation of human case data.

10.9.3.1 Currently available information

Low-resolution climate data with full geographical coverage are available as described above for other health impacts. Temperature and relative humidity are sometimes recorded locally as explanatory variables for variations in sandfly capture data.

Two- or three-year time series of sandfly abundance (and very rarely infection rate) have been published for a few specific study sites. Low-spatial-resolution distribution maps are available for sandfly species based on published records, although large unpublished databases are held by individual research groups, so that these could potentially be updated and increased in resolution in order to carry out geographical analyses of climate effects. Such unpublished databases have been used to generate preliminary statistical models of climate effects on sandfly distributions in Italy. Some endemic countries hold large databases of serological tests for infection in dogs, which can be used as a measure of zoonotic transmission intensity and potential for sandfly infection. Serological testing of foxes is very occasionally carried out, as a measure of the intensity of transmission away from the domestic environment.

Leishmaniasis is not a disease of compulsory notification in many endemic countries. However, local research networks, national health statistics office and the WHO Mediterranean Zoonosis control programme collate available data on notifications of numbers of cases, defined by clinical case definition and laboratory confirmation. Case notifications are often accompanied (at least at a local level) by data on age and HIV status.

10.9.4 Malaria

Malaria is characterized by relapsing febrile illness, and is caused by parasites of the genus *Plasmodium*, transmitted between humans by the bite of *Anopheles* mosquitoes. Malaria was highly endemic over much of Europe until the nineteenth century, transmitted by five members of the *Anopheles maculipennis* complex (*Anopheles atroparvus*, *A. claviger*, *A. labranchiae*, *A. messeae* and *A. sacharovi*) and *A. superpictus*. From the end of the nineteenth century, malaria began to disappear naturally from northern Europe due to a combination of changes in land-use and living conditions. Vector-control programmes (Jetten & Takken, 1994) later eradicated many southern foci. Transmission in Europe is currently limited to infections transmitted by mosquitoes accidentally imported from more endemic areas, and occasional transmission from infected travellers by local mosquitoes. However, malaria still remains a problem in the south-east part of Turkey. In some of the Newly Independent States (NIS) there has recently been a resurgence in malaria transmission (Armenia, Azerbaijan, Tadjikistan, Turkmenistan), apparently associated with civil unrest, migration and the breakdown of health systems (WHO, 1999b). The majority of locally transmitted cases in NIS are caused by *Plasmodium vivax*, although active transmission of the potentially fatal *Plasmodium falciparum* occurs in Tadjikistan (Pitt *et al.*, 1998).

Anopheles activity and malaria transmission are highly seasonal in current transmission foci, and show strong latitudinal and altitude associations in Europe and elsewhere. Variation in malaria incidence has been associated with El Niño events in Asia and South America (e.g. Bouma & Dye, 1997) and some field studies in Africa and Asia suggest consistent increases in malaria transmission associated with climate warming, due to local or global changes (Loevinsohn, 1994; Bouma *et al.*, 1996). Additionally, biological models have been generated which incorporate temperature effects on some of the components of vectorial capacity (extrinsic incubation period of the parasite, vector biting rate and survival, but excluding vector density) to generate a comparative index of transmission potential for local vector populations (e.g. Martin & Lefebre, 1995; Martens, 1998). Climate suitability models based on observed field data in Africa also successfully predict distributions outside of the original sample area (Craig *et al.*, 1999). Of course, nonclimatic factors also have strong effects on the distribution of malaria in space and time (Mouchet *et al.*, 1998).

The extrinsic incubation period of the malaria parasite relative to mosquito longevity is highly sensitive to temperature. The lower threshold for development of *P. vivax* is 16 °C (with development taking approximately 20 days) while optimum development occurs at 22–25 °C, requiring only 10–12 days. Climate warming will therefore increase parasite development rates and the viable period for malaria transmission in much of Europe. Temperature increases may also significantly favour the development of potential malaria vectors; mosquito development does not occur below 10 °C (*A. atroparvus*) or 15 °C (*A. labranchiae*) and is optimal at 22 °C (*A. atroparvus* and *A. messeae*) and 25 °C (*A. labranchiae* and *A. sacharovi*). Increasing temperatures will also tend to increase the feeding frequency of mosquitoes, favouring parasite transmission.

Temperature increases may, at the extreme, also increase mosquito mortality, and changes in temperature and rainfall are likely to affect the availability of standing water which provides larval breeding sites, so that the net effect on mosquito density remains to be defined. Adult habitats such as forests in northern Europe and shrubland in the Mediterranean are also likely to be affected by climate changes, altering the distribution of important species such as *A. labranchiae*. The contribution of changes in climate relative to land use, human migration, living conditions and malaria-control programmes has not been assessed, although historical data suggest that such nonclimatic factors have played a major role in determining malaria occurrence in Europe over the last 200 years.

10.9.4.1 Currently available information

Global climate data are available as above. Local temperature and humidity recordings sometimes accompany published data on vector population dynamics, and on the incidence of malaria cases.

Time series of mosquito abundance over a few years exist for a limited number of sites, with long time series (over 20 years) for *An. messeae* in Ekaterinburg,

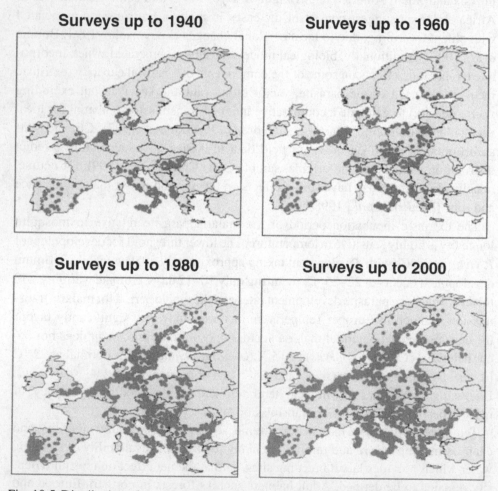

Fig. 10.5 Distribution of *Anopheles* mosquitoes in Western Europe according to published reports at different time points (data summarized from Kuhn, 1999b). Note that the apparently increased distributional range reflects greater sampling effort, rather than any real expansion.

Russia. Infection rate is rarely recorded, and relationships with climate are difficult to interpret due to control efforts. Low-spatial-resolution distribution maps for mosquito species have been published, and many unpublished records are held by individual research groups (summarized in Fig. 10.5). A climate-driven population dynamic model has been generated for *An. messeae*, and related to historical records of malaria cases in the Russian study site.

For health data, numbers of malaria cases defined by clinical case definition and/or parasitological diagnosis are reported to the Malaria Control Programme at WHO-EURO, through national health statistics office or local research networks. Case notifications differentiate between imported and locally transmitted cases, and now detail 5-year age bands and sex. A statistical model has been generated relating climate variation to malaria case notifications in Ekaterinburg over more than 100 years (Nikolaeva & Mazepa, 1999).

10.10 Conclusions

The primary objectives of climate and health monitoring are to support the early detection of climate effects on health, to help improve the prediction of future health impacts, and to provide evidence about how those health impacts may be ameliorated. Fundamental to meeting these objectives is research to improve our understanding of the complex relationships between climate, environment and health. In the context of Europe, the main health impacts of concern are: the direct effects of exposure to high and low temperatures; the health effects of extreme weather events (storms and floods); changes in the frequency of food- and water-borne disease; and changes in vector-borne illness, especially Lyme disease, tick-borne encephalitis, leishmaniasis and malaria. Many of the issues outlined above are relevant to the development of climate–health monitoring in other continents, although the relative importance of specific climate–related health outcomes will of course vary according to geographical location.

Climate change is a gradual process detectable only by monitoring meteorological conditions over decades. The impact of such climate change on health will therefore be similarly slow to evolve. Small changes in the underlying frequency of climate-sensitive diseases are observable only over a long time scale also because of their natural year-to-year fluctuation. Over the long time periods needed to show climate shift, changes also occur in nonclimatic risk factors and in disease detection/recording. These factors may render it difficult to assess the contribution of climate change to long-term trends in health. In contrast, short-term associations between climate and health are comparatively easy to assess. If knowledge of these associations is applied to recorded changes in climate, researchers can derive indirect estimates of the likely impacts of climate change on health (methods of quantitative health-impact assessment). The principal limitations of the indirect approach are: (i) that short-term associations may be modified over the longer term by factors such as physiological acclimatization, behavioural adaptation or public health interventions; (ii) with mortality from thermal extremes, there is some uncertainty about the degree to which short-term associations reflect minor shifts in the date of death (the so-called harvesting phenomenon).

High-quality data collected in the relatively near future (over a period of approximately ten years) could be used to assess short-term climate–disease relationships using time series methods. The suggested optimal time resolution of such data is: mortality – daily; extreme weather events – episodes; food- and water-borne disease – weekly; vector-borne disease – weekly (if feasible) or monthly.

Reliable temperature and daily mortality data are widely available and can provide long time series for multiple geographical locations to study the direct effects of climate variability (e.g. heat-waves) on health. Climate effects may potentially be confounded by the impact of diseases such as influenza, which follow a pattern that is seasonal, but not directly climate driven. Linkage needs to be made with (quality controlled) data on such illnesses from the various regional reporting systems.

A prime research need is to examine the modifying effect on the temperature–mortality relationship of individual, social and environmental factors, such as poverty, housing quality, pre-existing morbidity, and individual behaviours. This may be accomplished by comparison of climate–disease relationships in different geographical locations.

The most comprehensive (although not the only) source of data on effects of extreme weather events is the EM-DAT database, held by the University of Louvain. This contains a log of extreme weather events throughout the globe over the last hundred years, which should be a valuable resource for research on associated health impacts.

There are uncertainties over the consistency of notifications of food- and water-borne diseases over the long term, and recording of the source of infection is often incomplete. It is therefore problematic to assess impacts of climate change by direct measurement of case frequency. However, clear temperature relationships arc observable from analysis of weekly reports, even after allowing for cyclical seasonal fluctuations and long-term trends. Future research should use routine notifications to examine variation in the temperature–disease relationships between populations in different geographical areas. This should be done for a range of selected pathogens, e.g. *Salmonella, Campylobacter, Cryptosporidium* as well as for food-poisoning notifications as a whole.

For most vector-borne illnesses, it is difficult to assess the climate contribution to long-term trends in disease without reliable data on such factors as land use, host abundance and intervention measures. Detailed data on vector abundance, longevity and infection rate could be collected at high frequency (e.g. weekly) at a few locations in currently endemic areas, where few other factors are likely to alter over time. If resources allow, there would also be merit in having a low frequency of measurements (e.g. six to ten per year) at a much larger number of collection sites on transects that cut across the margins of current distributions. A comprehensive GIS database would be invaluable for linking existing and new data on climate, vector abundance and environmental conditions, to aid investigation of broad geographical patterns as well as local changes.

Monitoring is essential for the detection of early health impacts of climate change (and other global environmental changes), for improving our understanding of the mechanisms by which health may thus be affected, and for assessing the degree to which populations are able to adapt to climate change.

Acknowledgements

Many thanks are due to the following persons for helpful contributions during the preparation of the material in this chapter: Roger Aertgeerts, World Health Organization, Italy; Jamie Bartram, World Health Organization, Switzerland; Massimo Chiotti, World Health Organization, Denmark; Shakoor Hajat, London School of Hygiene and Tropical Medicine, United Kingdom; Marco Jermini, World Health Organization, Denmark; Elisabeth Lindgren, Stockholm University, Sweden; Eric Noji, World Health Organization, Switzerland; Sarah Randolph, University of Oxford, United Kingdom; Paul Reichert, Food and Agricultural Organization, Italy; Guido Sabatinelli, World Health Organization, Denmark; Rosalind Stanwell-Smith, CDSC, PHLS, United Kingdom; Jeff Tschirley, Food and Agricultural Organization, Italy.

References

Arnell, N. (1996). *Global Warming River Flows and Water Resources*. Chichester, UK: John Wiley and Sons Ltd.

Baird-Parker, A. C. (1994). Foods and microbiological risks. *Microbiology*, **140**, 687–95.

Balbus, J. M. (1998). Human health. In *Handbook on Methods of Climate Change Impact Assessment and Adaptation Strategies*, Version 2.0, ed. J. F. Feenstra, I. Burton, J. B. Smith & R. S. J. Tol. Amsterdam: UNEP, Nairobi/Institute for Environmental Studies.

Bartram, J., Carmichael, W. W., Chorus, I., Jones, G. & Skulberg, O. M. (1999). *Toxic Cyanobacteria in Water*, ed. I. Chorus & J. Bartram, pp. 1–13. London: Spon Press.

Beebee, T. J. C. (1995). Amphibian breeding and climate. *Nature*, **374**, 219–20.

Beniston, M. & Tol, R. S. J. (1998). Europe. In *The Regional Impacts of Climate Change: an Assessment of Vulnerability*. A Special Report of Working Group II, pp. 149–87. New York: Cambridge University Press.

Bentham, G. (1997). Health. In *Economic Impacts of the Hot Summer and Unusually Warm Year of 1995*, ed. J. P. Palutikof, S. Subak & M. D. Agnew. Norwich: University of East Anglia.

Bertollini, R., Dora, C. & Kryzanowski, M. (1996). *Environment and Health I: Overview and Main European Issues*. Rome: WHO European Centre for Environment and Health; Copenhagen: European Environment Agency.

Bouma, M. J. & Dye, C. (1997). Cycles of malaria associated with El Niño in Venezuela. *Journal of the American Medical Association*, **278**, 1772–4.

Bouma, M. J., Dye, C. & van-der-Kaay, H. J. (1996). Falciparum malaria and climate change in the northwest frontier province of Pakistan. *American Journal of Tropical Medicine and Hygiene*, **55**, 131–7.

Bouma, M. J., Kovats, R. S., Goubet, S. A., Cox, J. S. & Haines, A. (1997). Global assessment of El Nino's disaster burden. *Lancet*, **350**, 1435–8.

Checkley, W., Epstein, L. D., Gilman, R. H., Figueroa, D., Cama, R. I., Patz, J. A. & Black R. E. (2000). Effect of El Nino and ambient temperature on hospital admissions for diarrhoeal diseases in Peruvian children. *Lancet*, **355**, 442–50.

Colwell, R. R. (1996). Global climate and infectious disease: the cholera paradigm. *Science*, **274**, 2025–31.

Craig, M. H., Snow, R. W. & le Sueur, D. (1999). A climate-based distribution model of malaria transmission in sub-Saharan Africa. *Parasitology Today*, **15**, 105–11.

Crick, H. Q. P., Dudley, C., Glue, D. E. & Thompson, D. L. (1997). UK birds are laying eggs earlier. *Nature*, **388**, 526.

Davis, A. J., Lawton, J. H., Shorrocks, B. & Jenkinson, L. S. (1998). Individualistic species responses invalidate simple physiological models of community dynamics under global environmental change. *Journal of Animal Ecology*, **67**, 600–12.

DeGroot, R. S., Ketner, P. & Ovaa, A. H. (1995). Selection and use of bio-indicators to assess the possible effects of climate change in Europe. *Journal of Biogeography*, **22**, 935–43.

DETR (1999). *Indicators of Climate Change in the UK*, ed. M. G. R. Cannell, J. P. Palutikof & T. H. Sparks. London: DETR.

Epstein, P. R. (1999). Climate and health. *Science*, 347–8.

Garrett-Jones, C. (1964). Prognosis for interruption of malaria transmission through assessment of the mosquito's vectorial capacity. *Nature*, **204**, 1173–5.

Grabherr, G., Gottfried, M. & Pauli, H. (1994). Climate effects on mountain plants. *Nature*, **369**, 448.

Gray, J. S., Kahl, O., Robertson, J. N., Daniel, M., Estrada-Pena, A., Gettinby, G., Jaenson, T. G., Jensen, P., Jongejan, F., Korenberg, E., Kurtenbach, K. & Zeman, P. (1998). Lyme borreliosis habitat assessment. *Zentralblatt für Bakteriologie*, **287**, 211–28.

Gunn, S. W. A. (1990). *Multilingual Dictionary of Disaster Medicine and International Relief*. Dordrecht: Kluwer Academic Publishers.

Haines, A. (1999). Climate change and human health: the challenge of research and monitoring. In *Climate Change and Human Health*. London: Royal Society.

Haines, A. & McMichael, A. J. (1997). Climate change and health: implications for research, monitoring and policy. *British Medical Journal*, **325**, 870–4.

Haines, A., Epstein, P. R. & McMichael, A. J. (1993). Global Health Watch: monitoring impacts of environmental change. *Lancet*, **342**, 1464–9.

Hallegraeff, G. M. (1993). A review of harmful algal blooms and their apparent global increase. *Phycologia*, **32**, 279–99.

Hay, S. I., Tucker, C. J., Rogers, D. J. & Packer, M. J. (1996). Remotely sensed surrogates of meteorological data for the study of the distribution and abundance of arthropod vectors of disease. *Annals of Tropical Medicine and Parasitology*, **90**, 1–19.

Hennessy, K. J., Gregory, J. M. & Mitchell, J. F. B. (1997). Changes in daily precipitation under enhanced greenhouse conditions. *Climate Dynamics*, **13**, 667–80.

Hoogenboom-Verdegaal, A. M. M., de Jong, J. C., During, M., Hoogenveen, R. & Hoekstra, J. A. (1994). Community-based study of the incidence of gastrointestinal diseases in the Netherlands. *Epidemiology and Infection*, **114**, 41–4.

Jetten, T. H. & Takken, W. (1994). Anophelism without malaria in Europe: a review of the ecology and distribution of the genus *Anopheles* in Europe. Wageningen Agricultural University Papers 94–95. Wageningen Agricultural University.

Kahru, M. & Brown, C. W. (1997). *Monitoring Algal Blooms: New Techniques for Detecting Large Scale Environmental Change*. Berlin: Springer-Verlag.

Killick-Kendrick, R. & Killick-Kendrick, M. (1987). The laboratory colonization of *Phlebotomus ariasi* (Diptera: Psychodidae). Annals of Parasitology and Human. **62**, 354–6.

Kovats, R. S., Campbell-Lendrum, D. H., Reid, C. & Martens, P. (2000). Climate and vector-borne disease: an assessment of the role of climate in changing disease patterns. Maastricht: ICIS, Maastricht University.

Kuhn, K. G. (1999). Global warming and leishmaniasis in Italy. *Bulletin of Tropical Medicine and International Health*, **7**, 1–2.

Lacey, R. W. (1993). Food-borne bacterial infections. *Parasitology*, **107**, S75–93.

Last, J. M. (1995). *A Dictionary of Epidemiology*, 3rd edn. Oxford: Oxford University Press.

Lindgren, E. (1998). Climate and tickborne encephalitis. *Conservation Ecology Online*, **2**, 1–14. http://www.consecol.org/vol2/iss1/art5/

Lindgren, E., Talleklint, L. & Polfeldt, T. (2000). Impact of climatic change on the northern latitude limit and population density of the disease-transmitting European tick *Ixodes ricinus*. *Environmental Health Perspectives*, **108**, 119–23.

Lobitz, B., Beck, L., Huq, A., Wood, B., Fuchs, G., Faruque, A. S. & Colwell, R. (2000). Climate and infectious disease: use of remote sensing for detection of *Vibrio cholerae* by indirect measurement. *Proceedings of the National Academy of Science of the USA*, **97**, 1438–43.

Loevinsohn, M. E. (1994). Climate warming and increased malaria in Rwanda. *Lancet*, **343**, 714–18.

Lucientes-Curdi, J., Benito-de Martin, M. I., Castillo-Hernandez, J. A. & Orcajo-Teresa, J. (1991). Seasonal dynamics of Laroussius species in Aragon (N.E. Spain). *Parassitologia*, **33**, [Suppl], 381–6.

Malilay, J. (1997a). Floods. In *The Public Health Consequences of Disasters*, ed. E. K. Noji. Oxford: Oxford University Press.

Malilay, J. (1997b). Tropical cyclones. In *The Public Health Consequences of Disasters*. ed. E. K. Noji. Oxford: Oxford University Press.

Maroli, M., Bigliocchi, F. & Khoury, C. (1994). Sandflies in Italy: observations on their distribution and methods for control. *Parasitologia*, **36**, 251–64.

Martens, W. J. M. (1998). *Health and Climate Change: Modelling the Effects of Global Warming and Ozone Depletion*. London: EarthScan.

Martin, P. H. & Lefebre, M. G. (1995). Malaria and climate: sensitivity of malaria potential transmision to climate. *AMBIO*, **24**, 200–7.

McGregor, G. R. (1999). Climatic variability, extreme weather events and health. Setting an agenda for research on health and the environment. Workshop 1: Health and Climate Variability, 24–26 September 1999, Maastricht, The Netherlands.

McGuffie, K. A., Henderson-Sellers, A., Holbrook, N., Kothavala, Z., Balachova, O. & Hoekstra, J. (1999). Assessing simulations of daily temperature and precipitation variability with global climate models for present and enhanced greenhouse climates. *International Journal of Climatology*, **19**, 1–26.

McMichael, A. J., Haines, A., Sloof, R. & Kovats, S. (ed.) (1996). Climate change and human health. (WHO/EHG/96.7). Geneva: World Health Organization.

Mouchet, J., Manguin, S., Sircoulon, J., Laventure, S., Faye, O., Onapa, A. W., Carnevale, P., Julvez, J. & Fontenille, D. (1998). Evolution of malaria in Africa for the past 40 years: impact of climatic and human factors. *Journal of the American Mosquito Control Association*, **14**, 121–30.

NASA (2000). Center for Health Applications of Aerospace Related Technologies – Sensor Evaluation. http://geo.arc.nasa.gov/sge/health/sensor/

New, M., Hulme, M. & Jones, P. (1999). Representing twentieth-century space-time climate variability. Part I: development of a 1961–90 mean monthly terrestrial climatology. *Journal of Climate*, **12**, 829–56.

Nikolaeva, N. & Mazepa, V. (1999). Modelling of population dynamics of the malaria mosquitoes and incidence of malaria in the Middle Urals. *Proceedings of the XIIth European Meeting*, Society for Vector Ecology, Wageningen, September 1999.

Noji, E. K. (1997). *The Public Health Consequences of Disasters*. Oxford: Oxford University Press.

Parmesan, C. (1996). Climate and species' range. *Nature*, **382**, 765–6.

Parmesan, C., Ryrholm, N., Stefanescu, G., Hill, J. K., Thomas, C. D., Descimon, H., Huntley, B., Kaila, L., Kullberg, J., Tammaru, T., Tennent, W. J., Thomas, J. A. & Warren, M. (1999). Polewards shifts of butterfly species ranges associated with regional warming. *Nature*, **399**, 579–83.

Pitt, S., Pearcy, B. E., Stevens, R. H., Shapirov, A., Satarov, K. & Banatvala, N. (1998). War in Tajikistan and re-emergence of *Plasmodium falciparum*. *Lancet*, **352**, 1279.

Randolph, S. E., Gern, L. & Nuttall, P. A. (1996). Co-feeding ticks: epidemiological significance for tick-borne pathogen transmission. *Parasitology Today*, **12**, 472–9.

Randolph, S. E., Miklisová, D., Lysy, J., Rogers, D. J. & Labuda, M. (1999). Incidence from coincidence: patterns of tick infestations on rodents facilitate transmission of tick-borne encephalitis virus. *Parasitology*, **118**, 177–86.

Rioux, J. A., Golvan, Y. J., Croset, H., Houin, R., Juminer, B., Bain, O. & Tour, S. (1967). Ecology of leishmaniases in Southern France. 1. Phlebotomus. Sampling ethology. *Annales de Parasitologie Humaine et Comparée*, **42**, 561–603.

Rioux, J. A., Aboulker, J. P., Lanotte, G., Killick-Kendrick, R. & Martini-Dumas, A. (1985). Ecology of leishmaniasis in the south of France. 21. Influence of temperature on the development of *Leishmania infantum* Nicolle, 1908 in *Phlebotomus ariasi* Tonnoir, 1921. Experimental study. *Annales de Parasitologie Humaine et Comparée*, **60**, 221–9.

Roberts, J. A., Sockett, P. N. & Gill, O. N. (1989). Economic impact of a nationwide impact of salmonellosis: cost benefit of early intervention. *British Medical Journal*, **298**, 1227–30.

Sapir, D. G. & Misson, C. (1992). Development of a database on disasters. *Disasters*, **16**, 74–80.

Shih, C. M., Telford, S. R. III & Spielman, A. (1995). Effect of ambient temperature on competence of deer ticks as hosts for Lyme disease spirochetes. *Journal of Clinical Microbiology*, **33**, 958–61.

Sparks, T. H. & Yates, T. J. (1997). The effect of spring temperature on the appearance date of British butterflies 1883–1993. *Ecography*, **20**, 368–74.

Stanwell-Smith, R. (1998). Climate change: implications for European surveillance of infectious diseases. Paper presented at WHO meeting on early human health effects of climate change in Europe. Rome, May.

Swiss Federal Office of Public Health (1999). Reports of infectious diseases. http://www.admin.ch/bag/infreporting/

Thomas, C. D. & Lennon, J. J. (1999). Birds extend their ranges northwards. *Nature*, **399**, 213.

Todd, E. C. D. (1989). Preliminary estimates of costs of food-borne disease in the United States. *Journal of Food Protection*, **52**, 595–601.

Wheeler, J. G., Sethi, D., Cowden, J. M., Wall, P. G., Rodrigues, L. C., Tompkins, D. S., Hudson, M. J. & Roderick, P. J. (1999). Study of infectious intestinal disease in England: rates in the community, presenting to general practice, and reported to national surveillance. *British Medical Journal*, **318**, 1046–50.

WHO (1999a). Early human health effects of climate change and stratospheric ozone depletion in Europe. Geneva: WHO.

WHO (1999b). Strategy to roll back malaria in the WHO European region. EUR/ICP/CMDS 080302.

WHO (2000). World Health Report 2000. http://www.who.int/whr/2000/.

WHO (2001). Communicable disease surveillance and response. http://www.who.int/emc/.

WHO-ECEH (1999). WHO Surveillance Programme for Control of Foodborne Infections and Intoxications in Europe. http://www.who.it/docs/fdsaf/fs_survprog.htm

WHO/UNAIDS (1998). Leishmania and HIV in gridlock. WHO/CTD/LEISH/98.9, UNAIDS/98.23.

WMO (2000). World Meteorological Organization Home Page. http://www.wmo.ch/.

11

Epidemiology, environmental health and global change

ALISTAIR WOODWARD

11.1 Introduction

This chapter outlines the approach that is typically taken in epidemiology to the study of disease risk, gives examples of studies that currently serve as paradigms for epidemiological thinking, and compares these traditional approaches with what is needed to understand health problems associated with global environmental change. It is argued that in order to meet this new challenge, epidemiology must encompass three things: models of extended causation, multiple levels of analysis, and differential vulnerability to disease and injury. This chapter examines in detail the concept of vulnerability, the condition of individuals or groups that modifies the effect of environmental exposures on disease outcomes.

Modern epidemiology is both useful and wrong. By this I mean that epidemiology is an asset to public health because it points out ways to prevent disease and injury, but the foundations of the enterprise are shaky. "Wrong" is putting it strongly – "limited" may be closer to the mark. When you look closely you can see that the methods used by epidemiologists assume a world that does not exist. This is not a criticism of epidemiology, rather an observation on the way that science in general works. Like all scientific disciplines, epidemiology is a tool for creating knowledge. All tools provide partial and incomplete access to the external world because they themselves are imperfect reflections of the world. (Hammers assume there is not a plank that cannot be nailed; telescopes presuppose a universe in which all that matters can be seen.) In the case of science, all disciplines rely on short cuts, approximations and assumptions of convenience. No-one disputes the successes that are achieved nevertheless: satellites orbit, e-mail finds its mark and human lives are saved. Rather, the danger lies in assuming that a tool that has served us well in the past will do the job just as well when we are faced with a new task.

At the heart of the discussion is this point: that global change is a novel challenge. It differs in fundamental ways from the localized environmental perturbations that

290

have been bread and butter to epidemiology in the past. This means the old research tools may need to be reshaped or replaced. In this chapter I outline the approach that is typically taken in epidemiology to study the relation between environment and health, provide examples and explore some of the shortcomings. I will concentrate on two particular weak spots in the epidemiological foundations: the tendency to assume that a response to a particular exposure is uniform, and the belief that observations on a finer scale provide more certain information. This leads to a discussion of vulnerability (the state in which responses to exposures are not uniform): what it is, why it matters and what it means to epidemiology. In one sense the most important causes of vulnerability are those that are socially determined because these are, at least in theory, open to change while other categories of causes, such as biological traits or physical characteristics of particular locations, are more or less fixed. Clearly this provides a challenge to the reductionist model of epidemiology, because social factors are characteristics of groups and therefore require research at the level of populations not individuals. Finally, I propose ways in which epidemiology could be modified and enlarged to better fit the task in hand.

11.2 The epidemiological approach

11.2.1 General features of environmental epidemiology

"Environmental epidemiology" is shorthand for the use of epidemiology to investigate causes of disease that are found in the environment. In many ways, the application of epidemiology in this setting is little different to its use in other fields of public health. However, there are some features of environmental epidemiology that deserve mention, since these provide particular challenges to researchers.

In particular, environmental epidemiology is concerned generally with exposures that are, by definition, characteristics of the environment, not the individuals who live in that environment. Examples include infectious organisms in the water supply, features of the legislative environment (compulsory seat belt laws for example) and air pollutants both indoor and outdoor (see Box 11.1). This means that the exposures that are being studied are typically widespread and not readily controlled by the individuals who are directly affected.

What are the consequences for epidemiology? The fact that the exposures are widespread means that it may be difficult to find individuals who can act as an "unexposed" comparison group (for example, persons who are "not exposed" to air pollution). Sometimes the exposures are not only widespread, but also vary little within a given population (for example, air pollution levels in a neighbourhood) compared with the differences between populations. In these circumstances, ecological studies – in which the unit of comparison is the group not the individual – may

BOX 11.1
Second-hand smoke and heart disease. Reference: Law *et al.*, 1997.

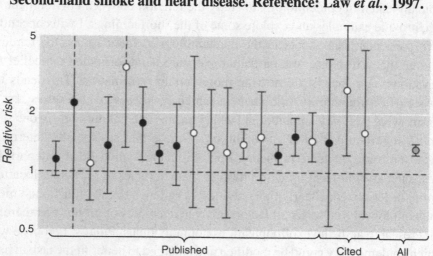

Relative risk estimates (with 95 % confidence intervals), adjusted for age and sex, from nine prospective studies (solid circles) and ten case–control studies (open circles) comparing ischaemic heart disease in life-long nonsmokers whose spouse currently smoked with those whose spouse had never smoked and three with results cited by others from abstracts or theses.

Second-hand cigarette smoke is typical of many environmental health problems in several respects, including:

- Localized exposures at low levels.
- Small risks for most individuals but high levels of concern.
- Exposures are common.
- Therefore potentially, high attributable risk.
- Smoke is a complex mixture; aetiological agents not well characterized.
- Difficult to mimic realistic exposure conditions.
- Hence laboratory studies have restricted application.
- But epidemiological studies limited also (e.g. by crude measures of exposure).

be particularly useful. An example of an ecological study of air pollution would be a study that compared mortality rates in different cities with the average levels of airborne particulates in those cities (Dockery *et al.*, 1993).

The particular character of environmental exposures also affects the types of intervention studies that can be carried out. Where the factor suspected of causing disease is more directly under the control of individuals (such as components of diet), it may be possible to carry out studies that mimic clinical trials. For example, studies have been conducted in which individuals were randomly allocated to different treatment groups to study the relationship between dietary vitamin A intake and the incidence of certain cancers (Omenn *et al.*, 1996). However, intervention studies of this kind are seldom applied in environmental epidemiology (for example, it is difficult to conceive of a study in which individuals' intake of air pollutants would be randomized). There are some examples of randomized trials applied to environmental health, such as a study in Melbourne, Australia in which effective and "sham" filters are being applied to household water supplies in a double-blinded, randomly allocated fashion, and the incidence of gastrointestinal illness is subsequently monitored. But most often information on the effect of interventions is gained from observing closely the results of "natural experiments" (such as reductions in air pollution resulting from altered industrial processes), in which the exposures are at the level of groups rather than individuals.

Policies and regulations concerning widespread exposures are generally framed in terms of average or peak levels in the environment rather than the quantities that are absorbed by individuals. Air pollution guidelines, for example, are expressed as time-weighted average concentrations in the ambient environment. Studies of the biological mechanisms by which environmental contaminants may cause disease must take into account variations between individuals in dose (the amount of an aetiological agent that is actually taken into the body). However, studies of the effectiveness of public health interventions are more appropriately based on exposures measured in the external environment.

Often environmental epidemiology is studying exposures at low levels. An example is radiofrequency radiation. Exposures in the general environment from sources such as mobile-phone base stations are orders of magnitude less than those that may be experienced in some occupational settings (such as workers operating plastic heat sealers). One consequence is that environmental epidemiology is frequently searching for risks on the margin of detectability. However, this does not mean that the risks presented by low-level exposures in the general environment are necessarily unimportant. First, these exposures are typically involuntary (people who live close to mobile phone cell-sites, for example, have little choice over whether they are exposed to radiowaves from the sites), and the public is far more sensitive to potential dangers of this kind than exposures that are seen to have an element of discretion about them. Second, the increase in risk for an individual may be small, but if exposures affect large numbers of people then the overall burden of illness attributable to the exposure will be substantial. Relatively few people are exposed to radiofrequency radiation at work, so although this may be an important personal health issue, the impact on the health of the population overall will be relatively small. On the other hand, if low-level exposures from mobile phone cell-sites do have health effects, then this would be a significant public health issue since the number of people at risk would be very large. Second-hand tobacco smoke provides another example. Although the increase in risk of respiratory illness for an individual is only moderate (roughly 30–50 % for most conditions), second-hand smoke is widespread and many people are exposed. In Australia, it was estimated in 1997 that about one-third of the population was regularly exposed to second-hand smoke and the numbers of cases of illness that resulted each year amounted to tens of thousands (National Health and Medical Research Council, 1997).

Finally, studies of environmental causes of disease and injury tend to involve dispersed, heterogeneous populations. This provides particular challenges in recruitment of study participants and in the analysis of findings. By contrast, epidemiological studies that are based on workplaces or health-care settings tend to have less difficulty identifying and gaining access to the potential study population. In the case of environmental epidemiology it may be difficult to define the exposed population, weakening the confidence with which results can be extrapolated to other groups. The very mixed nature of the general population (in terms of age, health status and co-exposures) means that an overall average risk estimate may mask considerable variations in the strength of effect in different subgroups. Consider for example a study of hospitalization rates in relation to ambient levels of air pollution in a major city. The population of the city will include a number of groups that are likely to be more susceptible than the "average" city inhabitant to the effects of pollution (such as the elderly, people with pre-existing chest disease, outdoor workers). Typically the numbers of susceptible individuals and the exposures they receive are not known, and caution must be applied in interpreting the relationship observed between pollution levels and health outcome (e.g. numbers of hospital admissions per day). Exposure guidelines based on studies of this kind may not provide adequate protection to the most sensitive groups in the population, and this must be taken into account when epidemiological results are translated into public health policy (Woodward *et al.*, 1995).

11.2.2 Some shortcomings

The conventional epidemiological approach assumes multiple, independent exposures. In practice the situation is more complex. Even with an exposure as apparently straightforward as outdoor air pollution, it appears that the different components interact so that the health effects of the total package of air pollution are greater than would be anticipated (Woodward *et al.*, 1995). At the level of ecosystems, the complexities are even greater, and the limitations of single-exposure risk assessment become more apparent.

In epidemiology, partial and incomplete answers about the causes of the disease may be "right", in the sense that they provide sufficient information for action. For example, epidemiology provided in the middle of the last century a "right" answer to the question of why lung cancer rates had risen in Britain, the U.S. and many other countries (Doll & Hill, 1950). But the methods developed to study the immediate causes of noncommunicable diseases, such as lung cancer, have been less effective in tackling other problems. We have made less progress, for instance, explaining variations in smoking rates than in establishing the link between smoking and lung cancer. One reason for this is that the epidemiological tool kit has lacked instruments to engage social and cultural phenomena, especially those that are not well-suited to quantification (Chapman, 1986).

When we consider the study of global environmental change, two weak spots are apparent in the foundations of epidemiology: homogeneity and reductionism. Homogeneity means that the relationship between exposure and outcome is assumed to be the same across all units of observation. Reductionism is the view that the relationship between environmental exposures and disease outcomes is understood best when the unit of observation is reduced to the smallest scale possible. The first is an assumption of convenience: homogeneity is often assumed to bridge gaps in the data. A statistical test can be applied to check whether the assumption is warranted, but this is a relatively insensitive measure of variations in effect across study populations (Rothman, 1996). Reductionism, on the other hand, is principally a matter of belief. The view that the finer the scale of understanding the more certain the knowledge is widely held. Although epidemiology is literally the study of populations, most epidemiological research seeks to understand the causes of disease exclusively at the level of individuals. "Molecular epidemiology" provides an opportunity to observe the causes of disease at an even smaller scale, and some assume that this means the research findings will (necessarily) be more informative than those provided by "classic" research (Wilcox, 1995). Others have argued that measuring maladies in molecules may be fine for some purposes, but this approach should be seen as complementing rather than replacing coarser-scale epidemiology (McMichael, 1994).

The assumptions of homogeneity and reductionism are short cuts, necessary short cuts, in research relating disease incidence to aspects of personal lifestyles. However, they are poorly suited to the study of global environmental change (McMichael, 1993). In the next section I introduce the concept of vulnerability, which is the antithesis of the "one size fits all" research approach. In thinking about the causes of vulnerability and what it means for research on global change, problems arising from the second assumption (reductionism) become apparent.

11.3 Vulnerability

11.3.1 Definitions

Vulnerability "is the state of individuals, groups or communities, defined in terms of their ability to cope with and adapt to any external stress placed on their livelihoods and well-being" (Adger & Kelly, 1999). A shorter definition is "capacity for loss". What this means is that the effect on an organism of a change in the environment does not depend solely on the nature of the change or its magnitude. In this there is an echo of the evolution of germ theory thinking in the early twentieth century, when simplistic notions of risks determined by the presence or absence of the infectious agent were modified to incorporate the modulating effects of host and environmental characteristics. The impact depends also on how well the organism or population involved responds to the challenge.

Vulnerability implies being open to harm (Jones *et al.*, 1990) (see Box 11.2). In this sense it is a one-tailed version of what the Intergovernmental Panel on Climate Change calls "sensitivity" (the capacity to respond, in either a negative *or* a positive fashion, to changed circumstances) (Adger, 1996). Sometimes the term includes exposure (a vulnerable system being one that is exposed to challenge *and* is unable to respond in a positive manner). For the purposes of this chapter, I choose to define vulnerability solely in terms of negative responsiveness, that is, the amount of harm suffered for a fixed level of exposure.

BOX 11.2
The definition of vulnerability

Vulnerability is:

- Undesirable (adaptability is positive).
- Relative (everyone is vulnerable, to some extent).
- Particular (vulnerability refers to specific hazards or specific outcomes).
- Multi-levelled (it may apply to individuals, households or regions).
- Being at high risk of losses in the future (the most vulnerable may not be those most severely affected at present).

The consequences of threatening environmental change depend (among other things) on the qualities of resistance (ability to adhere to a pre-existing condition in the face of a force for change) and resilience (ability to recover or adapt after change has occurred) (Begon *et al.*, 1990). There may be a trade-off between these components of vulnerability. For example, surveys of insect populations have shown that in extreme climates there is typically a wide fluctuation in insect numbers (that is, low resistance) but as a general rule populations restore themselves after a very severe season (high resilience). In contrast, in temperate areas populations tend to be more stable, but when they are perturbed numbers recover slowly. Ecologists have suggested that under relatively constant environmental conditions, organisms are selected for qualities that maximize stability, but these qualities may not be adaptive in times of rapid change (Berkes & Folke, 1998). A high degree of specialization, for example, may enable an organism to do well in an unvarying environment, but this characteristic may be a liability in changed circumstances. There may be parallels in economics: Levin *et al.* (1998) suggest that "companies that live in a very stable, protected environment will not have incentives to develop the flexibility and competitiveness that would be needed if a major shock was to hit the market". This is not to say that stability is a bad thing, rather that the qualities required for success under stable conditions may be different from those that are desirable when the world changes.

11.3.2 Manifestations

The concept of vulnerability may be ascribed to systems at any level of aggregation, from subcellular to the complete organism to populations of organisms. It is a concept that may be more familiar to ecologists, economists and entomologists than it is to epidemiologists. However, vulnerability and its twin, adaptive capacity (being able to take action to reduce the capacity for loss), are just as applicable to human populations as to ecosystems or financial markets.

As the predominant environmental issues shift scale, from localized hazards with immediate effects to widespread exposures with impacts on global systems that stretch into the long-term future, the concept of vulnerability will become even more important. There are theoretical reasons for this: with up-scaling of environmental disturbances, there is a greater risk of crossing ecological thresholds, obstructing recovery pathways, depleting vital stocks and in these and other ways increasing the chances of sudden and unexpected outcomes (Levin, 1998). Practical considerations apply also.

First, vulnerability has to be taken into account in any calculus of impacts. The reason is that indirect or mediated effects are a significant feature of macro-scale environmental disturbances. Climate change, for example, is likely to have relatively direct effects on human health such as increasing death rates due to more frequent extremes of heat. But there will also be "knock on" effects when climate change acts by way of other systems. Agriculture for instance will be affected by changing temperatures and rainfall patterns and in some parts of the world food production could be severely impaired (Parry & Rosenzweig, 1993). If producers in these parts of the world can't buy food from elsewhere they will go hungry. According to one estimate, the number of people at risk of hunger globally in 2060 will increase from a baseline figure of about 400 million to about 700 million as a consequence of climate change (Rosenzweig & Parry, 1994). Food shortages on this scale would bring with them a variety of health problems that are likely to be far more serious and more widespread than increases in mortality due to extreme heat. To estimate the potential size, distribution and timing of these "knock on" effects, researchers have to follow the causal pathway and consider at each step the factors that modify the response to an environmental stimulus. Which farming practices influence crop responses to altered soil moisture conditions? How do social structures and political events construct food shortages, or work to avoid them? If food supplies are limited, who will be most susceptible to poor health as a consequence? These questions are particularly important because the causal pathway stretching from global change to human health is so elongated. The ultimate impact is a function of compounding adaptive responses. This means that any model relating global environmental change to health outcomes must include vulnerability in the equation.

Second, it helps to understand the causes of vulnerability since this means it may be possible to reduce the harmful consequences of environmental changes, even if these changes cannot be prevented. In some circumstances, environmental change in the future is largely determined by past events or changes may be avoidable in principle, but at great cost. Such considerations are particularly relevant to global environmental change because there is a lengthy gap between "exposures" such as diminishing biodiversity and human health outcomes. A good example of pre-determined or "committed" change is warming due to accumulation of greenhouse gases (see Box 11.3). As a result of past emissions of CO_2 and other greenhouse gases, the biophysical systems that regulate the composition of the Earth's atmosphere are out of balance. Sinks and stores take up CO_2 in the short-term, but the complete planetary carbon cycle has a long time course. Even if emissions of CO_2 were to be held constant from now onwards, atmospheric concentrations of CO_2 would continue to rise for many years, as a result of the carbon load that has already been applied. The same will be true for global temperature. It is thought that the warming effect of a thickening greenhouse blanket takes hundreds of years to be fully expressed as the processes of establishing a new equilibrium gradually unfold (Houghton *et al.*, 1996).

We are committed to change in the future in other ways also. Emerging, re-emerging and changing infections will be a feature of the next 20, 50 and 100 years. The success of antibiotics in the 1940s and 1950s and immunization programmes in the 1950s and 1960s led some to conclude that infectious diseases were conquered or, at least, were conquerable. Since then almost 20 million people have died of HIV infection and the epidemic is still gathering momentum: it is expected that half the

BOX 11.3
Vulnerability and climate change

The Intergovernmental Panel on Climate Change (IPCC) has produced three major reports between 1992 and 2001 (McCarthy *et al.*, 2001). Over this time the emphasis in the reports has shifted from "impacts" alone to "impacts" and "vulnerability and adaptation". Vulnerability, in IPCC terms includes degree of exposure, sensitivity (the extent to which a system affected by climate change) and adaptability. Apart from this attempt to standardize the definition, the assessment of vulnerability in IPCC reports varies between sectors and discipline-specific chapters. There are common themes. For example, sensitivity and adaptability are consistently related to existing pressures upon physical and social systems. Environments that are currently stressed as a result of growth in human numbers, intensity of resource use, fragmentation of habitats, invasion by exotic species and pollution will be particularly susceptible to losses due to climate change. The same principle applies to human systems: populations that are divided, lack access to resources and effective and accepted institutions of governance will struggle to adapt to rapid changes in climate. This applies at the national level, and also to subgroups within countries (for example, rural communities, indigenous peoples, urban poor).

The effect of climate change on food security and hunger will be greater in countries that are buyers than those that are sellers, and will be most serious for populations that lack access to markets altogether. Whether or not intensively managed ecosystems (such as commercial forests and fish farming) adapt better to change than "wild" systems will depend on the quality of the managers and the resources at hand. Climate thresholds are important also. Populations in mid and high latitudes, for example, will generally benefit from increased crop yields in the early stages of global warming. Countries in lower latitudes, where temperatures are already close to critical values for crop performance, will suffer earlier under climate change. Species growing near the upper limit of their "temperature envelope" will be endangered first. Ironically, there is a marked trend in many countries to more intensive human settlement of exposed coastal regions that are affected by tropical cyclones and storm surges. This pattern increases vulnerability under climate scenarios that indicate more frequent and/or more severe extreme events in the future.

Diversity will be an important element of adaptive capacity in several sectors. Access to genetically diverse crop types and risk spreading through diversification of business activities are two examples. Access to markets is a positive feature of adaptation, as long as this goes with the ability to influence the rules of the market. Openness to trade in goods and human skills can also be a cause of vulnerability, for countries that lack political clout internationally (for example, the small island states of the Pacific).

The IPCC reports acknowledge that there are different views on the best means of ensuring that adaptation is pro-active and equitable, but emphasize that many adaptation strategies could provide multiple benefits, in the short and longer terms, whatever the trajectory of climate change in the future.

teenagers in southern Africa will die from AIDS-related illnesses before the age of 50 (Anonymous, 2000). As the smallpox elimination programme closed in on its quarry, a related pathogen (but fortunately one much less virulent), the monkey pox virus, was detected for the first time in Zaire (Wenner *et al.*, 1968). Elsewhere long-standing infections such as tuberculosis are making new ground (Balinska, 2000). We should not overlook the "irresistible evolutionary opportunism" (McMichael, 1995) of our microbiological travelling companions. One could almost say that change and challenge are "hard-wired" ecologically. At the very least it would be foolish not to expect unexpected threats to human health from infectious disease to be a feature of future epidemiological landscapes.

11.3.3 Causes of vulnerability

Where primary prevention is not entirely within our grasp, then complementary strategies are needed. We have to ask: what can be done to reduce the harmful effects of unforeseen or unavoidable hazards? This leads to a more basic question: what are the causes of vulnerability?

Ecologists point out that there are many factors that influence the vulnerability of species and ecosystems. These include location, resources, diversity and "connectivity" or complexity. Some of these variables have obvious application to human populations; others are less straightforward to apply. Human populations that reside on the margins are obviously at greatest risk when faced with environmental threats. This is true in a physical sense (Indonesian transmigration settlements squeezed between inhospitable swamps and storm-prone coasts, for example), and also in terms of economic resources. Mortality due to tropical cyclones in Japan is considerably less now than it was 50 years ago (although economic losses have increased). The storms are no less severe, but the health of the population is better protected as a result of the economic and social development that occurred after 1950 (Nimura, 1999).

Vulnerability is not only a function of the level of resources available to a population: the ways in which these resources are organized and distributed are important also. A neighbour of Japan's, North Korea, was affected by strong storms with torrential rain, in 1995 and 1996 (Fig. 11.1). The storms were not unusual for the region, but the impact in North Korea was unexpectedly severe. Crops failed, the arrangements for distribution of stored food failed also and famine occurred. World aid agencies were caught by surprise, and it took some time to appreciate the magnitude of the disaster (International Federation of Red Cross and Red Crescent Societies, 1996). In 1996–97, it was estimated that 15–20 % of emergency food stores, worldwide, were required to meet immediate needs. Five years later, it is thought that a substantial proportion of the North Korean population is still seriously

Fig. 11.1 Floods in Democratic People's Republic of Korea, 1995.

under-fed. A survey by the World Food Programme in 1998 suggested that the rate of acute malnutrition among children six months to seven years was 16 %, one of the highest rates of wasting in the world (FAO/WFP, 1999). As conditions have deteriorated, tuberculosis, malaria and diarrhoeal diseases of childhood have re-emerged as major causes of mortality. A vicious circle of poor nutrition compounding poor health leading to low productivity and further reduced food supplies has become deeply entrenched.

Why did the floods have such critical effects? Apart from the double hit of serious flooding in two consecutive years, there were a number of features of North Korean society that rendered it susceptible to losses due to natural disasters (FAO/WFP, 1999). These included its economic isolation, lack of reserves, the highly centralized arrangements for storage and redistribution of foods, the lack of variety in agricultural practice, and a strictly hierarchical political system that concentrated decision-making. Such characteristics, typical of military rule all over the world, sustained North Korea during civil war and post-war reconstruction. A strong element of central planning may be protective with certain types of risk. For example, comparisons of responses to flood hazards in Vietnam and Bangladesh in the early 1970s suggested that coordinated and well-directed civil defence measures in Vietnam reduced mortality. In Bangladesh, on the other hand, absentee landlordism, insecure land tenure and lack of collective action resulted in many more deaths for similar exposures (Adger, 1999). But a high degree of central direction may be a liability in other circumstances, and in the case of North Korea the combination

Table 11.1. *Some characteristics of social vulnerability and approaches to measurement*

Indicator of vulnerability	Proxy for	Mechanism of action	Measured by
Poverty	Marginalization	Narrowing coping strategies, less diversified and restricted social entitlements, lack of power	Poverty measures (either material or experiential)
Inequality	Degree of collective responsibility, risk-sharing	Direct: concentration of resources in hands of a few Indirect: entitlements not available to many	The quantitative distribution of assets and entitlements
Institutional adaptation	Context that determines exposure; institutions as conduits for collective perceptions of vulnerability; political institutions shape adaptive responses	Responsiveness, evolution and adaptability of institutions	Study of institutions through decision-making and social learning

Adapted from Adger & Kelly (1999).

of a command and control model and severe political rigidities (and a declining resource base) meant that adaptive capacity was seriously compromised.

Circumstances in North Korea in the 1990s were unusual, but the underlying causes of vulnerability are repeated elsewhere. In 1998 Hurricane Mitch destroyed the homes of more than a million people in Central America. It was not a particularly severe storm but its effects were amplified by loss of forests, replaced with intensive farming even on steep slopes, crowded settlements on flood plains and the lack of any systems for early warning, protection and evacuation (Copley, 1998). Work in other parts of the same region has shown that groups with little influence over decision-making processes are more susceptible to loss caused by earthquakes, floods and other disasters, and also have greater difficulty recovering from these events (Nigg, 1995). In general terms, these examples serve to make the point that adaptation is essentially a social phenomenon (see Table 11.1). Individuals make decisions that are shaped by their social environment and institutions play important roles over and above the actions of individuals. In Vietnam, for instance, Adger and Kelly (1999) found that the responses of farmers to the threat of typhoons were strongly conditioned by prevailing social and economic hierarchies and the actions of local and central government. In his account of the late nineteenth century Indian famines associated with El Niño events, Mike Davis argues that the extremely high mortality rates were largely caused by the exploitative land-use and tax regimes imposed by the British (Davis, 2001). These conditions favoured cash cropping, exports and the needs of "home" (i.e. British) markets, but seriously handicapped Indian farmers in times of drought.

Diversity is generally associated with resilience in ecosystems, although the relationship is not entirely straightforward: the nature of the environmental disturbance and the context are important. For example, simple systems tend to do better in very harsh conditions when "doing better" is defined simply as survival. There is fierce debate amongst ecologists over the mechanisms by which diversity may act (Kaiser, 2000), but in most circumstances variety would be seen as an advantage. It is plausible that the same applies to human communities, where diversity means that risks are shared and a range of resources are available to cope with a range of exigencies. There is relatively little written about this aspect of vulnerability and adaptation, but measles provides an interesting model.

A puzzling aspect of the history of measles is the extremely high mortality suffered by populations in North and South America and the Pacific on first contact with the virus. Lifestyles and health status could not explain the death rates – at that time life expectancy, measles aside, was probably greater in much of the New World than in Europe. The possibility of a genetic cause of susceptibility to measles has been explored closely, without uncovering any feature of individual Americans or Pacific Islanders that could explain the history. This is hardly surprising, since at an individual level the variation in genetic make-up between ethnic groups tends to be much less than the variation within populations. But the variability of genetic traits is not the same in different ethnic groups, and it has been suggested that heterogeneity may be a significant factor in the transmission of measles and possibly other infections also.

There are two lines of evidence that support this proposition: studies of the virus show that it has the capacity to "pre-adapt" to the immune defences of a new host as a consequence of lability in replication, and epidemiological research reporting that case fatality rates are higher when the infection is contracted from a close family relative than a contact outside the family (Garenne & Aaby, 1990). New World populations that began with small settler groups display less genetic variability than populations in Europe, which in turn are less genetically diverse than African populations. Putting all this together, it has been suggested that the lesser variety of genotypes within indigenous populations in the Pacific and the

Table 11.2. *Areas vulnerable to water-related conflict around 2025.*
Ratio of water use to water availability estimated from current
consumption, population trends and forecast regional climate
changes (rainfall and runoff)

Inflow of country's total water supply	Use–availability ratio		
	<40 %	40–60 %	>60 %
<25 %	Poland	Algeria	Israel
25–50 %		Nigeria Pakistan	Tunisia
50–75 %	Iraq	Thailand	
>75 %		Germany Mauritiana Sudan	Egypt Niger Senegal

Adapted from Kulshreshta (1998)

Americas caused these populations to be susceptible to measles (Black, 1992). The reason: when it spread the virus was more likely to encounter a new host with a familiar immune environment than was the case in more diverse Old World populations. No doubt there are other factors influencing case fatality from measles, but this example is helpful to make two points: causes of vulnerability that operate at the level of populations (such as a limited range of genotypes) require study designs that make the population the unit of observation, and diversity of resources (in this case, stocks of DNA) is likely to be protective in human populations, as it is in ecosystems generally.

11.3.4 Implications for epidemiology

How can epidemiology contribute to a better understanding of vulnerability and adaptation? A first step would be to develop ways of measuring these qualities that are relevant to human health. Other disciplines have their own approaches to conceptualizing vulnerability, and techniques for turning the concepts into empirical data. An example from the water resources area is shown in Table 11.2. Social scientists have proposed measures that combine the present incidence of poverty, the riskiness of sources of income (i.e. the likelihood of future poverty), equality in the distribution of resources and entitlements and the responsiveness of institutions (Adger & Kelly, 1999). In the environmental sciences, scoring systems have been devised to reflect the fragility of national or regional environments. One of these, the Environmental Vulnerability Index created by the South Pacific Agency for

Geosciences (SOPAC), summarizes 57 indicators from three domains (risks to the environment, intrinsic resilience and ecosystem integrity) (Kaly & Pratt, 2000). In general, these scoring systems are based either on an average of potential determinants of future vulnerability (the Environmental Vulnerability Index is an example of this type), or they summarize the stability of key outcomes in the past. Investors, for instance, may use the historical variability of stocks (in relation to the variability of the market overall) as an indication of vulnerability to adverse economic circumstances in the future. Both approaches could be applied epidemiologically.

Another step that epidemiologists need to take is to apply research techniques that better capture the emergent properties of populations. Epidemiology has a long tradition of ecological research – research that makes the population the unit of observation and measurement. However, in the latter part of the twentieth century the ecological study came to be regarded as the poor cousin in the epidemiological family. This was a consequence of the reductionist thinking that followed the early triumphs of the germ theory and led to a pre-occupation with studying individual-level risk factors for disease.

The risk factor version of epidemiology has been successful in many areas (smoking and lung cancer has already been mentioned as an example). But this particular view on the world has limitations (Pearce, 1996). In the previous section I described an approach to genetic susceptibility that focused on one individual after another, without stepping back and considering how these individuals related to one another. In the same way, risk factor epidemiology runs the risk of overlooking important aspects of the structure of populations. For example, it is commonly assumed that the frequency of disease is determined by the proportion of the population that experiences the relevant "exposures". But this is only part of the story: the incidence of disease is determined also by the relationships between individuals or, in other words, the arrangement or structure of the population. This is evident in the case of infectious diseases. The spread of infection with HIV-1, for example, is a function of the number of individuals who are infected and the average number of sexual partners. However, the pattern of sexual relationships is also a key variable. Simulations show that the magnitude and time course of an epidemic of HIV infection are sensitive to the degree of mixing between different risk categories (Anderson, 1994). If all other variables are held constant, the eventual number of cases is much greater if there is a high degree of mixing between low- and high-risk categories than if sexual relationships predominantly occur within categories, although the rate of increase in the frequency of infection is slower initially. Here is an example of a population characteristic (the degree of compartmentalization of risk behaviours) that shapes vulnerability to the spread of infection. It would not be detected by a study that looked only at the characteristics of individuals with and without disease.

Connections may be harmful or protective. Social scientists have long been familiar with concepts of cohesion, connections and trust, expressed recently by the term "social capital". Epidemiologists have been slower to recognize the importance of these ideas, and to appreciate the implications for study design. Conventional, individual-level studies are suitable for investigating the association between health outcomes and individuals' experiences of social networks. However, research to do with the effects on health of phenomena that are a property of the group, not the individual, such as social capital, requires comparisons between rather than within populations. For example, states make a natural unit of observation in the U.S. for investigations of social cohesion and health (Kawachi *et al.*, 1997). In another setting, social causes of vulnerability to storm losses were studied in the Caribbean by comparing the experiences of island states (Rosenzweig & Parry, 1994).

Leading the development of new population-level methods in epidemiology are researchers in the fields of infectious disease and social determinants of health. For example, infection transmission models have been developed that deal explicitly with interactions between individuals and the powerful modifying effects of time (Koopman & Lynch, 1999). It is seldom the case that the important causes of disease operate exclusively at either the population level or the individual level. More often there are important influences on disease incidence that operate at a variety of levels of complexity, and it would be ideal to include in the one analysis characteristics of both individuals and populations. This is the purpose of multi-level techniques, as developed and applied in areas such as educational research and psychology and now increasingly used in epidemiology. In the social determinants field, researchers are wrestling with the best ways of applying these conceptually and computationally demanding techniques to help answer public health questions (Blakely & Woodward, 2000). Challenges include specification of variables (such as clarifying the distinction between summaries of individual-level variables and variables that describe emergent qualities of groups), identifying sources of error and untangling causal pathways (distinguishing, for example, between variables that confound and those that lie along the causal chain) (see Box 11.4).

Techniques such as infection transmission modelling and multi-level analyses have been developed in data-rich environments that support a variety of experimental and observational research. Global environmental change is a different kind of research setting, as pointed out many times in this book. Often the outcomes (such as climate change) have not yet been experienced – they lie in the future. This means analyses take the form of predictive models, drawing on data from climate variability "analogues". Epidemiologists face a challenge: to make the best use of methods such as infection transmission models while paying due regard to the very large uncertainties that are inherent in the study of large-scale phenomena that are geographically and temporally diffuse. Modelling the potential spread of

---

BOX 11.4
Multi-level research

Multi-level research aims to combine analyses of ecological or group-level effects with causal factors that operate at the level of individuals. In such studies it is desirable to:

- Maximize variation between groups (in general, increasing the number of study groups will improve precision more than increasing the number of individuals within each group).
- Consider the different mechanisms by which ecological (group-level) variables may operate:
  — Direct cross-level effects.
  — Cross-level effect modification (ecological variable modifies effect of individual-level exposure).
  — Indirect cross-level effects (ecological variable acts through intermediate individual-level causal factor).
- Handle potential intervening variables with care (ideally, present results with and without adjustment for individual-level co-variates that may be confounders or may be part of the causal chain).
- Consider whether the ecological units are the most appropriate groupings (effects may depend on the level of aggregation).
- Explore the effects of choosing different models and datasets (testing, for example, possible influences of nonlinearity and collinearity).

Reference: Blakely & Woodward (2000).

---

malaria in a warmer and wetter world provides a good example (Martens *et al.*, 1995). The work of Martens and colleagues developed and expanded during the 1990s, with increasingly sophisticated models of the interactions between vector, pathogen and climate. Their approach could be extended to include projections of host numbers, lifestyles and susceptibility. However, these projections are based on many assumptions, and investigators are in a bind. Unless we can reduce the underlying uncertainties, the more comprehensive these models become the more difficult it may be to interpret the results.

11.4 Conclusion

Koopman has written "epidemiology is in transition from a science that identifies risk factors to one that analyses the systems that generate patterns of disuse" (Koopman, 1996). Such a change is needed to obtain a better understanding of

global environmental change, vulnerability and human health. But it will require
rethinking the methods used by epidemiologists, with particular attention to two
points: recognizing that populations are more than the sum of the individuals included, and incorporating and elaborating variations in susceptibility to ill-health.
We should not lose sight of the reason for this rethink. A double-barrelled response
is needed to the new generation of long-term, widely disposed environmental hazards. It makes sense to pursue both mitigation and adaptation options.

References

Adger, W. N. (1996). Approaches to vulnerability to climate change. *CSERGE Working
Paper GEC 96-05*. Norwich: Centre for Social and Economic Research on the Global
Environment, University of East Anglia.

Adger, W. N. (1999). Social vulnerability to climate change and extremes in coastal
Vietnam. *World Development*, **27**, 249–69.

Adger, W. N. & Kelly, P. M. (1999). Social vulnerability to climate change and the
architecture of entitlements. *Mitigation and Adaptation Strategies for Global
Change*, **4**, 253–66.

Anderson, R. M. (1994). The Croonian Lecture, 1994. Populations, infectious disease and
immunity: a very nonlinear world. *Philosophical Transactions of the Royal Society of
London Series B*, **346**, 457–505.

Anonymous (2000). UN predicts half the teenagers in Africa will die of AIDS. *British
Medical Journal*, **321** (7253), 72.

Balinska, M. (2000). Tuberculosis is spreading in central and eastern Europe. *British
Medical Journal*, **320** (7240), 959.

Begon, M., Harper, J. & Townsend, C. (1990). *Ecology, Individuals, Populations and
Communities*. London: Blackwell.

Berkes, F., Folke, C. (eds). (1998). *Linking Social and Ecological Systems: Management
Practices and Social Mechanisms for Building Resilience*. Cambridge: Cambridge
University Press, 1998.

Black, F. L. (1992). Why did they die? *Science*, 258, 1739–40.

Blakely, T. & Woodward, A. (2000). Ecological effects in multi-level studies. *Journal of
Epidemiology and Community Health*, **54**, 367–74.

Chapman, S. (1986). If your only tool is a hammer, then all your problems appear like
nails. Unpublished Paper. Presented to *Australian Community Health Association
Conference*, Adelaide 1986.

Copley, J. (1998). Recipe for disaster. Why a weakening hurricane wrought such havoc.
New Scientist, 14 November, p. 5.

Davis, M. (2001). *Late Victorian Apocolysis*. London: Verso.

Dockery, D. W., Pope, C. A., Xu, X. *et al.* (1993). An association between air
pollution and mortality in six US cities. *New England Journal of Medicine*, **329**,
1753–9.

Doll, R. & Hill, A. B. (1950). Smoking and carcinoma of the lung. *British Medical
Journal*, **ii**, 739–48.

FAO/WFP (1999). *Crop and Food Supply Assessment Mission to the Democratic
People's Republic of Korea* [special report]. Geneva: Food and Agriculture
Organization.

Garenne, M. & Aaby, P. (1990). Pattern of exposure and measles mortality in Senegal. *Journal of Infectious Diseases*, **161**, 1088–94.

Houghton, J. J., Meiro Filho, L. G., Callander, B. A., Harris, N., Kattenberg, A. & Maskell, K. (eds). (1996). Climate change 1995 – the science of climate change. Contribution of Working Group I to the *Second Assessment Report of the Intergovernmental Panel on Climate Change*. Cambridge: Cambridge University Press.

International Federation of Red Cross and Red Crescent Societies. (1996). *World Disasters Report 1996*. Oxford: Oxford University Press.

Jones, G., Robertson, A., Forbes, J. & Hollier, G. (1990). *Environmental Science*. London: Collins.

Kaiser, J. (2000). Rift over biodiversity divides ecologists. *Science*, **289**, 1282–3.

Kaly, U. & Pratt, C. (2000). Environmental vulnerability index: Development and provisional indices for Fiji, Samoa, Tuvalu and Vanuatu, EVI Phase II Report. *SOPAC Technical Report 306*. Fiji: South Pacific Applied Geoscience Commission.

Kawachi, I., Kennedy, B. P., Lochner, K. & Prothrow-Stith, D. (1997). Social capital, income inequality and mortality. *American Journal of Public Health*, **87**, 1491–8.

Koopman, J. S. (1996). Emerging objectives and methods in epidemiology [comment]. *American Journal of Public Health*, **86**, 630–2.

Koopman, J. S. & Lynch, J. W. (1999). Individual causal models and population system models in epidemiology. *American Journal of Public Health*, **89**, 1170–4.

Kulshreshta, S. N. (1998). A global outlook for water resources to the year 2025. *Water Resources Management*, **12**, 167–84.

Law, M. R, Morris, J. K. & Wald, N. J. (1997). Environmental tobacco smoke exposure and ischaemic heart disease: an evaluation of the evidence. *British Medical Journal*, **315**, 973–80.

Levin, S. A. (1998). Ecosystems and the biosphere as complex adaptive systems. *Ecosystems*, **1**, 431–6.

Levin, S. A., Barrett, S., Aniyar, S., Baumol, W., Bliss, C. *et al.* (1998). Resilience in natural and socioeconomic systems. *Environment and Developmental Economics*, **3**, 225–36.

Martens, W. J. M., Niessen, L. W., Rotmans, J., Jetten, T. H. & McMichael, A. J. (1995). Potential risk of global climate change on malaria risk. *Environmental Health Perspectives*, **103**, 458–64.

McCarthy, J. J., Canziani, O. F., Leary, N. A., Dokken, D. J. & White, K. S. (2001). Climate change 2001: Impacts, adaptation, and vulnerability. Contribution of Working Group II to the Third Assessment Report of the Intergovernmental Panel on Climate Change. Cambridge: Cambridge University Press.

McMichael, A. J. (1993). Global environmental change and human population health: a conceptual and scientific challenge for epidemiology. *International Journal of Epidemiology*, **22**, 1–8.

McMichael, A. J. (1994). Invited commentary – molecular epidemiology : new pathway or new travelling companion? [see comments]. *American Journal of Epidemiology*, **140**, 1–11.

McMichael, A. J. (1995). The health of persons, populations and planets: epidemiology comes full circle. *Epidemiology*, **6**, 633–6.

National Health and Medical Research Council (1997). *The Health Effects of Passive Smoking. A Scientific Information Paper*. Canberra, Australia: Commonwealth of Australia.

Nigg, J. M. (1995). Disaster recovery as a social process. In *Wellington after the Quake: the Challenge of Rebuilding*, pp. 81–92. Wellington: The New Zealand Earthquake Commission.

Nimura, N. (1999). Vulnerability of island countries in the South Pacific to sea level rise and climate change. *Climate Research*, **12**, 137–43.

Omenn, G. S., Goodman, G. E., Thornquist, M. D., Balmes, J., Cullen, M. R., Glass, A. *et al.* (1996). Effects of a combination of beta carotene and vitamin A on lung cancer and cardiovascular disease. *New England Journal of Medicine*, **334**, 1150–5.

Parry, M. L. & Rosenzweig, C. (1993). Food supply and risk of hunger. *Lancet*, **342**, 1345–7.

Pearce, N. (1996). Traditional epidemiology, modern epidemiology and public health. *American Journal of Public Health*, **86**, 668–73.

Rothman, K. J. (1986). *Modern Epidemiology*. Boston, MA: Little, Brown and Company.

Rosenzweig, C. & Parry, M. (1994). Potential impact of climate change on food supply. *Nature*, **367**, 1933–8.

Wenner, H. A., Macasaet, F. D., Kamitsuka, P. S. & Kidd, P. (1968). Monkey pox. I. Clinical, virologic and immunologic studies. *American Journal of Epidemiology*, **87**, 551–66.

Wilcox, A. J. (1995). Molecular epidemiology: collision of two cultures. *Epidemiology*, **6**, 561–2.

Woodward, A., Guest, C., Steer, K. *et al.* (1995). Tropospheric ozone: respiratory effects and Australian air quality goals. *Journal of Epidemiology and Community Health*, **49**, 401–7.

12

Dealing with scientific uncertainties

TIM O'RIORDAN & ANTHONY J. McMICHAEL

> That the unknown is like the known makes science possible; that it is also
> unlike the known makes science necessary. This conflict is the reason
> that all theories are eventually proven to be wrong, limited, irrelevant, or
> inadequate.
>
> *Levins, 1995*

12.1 Introduction

The study of global environmental change and its current and potential health impacts encounters uncertainties at many levels. These have profound implications for the form and content of science, scientific communication and policy-making. Indeed, it was in the course of the first substantive formal international discussion of these large-scale environmental issues, at the 1992 UN Conference on Environment and Development (the "Rio Conference"), that the Precautionary Principle was clearly enunciated and endorsed. That principle states, in essence, that scientific uncertainty in relation to a phenomenon with potentially serious, perhaps irreversible, consequences does not justify lack of preventive action.

During the past decade, the important role of uncertainty has become especially evident in the scientific and public debate around global climate change and its impacts upon human societies and population health. In each such discourse, pertaining to global environmental changes, there is continuing debate about the nature and quality of the scientific evidence for both the process and its impacts; about the relevance and legitimacy of multi-decadal-length modelling predictions; about the assumptions that can or should be made about future human societies and their capacity to handle changed circumstances; about the extent to which the medium to distant future should be discounted; about moral obligations between rich and poor nations and between present and future generations; about the extent to which decisions can be made within an orthodox economic framework by assigning money values to all present and future variables in the equations; and about the decision-making structures and forms of international governance appropriate to this type of large-scale phenomenon.

Consider, first, the scientific evidence. Very often, the familiar intellectual comforts of steady-as-she-goes empirical scientific research and the apparent precision that comes from reductionist analyses (of well-specified, discrete variables) and the attendant specific risk assessments do not apply in this domain. Many of the ecological, biophysical and social processes are too inherently complex, dynamic and feedback-rich to be sensibly reducible to conventional bivariate or multivariate models or even to be represented by directed acyclic graphs. In the words of Karl Weick (1999): "Global issues that involve organizing on a massive scale have been described as contested, nonlinear metaproblems with long lead times, unintended side effects, unclear cause–effect structures, and consequences that are often irreversible." For the classical scientist, denied the opportunity of experimentation, the lack of clarity of cause–effect structures is sufficient to preclude orthodox reductionist approaches to modelling and analysis.

A particularly severe form of uncertainty exists when information is insufficient to allow the probabilities of occurrence to be estimated. Whenever a probability can be estimated, then a statement about the level of risk can be made – either the absolute or relative risk, albeit with varying degrees of confidence. In some other situations, however, while environmental change and indeed its direction may be foreseeable, if the actual probabilities of occurrence and magnitude of change (and its impacts) are not estimable then the situation is inherently uncertain. This comparison was well summed up by Frank Knight in 1921, who said, "If you don't know for sure what will happen, but you know the odds, that is risk; if you don't even know the odds, that's uncertainty." Nevertheless, there is no simple dichotomy here. Rather, we are spanning a continuum that extends from situations of high information and quantitative precision, with some uncertainty, to situations of great uncertainty, beset by complex processes, often displaced into the future, and with imprecise, perhaps qualitative-only, estimates of future outcomes.

In this chapter we explore the nature of the myriad uncertainties that apply in this domain of research and policy-making. We then examine the implications for risk assessment and risk management, particularly considering the need for a broadly based consultative process that embeds the responsibility for appraising and using the scientific "knowledge" within the community at large.

12.2 Classifying "uncertainties"

There are many ways in which the uncertainties (or "incertitude") in science can be defined. van Asselt (2000: pp. 82–90) offers a useful overview of this topic. She notes the pioneering role of Weinberg (1972) who considered that certain generic aspects of scientific enquiry are "trans-scientific", i.e. beyond the normal rules of conventional scientific method. These related to four categories of question in scientific research:

- Questions that require very lengthy experiments or prolonged monitoring.
- Questions that depend on predicting human behaviour.
- Questions pertaining to the future.
- Questions that involve substantive value judgements.

van Asselt (2000) describes two primary sources of uncertainty (see Fig. 12.1). The first results from the inherent variability of the events or phenomena being studied, reflecting the natural rhythms of chaos or the equivalent. The second type of uncertainty results from quixotic or partisan human values and behaviour – characteristics that are unreliable (i.e. inexact or nonmeasurable) and inherently structural (i.e. reflecting ignorance or conflicting evidence). The types of uncertainty that result are, according to van Asselt, of a technical, methodological or epistemological kind. These in turn yield quantitative and qualitative, or structural, uncertainties.

Perhaps it is unavoidable that the characterization of "uncertainty" should itself be somewhat uncertain. Along with van Asselt's helpful typology of uncertainty, various other formulations have been published by others. The formulation portrayed in Table 12.1, for example, is based on the work of Hexham and Sumner (2000), and that for Table 12.2 on the work of Andrew Stirling (see O'Riordan *et al.*, 2001).

Table 12.2 proposes four distinctive "zones" of incertitude. These four categories are directly relevant to assessing the impacts of global environmental changes on human health, and they therefore have a bearing on the case studies that follow. The table characterizes the type of incertitude as a function of knowledge of outcomes (i.e. consequences) and knowledge of likelihood (i.e. probabilities).

Risk is the zone in which both the causally attributable outcomes and the likelihoods of exposure are reasonably well known. This enables formal risk assessment to be conducted, to estimate the dangers posed by each of the outcomes known to be caused by the exposure. Typical risk valuations in the engineering and chemical industries follow this procedure, sometimes referred to as hazard operational studies. This entails the use of both fault tree models (i.e. "what if?" scenarios) and frequency curves of numbers of possible fatalities for particular kinds of hazards. Note that "risk" is defined as the combination of causal processes with their likelihoods of occurrence (and not just the likelihood alone, as is widely believed). Therefore, we need to look at both the degree of certainty about cause and effect (outcome) and the probability of occurrence (likelihood and distribution). That dimension of certainty, or lack of it, is integral to both risk estimation and precautionary policies in the modern age.

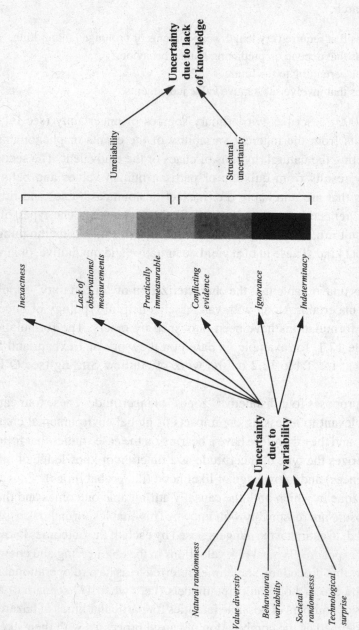

Fig. 12.1 Typology of sources of uncertainty (source: van Asselt, 2000).

Table 12.1. *Categories of scientific uncertainties (based on Hexham & Sumner, 2000)*

	Predictive uncertainty	Inductive uncertainty
Measurement uncertainty	Sampling error and limited monitoring capability	Evolving models of best-guess linkages
Process uncertainty	Deficient understanding of cause and effect	Creating scenarios of best-guess assumptions

This particular typology contrasts uncertainties that result from uncertain measurement, monitoring and data collection (top row within table) to those resulting from more fundamental uncertainties in human interpretive capacity and modelling abilities (bottom row).

Uncertainty applies where there is no firm basis for probabilities, yet there is some reasonably clear idea as to outcomes should an adverse probability come into play. An example of this is the possible health effects on young brains associated with the use of certain kinds of mobile phone. Under conditions of uncertainty, and especially where targeted groups or individuals could be vulnerable (e.g. young, elderly, weakened immunity), then the Precautionary Principle should be invoked. This means:

- Improving the use of cautionary advice about exposure.
- Opting to "over regulate" in the absence of better information.
- Delaying the use of a substance or procedure until the evidence becomes clearer.

Ambiguity applies to circumstances where the outcomes are not clear, but there is evidence of likelihood that is reasonably well developed. A good example is the effect of chemical prophylatics on lice-infested farmed salmon and their consequences for the marine environment. A further example in the case of farmed genetically modified salmon is the possibility of their interbreeding with wild stock. This entails no certain outcome for the health and vitality of the wild species, even though the likelihood of some interbreeding is quite high.

Table 12.2. *Four zones of incertitude: risk, ambiguity, uncertainty and ignorance*

	Knowledge about outcomes	
Knowledge about likelihoods	Outcomes well defined	Outcomes poorly defined
Some basis for estimating probabilities	Risk zone	Ambiguity zone
No basis for estimating probabilities	Uncertainty zone	Ignorance zone

Ambiguity inevitably poses difficulties for government, business and regulatory authorities: the evidence is never sufficiently well defined for authoritative judgements to be made on the basis of science alone. In such ambiguous situations, there is a strong case for sharing the responsibility for information and judgement with a wide range of interested parties. Hence, the key needs arising from the existence of ambiguity are to:

- ensure that all relevant information is shared with a representative array of interests;
- minimize using business confidentiality clauses as a reason for withholding information;
- seek alliances in advocacy with key public voluntary sector bodies based on precaution and prevention of harm; and
- share with the media the ambiguity of the evidence, and indicate a continuing willingness to be published or to be broadcast, preferably in situations of shared advocacy.

Ignorance prevails when the exposure is to an innovative or non-natural product or substance, and there is no prior information about cause-and-effect relationships to enable estimation of consequences. In this situation, science is ignorant – it cannot predict either likelihood or outcome. For example, the genetic modification of crops carries with it possible unintended genetic consequences, via inadvertent hybridization, for both wild species and commercial nonmodified varieties. We simply do not know all possible consequences, and science cannot provide a comprehensive answer via field trials or limited-term observation. Furthermore, the intrinsic variability of ecosystem processes creates a background "noise" that tends to obscure any discernible "signal" of cause and consequence.

Ignorance extends to areas of possible changes to social morality and social futures where people themselves have a legitimate interest in being part of the analysis. Thus, to eat genetically modified food, for some people, is a personal anathema – even in amounts that are regarded in regulatory terms as below the threshold of observation. The future planting of genetically modified plants depends hugely on how well they are regulated, and general public interpretations of the human nutritional benefit and environment-protecting values of the crops themselves. These are all arenas where many voices clamour for attention.

12.3 Risk, duty-of-care and adaptive action: the policy-making landscape

In recent years, these concepts of a duty-of-care, risk minimization or avoidance, and anticipatory and adaptive action have evolved into many international treaties and laws (see O'Riordan *et al.*, 2001). Indeed, the application of the Precautionary

Principle has been extended from the scientific to the political sphere as various parties have sought to use it to promote wider, sometimes illegitimate, political ends. (For example, recent efforts by the European Union to control the introduction of commercial genetically modified crops, ostensibly because of the uncertain biological and ecological risks involved, appear to have invoked the Precautionary Principle largely as a way of controlling international trade.)

This genre of scientific thinking, and its connections to policy-making, has particular relevance to today's large, complex and uncertain processes of environmental (and social) change – where the attendant risks may be rather low but of great potential severity, and where there are diverse points-of-view and contending values. In such situations the lines of risk management are neither clear nor readily consensual. Therefore, prudent and accommodating decisions will need to be made. This is the essence of what has been called "post-normal" science (Funtowicz & Ravetz, 1993). Normal science, grounded in classical Cartesian–Newtonian reductionism, envisages specific, material, constant processes (Tarnas, 1991). Scientists and policy-makers are confident of reality, there is an expectation of predictability, and research serves to reinforce and fine-tune hypotheses. Paradigms remain unshaken. In policy-making, the applicable social values are, seemingly, clearcut. However, as has been pointed out by Woodward in Chapter 11, epidemiological research, faced by the advent of global environmental change, "must encompass . . . models of extended causation, multiple levels of analysis, and differential vulnerability to disease and injury".

Meanwhile, there is a growing recognition of the importance of analysing patterns of health and disease at the population level, thereby attaining a more integrated basis for understanding and for preventive strategies (McMichael & Beaglehole, 2000). This recognition is part of the emerging "ecological" perspective that will become increasingly important in a world undergoing scalar shifts in social, economic and environmental influences on health. The various radical processes of globalization, urbanization, population ageing and global environmental changes are bringing larger and more complex influences to bear on the health of human populations. Increasingly, the perturbation of large complex systems – the climate system, the hydrological system, ecosystems and urban social systems – is posing risks to human health. Interestingly, recognition of this ecological, systems-based dimension to human population health accords with the ongoing broadly based rise of post-modernist ideas in science, acknowledging the complexity, uncertainty, relativism and conditionality of scientific knowledge. Scientists and policy-makers must learn to communicate in these terms, and to understand that while uncertainties (especially about the future) abound, where the stakes are high the need for enlightened and precautionary policies is paramount.

12.3.1 The example of climate change and its impacts

These issues are well illustrated by the steadily maturing discourse about the sources, characterization and communication of uncertainty in relation to global climate change and its impacts.

Climatologists have recognized the irreducible nature of most of the uncertainty that attaches to modelling the world's future climate, especially the variability due to uncertainties about the future shape, form, values and technological practices of human societies. They have therefore deliberately chosen to use the term "scenarios", as opposed to climatic predictions. This term makes explicit that the description of a future climatic configuration is based on various assumptions – and that there are many alternative future scenarios. Here, of course, a difficulty arises for policy-makers. How likely are these alternative scenarios? Which should be used? Should the presumptive worst-case and best-case scenarios be used to assess the range of potential consequences? In fact, because of the uncertainties in the situation, it is not clear which scenarios are more likely and which are less likely – the range of plausible future climate scenarios is *not* a probability distribution. Strictly then the scenario at the centre of the range is not more likely than those scenarios towards the ends of the range. To call a risk assessment a "scenario-based risk assessment" is to recognize that it is, in fact, an uncertainty-based risk assessment. Although the estimated impact provides an indicative forecast, the lack of an associated formal statistical probability precludes an unconditional risk-based policy decision.

In the first two Assessment Reports of the Intergovernmental Panel on Climate Change (IPCC), published in 1990 and 1996, uncertainties attaching to climate-change scenarios were primarily handled by presenting an array of temperature ranges, sea-level rises and precipitation charts (IPCC, 1990, 1996). These uncertainties reflected the dual weaknesses of the inaccuracy and incompleteness of the input data and the approximations, assumptions and untested nature of the models used to then process the data. The IPCC's Third Assessment Report, published in 2001, not only takes into account improvements in the estimated values of these parameters, but also includes a range of plausible future-world scenarios of socio-economic and political circumstances, translated into greenhouse gas emissions and, through those, into climate-change parameters (IPCC, 2000). This is the first time that the IPCC's climate scientists have collaborated with the social–economic scientists to create coherent linkages between socioeconomic behaviour and climate change.

Human societies, of course, have long experience of uncertainty-based policy-making. For example, we avoid locating housing developments around nuclear power plants because of the recognized finite but nigh-unquantifiable risk of serious accident. We have taken various actions to prevent the final extinction of many species of plants and animals, in part because of concerns about likely, though uncertain, knock-on consequences for the functioning of ecological systems. Yet it is also clear that many such precautionary decisions are delayed or otherwise hampered by a lack of information about quantifiable risks, and hence also a lack of information about the likely economic costs to society.

There is a need to clarify the relationship between these two domains, of risk-based and uncertainty-based policy-making. That will help to diminish our discomfort as we come to terms with the as-yet unfamiliar inevitability of a substantial amount of uncertainty, as a property of the systems and processes in which changes are occurring. Hulme (2000) has claimed that, over recent years, many climate scientists have embraced the notion of probability-based statements about how the climate in future may change. These statements, which include explicit caveats (referring to necessary assumptions), fit more readily into the conventional risk assessment framework. He states, "This development in the way we talk about future climate change is therefore both possible and necessary. Possible because of improved techniques for representing uncertainty in climate change scenarios; and necessary because resource managers and social planners are better able to incorporate such information into decision-making frameworks." (Hulme, 2000).

In the popular mind nowadays, science is judged more by what it does for the good of humanity than by its purpose or method. A survey of MORI polls found that more than half of a representative sample of the British public judged science by its social value, or possible environmental harm, rather than by its role of looking at and predicting the workings of the world (Worcester, 2000: pp. 9–11). In general, science is also valued according to views on the credibility and reliability of its organizational setting and funding. Scientists working for government and industry are trusted by only half of the population. Scientists operating through universities are deemed credible by 60%, while scientists working for environmental groups are trusted by over three-quarters of the population. These belief ratings relate to the general faith in the supporting and surrounding institutions. They have very little to do with the qualities of the actual science.

12.4 Dealing with risks

Risk research has reached the important stage where it is now generally agreed that risk perception is a process locked into cultural discourse and personal judgement, and this is the framework within which science is interpreted. Science is itself subject to cultural norms and institutional framing. All this is fairly familiar to social scientists interested in the sociology of science (see, for example, the excellent summary by Jasanoff & Wynne, 1998). But it has taken some time for the scientific community as a whole to come to terms with the fact that they have a partnership role to play in risk perception and management. Moreover, the way in which science handles uncertainty has a powerful influence on the relationships formed with the various interested parties who make up the "public interest" in a reasonably well functioning democracy.

How these interested parties, or stakeholders, view the effectiveness of the handling of uncertainty is a function of trust, transparency, inclusiveness of negotiation and a communicative process that embraces civic science within traditional scientific procedures. The handling of scientific uncertainty in the modern age is as much a function of democratic participatory style as it is a matter of scientific integrity. This development is rewarding for science, as it extends the finest of scientific traditions into a coupled relationship with the active civil society. Handling global environmental change has arrived at a time when a more negotiated approach to achieving social consensus is in vogue.

12.4.1 The changing interpretation of risk

Three volumes published by the Royal Society (1983, 1992, 1997) reveal the progression in the relationship between the natural and social sciences in the understanding of the processes underlying the perception of risk. A valuable summary of this evolution and the consequences for climate change policy analysis is provided by Thompson and Rayner (1998a). In the same volume, Jasanoff and Wynne (1998) provide an extensive review of social studies of scientific knowledge, and Thompson and Rayner (1998b) look widely at cultural discourses underlying science, notably with respect to climate change. The text that follows draws on all three of these excellent chapters.

The early Royal Society volume indicated that there was no meeting of the minds between statisticians, natural scientists and those social scientists who argued that risk perception was a process that was embedded in the very ways in which science was constructed. The more scientists argued that the matter of probability was a statistical exercise, located in burdens of proof, falsification and informal judgement in which "the public" played little part, the more the social scientists claimed that these very attitudes actually altered the popular perception of risk.

In the two decades since the first Royal Society study, the interpretation of risk has moved from being a technical and essentially specialist exercise into a coupled combination of scientific analysis and social discourse. The phrase "social discourse" covers the means through which individuals take their cue as to danger, trust, ability to cope and accountability amongst those responsible for creating and managing risk. Jasanoff and Wynne (1998: pp. 16–20) survey the literature on the social construction of science, in which risk discourses are framed. They point out that knowledge and technological advance are part of a series of institutional contexts in which scientists and technical experts engage with multiple audiences of sponsors, specialist peers, other scientists, consumers, interpreters and users. Shared meanings, joint expectations, rules of behaviour and evaluation, and agreed goals all help to influence the ways in which knowledge is constructed and particular technologies interpreted and promoted. For example, the current debate over the appropriate role of genetically modified organisms in food technology is a living demonstration of both commercially influenced technological drives and a societal response that is, in part, a function of the apparent "nonaccountability" of that knowledge construction, at least as it appears to operate within certain global chemical companies.

It is worth reflecting on why this widening of the nature of risk perception has come about. One reason is the continued resistance by consumer and environmental groups to buying into a scientific and psychometric approach to risk perception. This is termed the "objectivist" approach by Thompson and Rayner (1998a: p. 165) where the problem is predetermined by a single definition. This approach assumed that there was a scientifically grounded "objective" risk, and that "lay" people distorted this because of their ignorance and emotional prejudice. Consumer opposition to nuclear power, endocrine disrupters, food safety hazards and to genetically modified organisms (GMOs) – which occurred despite scientific assurances of negligible risk – shook the scientific establishment, but softened up the politicians. The psychometricians also began to realize that the "numbers game" of dread and involuntariness that was often used to characterize risk adverse behaviour simply did not provide sufficient insight into the subtle social, cultural and political processes at work – processes that linked individuals to socially influential groupings, and to institutional criticism generally (see Marris *et al.*, 1997 for a review).

There is also a socio-psychological element that reinforces these belief structures through social bonds. Eiser (1994: p. 216) looks to social networks as a source for re-examining social bonds. The relevant theory here is social impact theory which is based on:

- The number of sources advocating a particular position.
- Their immediacy, or closeness, in space or time.
- Their strength in the form of authority or persuasiveness.

According to this analysis, individuals are responsive to how strongly others' opinions are held, and are most susceptible to the opinions of those close to hand. Social networks of varying configurations thus reinforce the individual patterning and connectivity of attitude that make up any belief structure. In this light, it is not surprising that Marris *et al.* (1997) in the Norwich study found that family, friends and nongovernmental organizations are the most trusted sources of knowledge for risks whose scientific interpretations are shrouded in uncertainty. One should bear in mind, however, that such social influences also have origins in general outlooks, and in the interpretation of media coverage.

In all this the "risk" itself is only part of the picture. Much more of the image is coloured by an individual's attitudinal history, experiences with danger, views on fairness and interpretations of truthfulness for those in authority, together with judgements about the naturalness or non-naturalness of the various processes or products that are being assessed.

Another angle to this is the scope for relating outlooks on trust and fairness to the ways in which people bond to each other and judge prescriptions determined for the public good by regulatory and political bodies. This is known as *cultural theory* (see Tansey & O'Riordan, 2000 for a detailed analysis). The reasoning behind this theory is that groups form constellations of outlook as determined by the nature of their social attachments. Four "cultural types" have been singled out for analysis. *Fatalists* are loosely connected and generally unhappy about social regulation. They tend to see risk and uncertainty as a lottery. *Market libertarians* are also disconnected, but like regulation so long as it ensures competitive advantage and encourages individual responsibility. *Hierarchists* like regulation and feel connected to society. They reach and accept a degree of order in risk regulation. *Egalitarians* are also bonded to each other, but do not like prescription. They seek much more fair and participatory means for managing uncertainty.

In general, these typologies do not hold firmly in reality. These categories are not fixed; for example, people move in and out of social relationships and they experience risk and uncertainty in different ways at different times of their lives. Nevertheless, an analysis of this kind does suggest that different kinds of messages should be created for different patterns of social outlook, building on credibility, openness and fairness.

12.4.2 Implications for science

In a valuable assessment of how scientists handle uncertainty, Schrader-Frechette (1996: p. 13) concludes that scientists make use of two frames for structuring their data and for solving problems. One is the Popperian *falsification* frame, in which, if rigorous replication fails to falsify a provisional hypothesis, then for the moment the hypothesis continues to stand. In the second frame, the study hypothesis is *provisionally accepted* as an apparently reasonable proposition. Whichever the frame, scientific uncertainty or certainty is presumed to be in a state of constant review and critical enquiry. However, where major political and/or commercial interests are involved, if the initial provisional acceptance of "no evidence of a health risk" is to be reversed, there are very great peer pressures on scientists not to make a type I error.

Because [scientists] are more interested in avoiding false positions (type I errors, where to err on the side of uncertainty may result in heavy consequences which are subsequently not shown to be justified), rather than false negatives (type II errors, where the presumption that there is no health risk is shown to be subsequently wrong), scientists place the greater burden of proof on the person who postulates some, rather than no, effect.

(Schrader-Frechette, 1996: p. 13)

Schrader-Frechette argues for an alternative approach, based on the ethical norm that those least able to defend themselves should be protected by greater use of a prudent, false-negative avoidance, approach – that is, an assumption that real risks exist but have not yet been detected. This is a tricky area, for it explicitly places science in the frame of more judgemental processes of risk assessment, where social norms, distributional fairness and the Precautionary Principle are brought into play. Various recent food health scares suggest that such a shift has begun to take place. The decisions in the U.K. to ban beef on the bone and to exclude blood donors who may conceivably have been exposed to the agents responsible for nvCJD (new-variant Creutzfeldt-Jacob Disease) are suggestive of this kind of response. The point here is that though the final decisions were clearly political, that is to favour a political party advantage in the voter's mind, they were guided by scientific advice that to err on the side of caution in such cases was appropriate.

This so-called maximin approach to uncertainty is increasingly followed where the risk is involuntary at the point of impact, unavoidable and likely to hurt those most removed from the cause and least knowledgeable of the possible consequences. This lies at the heart of the current treatment of scientific uncertainty. Where a class of risk is seen to be societal in scope, where individual freedom of choice to avoid the risk is not possible, or at least not guaranteed, and where judgements about acceptable exposure relate to ethical norms about communicating reliable and intelligible information and fair treatment, then the utility maximization rules, commonly associated with Bayesian statistics, are successfully challenged by the egalitarians and the maximin sympathizers.

What we really need for a future science is a coupling of the cognitive and the analytical, of the data acquisition and the ethical framing, of the possible outcomes and the social conditions that may alter each possible outcome as stakeholders negotiate their way through a maze of uncertainty. In short, we require – and are seemingly now ready for – a participatory science.

12.4.3 Deliberative and inclusionary science

To pursue participatory science is more easily proposed than achieved. Once science enters a realm of direct democracy it becomes imbued with the failings as well as the triumphs of democracy. Countless studies of democracy (for example, in Lafferty & Meadowcroft, 1996) have shown that power is not distributed evenly, that some interests are more likely to get their way than others in particular classes of policy and decision, that information is manipulated by bias and motivation, and that powerlessness is endemic to particular groups by virtue of their class, colour, creed, culture or their choosing to be peripheral to the currents of social change.

. All this is well analysed and, for the most part, well understood. What is less well understood is how to incorporate deliberative and inclusive procedures into the scientific discourse. Some definitions are appropriate here. The notion of deliberation is essentially the public discussion of issues that can only be explored and understood in relation to the public interest (Stewart, 1996: p. 2). Fishkin (1992: p. 12) puts it thus:

The distinction between the inclinations of the moment and public opinions that are refined by sedate reflection is an essential part of any adequate theory of democracy. Political equality without deliberation is not of much use, for it amounts to nothing more than the power without the opportunity to think about how that power should be exercised.

Ortwin Renn and his colleagues (1995) have examined how democratic procedures should be based on the building of trust, including representativeness, generating nondistorting communication and reaching open consensus. The key here is inclusiveness and consensus-building. This is a complicated literature, but the nub of the issue is as follows:

- *Trust* is only possible if interests feel connected, respected, listened to and inclusive.
- *Representativeness* is only possible if all participants are networked to all their constituent interests. This can best be achieved by the kinds of social impact connectors outlined earlier, and undertaken by local "facilitators". These are invaluable people who bridge individuals to their respective interest groupings, as much by awareness raising as by direct communication.
- *Inclusiveness* is only possible when all the relevant parties are connected to and trusting of one another. Inclusiveness is essentially an outcome, an emergent property, of a successful participatory process. It cannot be simply specified as an itemized input.
- *Fairness* is difficult to ensure in a democracy. Fairness comes out of empowerment, which in turn is a product of genuine respect, and the revelation that others' interests are part of one's own self interest. To achieve fairness, therefore, there needs to be agreement about what principles underlie justice and appropriate

treatment amongst the various social groupings involved. This is unlikely to be reached without a process of consensus-building, the result of confidence in the process and in one another, and appropriate use of compensation or liability rules.

BOX 12.1
Innovations in public participation

Attempts to widen the basis for consultation are becoming more confidently pursued in an age of more active civic involvement, and a recognition that expertise has to be linked to interest-group judgements and biases if politically acceptable decisions are to be reached. Some of the ways for incorporating inclusive and deliberative participation are listed below, but the examples are by no means exhaustive.

Judgements informed by deliberation

- *Citizens' juries*: a group of representative citizens listen to presentations and argument amongst experts and informed public opinion, and come to a consensus conclusion. They may be assisted by a facilitator, or a "judge", though this is not always required.
- *Consensus conferences*: a large group of representative citizens hear argument from a range of opinion on a matter of contested science. They vote, usually electronically, on a series of points raised by a moderator, but according to an agreed agenda.
- *Referendums*: a voter-wide poll, triggered by a "question" and informed by a variety of debating arrangements. In some countries (e.g. Switzerland, Sweden) and some states (e.g. in the U.S.) these have the force of law. Elsewhere, they are advisory to a legislature, but, in effect, politically binding. The level of discussion is variable.

Discourse techniques

- *Issues forums*: local forums that may meet over many months to discuss a range of issues, often hearing evidence. Many of these become round tables for particular themes, for example, Local Agenda 21.
- *Study circles*: groups of people in social settings aiming to be informed and to reach consensus on a range of issues.
- *Citizens' panels*: a mix of the two above, more formal and representative than the circles, but not so coherent as the forums generally, these have value at the local level, and are the focus of a variety of participatory techniques such as Future Search, Visioning Exercises, Integrated Assessment modelling.

Source: Stewart, 1996: pp. 3–14.

Inclusive deliberation for a participatory science requires a set of conditions that are hard to fill. There are various types of procedures as indicated in Box 12.1, but all require the following underlying commitments:

(i) *That elected politicians agree to share power.* In so doing, politicians should actually enhance their power finally to act. But that assumption is speculative in an uncertain new patterning of democracy, so most politicians remain deeply sceptical about real community democracy.

(ii) *That regulatory and executive agencies become fully transparent.* This means that they agree to allow interests to enter fully and openly into their deliberations. Again this is not easy, for most senior executive regulators are chary of relinquishing their expected and accessional decision rules. But in many areas of food safety, including GMO regulation, the entry of consumer interests and other public-interested individuals is an important innovation. So too is the introduction of the use of the World Wide Web for communicating agendas and minutes of deliberations, as is the principle of arriving at all decisions by consensus. This latter principle means that there is an obligation on all parties to move to a common position or jeopardize the outcome by the exercise of a voluntary veto.

(iii) *That participants accept responsibilities to reach agreement.* Rights to participate are not absolute. There is an obligation to incorporate the interests of other parties when inclusive procedures are invoked. Social networks usually throw up elites who do not always want to share power that they have recently attained. Inclusive procedures suggest they must do so, but in a manner that is dignified and fair.

Such conditions are difficult to find. Democracy may be well developed but it is not fine-tuned to inclusive deliberation. We need to find many case studies of extended and learning democracies to discover just how possible it is to share power, and to reach maximum outcomes. The advent of a more active civil society in recent years, well beyond the formal pressure groups, is a hopeful sign. So, too, is the emergence of Local Agenda 21 and civic involvement in sustainability indicators. These are two initiatives of the late 1990s that show the scope for more inclusive deliberation and a social and political norm.

However, the crucial issue is how seriously and purposefully these procedures are treated by those actually in power. Citizens' juries and consensus conferencing are well and good as deliberate procedures. The point is how they relate to other arrangements for scientific debate, such as advisory committees, specialist panels and technology assessments.

This is an arena for the most searching analysis. If business and science open up, as most are now doing, to a genuine debate, then it is desirable to find new institutional arrangements to promote the rules of transparency, inclusiveness, accountability and consensus via appropriate negotiation, following maximin objectives.

BOX 12.2
Decision-making under uncertainty

Global climate change provides an instructive example of decision-making under uncertainty. The role of the Precautionary Principle is central to this process. Here, in summary, is the strategic situation for dealing with greenhouse gas emissions and climate change.

Policymakers are obliged to respond to the risks posed by anthropogenic greenhouse gas emissions in the face of great scientific uncertainties. Climate-induced environmental changes cannot be reversed quickly, if at all, because of the long time-scales over which changes in the climate system occur. Decisions taken now may limit policy options in future because high near-term emissions would require deeper reductions in the future to meet any given target. Delayed action might reduce the overall costs of mitigation because of future technological advances but could increase the rate and eventual magnitude of climate change, and hence the adaptation and damage costs.

Policy-makers must decide to what extent they want to take precautionary measures to limit anthropogenic climate change by mitigating greenhouse gas emissions and enhancing the resilience of vulnerable systems by means of adaptation. Uncertainty does not mean that human society cannot position itself better to cope with the broad range of possible climate changes or protect against potentially costly future outcomes. Delay may impair the capacity to deal with adverse changes and may increase the possibility of irreversible or costly consequences. Options for mitigating change or adapting to change that can be justified for other reasons today and which make society more resilient to anticipated adverse effects of climate change are particularly desirable.

Without action to limit greenhouse gas emissions the Earth's climate will warm at an unprecedented rate. This would have adverse consequences for the natural environment and human society, undermining the very foundation of sustainable development. Proactive adaptive strategies to deal with this issue need to be developed, recognizing issues of equity and cost-effectiveness.

While protection of the climate system will eventually require all countries to limit their greenhouse gas emissions, the UN Framework Convention on Climate Change (UNFCC) recognizes the principle of differentiated responsibilities. The UNFCC also recognizes that developed countries and countries with economies in transition should take the lead in limiting their greenhouse gas emissions, in view of their historical and current emissions of greenhouse gases and their current financial, technical and institutional capabilities. Current emissions in developing countries are much lower, both in absolute and per capita terms. Even though it is well recognized that emissions from developing countries are increasing rapidly and are likely to surpass those from developed countries within a few decades (absolute terms, not per capita), their contribution to global warming will not equal that of developed countries until nearly 2100 – since the climate system responds to the cumulative, not annual, emissions of greenhouse gases.

Energy services are critical to poverty alleviation and economic development. Increased energy provision in developing countries is essential to alleviate poverty and underdevelopment, in a world in which two billion extremely poor people are without electricity. Hence the challenge is to assist developing countries, via various international financial instruments, expand their production and consumption of energy in the most efficient and environmentally benign manner. Furthermore, enhanced energy research and development for energy-efficient technologies and low-carbon technologies would allow the world to meet its energy needs in a more climate-friendly and air-quality-friendly manner.

Based on a presentation by Dr. Robert T. Watson to the Sixth Conference of the Parties, under the UN Framework Convention on Climate Change, The Hague, November 2000.

One arena worthy of closer examination is the emergence of wide-ranging scientific commissions of review and advice for both governments and citizens alike. The IPCC is an extensive example of this. The application of precautionary approaches and the handling of unavoidable scientific uncertainties is well illustrated in the argument propounded by the chairman of the IPCC, Dr. Robert Watson (see Box 12.2). Another example of such a science/policy interface body is the relatively new Agriculture and Environment Biotechnology Commission (AEBC) in the U.K. This body, with wide-ranging membership, encompasses science, biotechnology businesses, nonorganic and organic farming interests, consumers, healthcare, media and ethics. It is also distinguished by a very open method of working (www.aebc.gov.uk), holding open public meetings for direct involvement and observation, with fully published minutes, working documents, focus groups and other interactive procedures. It is too soon to evaluate just how successful these approaches will prove to be in the building of public trust. However, it is clear that the Commission will be authoritative because of its particular style, and because the national and devolved governments look to its particular method for working as a model for science policy advice in the future.

12.5 Conclusion

An integral part of the wider sea-change that characterizes contemporary science is the recognition that the world is less certain than our immediate predecessors thought. During the first third of the twentieth century, classical materialist physics was confronted by the several unsettling insights from relativity theory, quantum mechanics and the uncertainty principle. Space-time was curved, energy and matter were interconvertible, atoms were mostly empty, both the position and speed of an electron could not be known simultaneously, and so on. While the rise of molecular biology has seemed to provide a new dimension of specificity and detail, the concurrent rise of ecology over the past half-century has provided a counterpoint to that reinforced reductionism. We have learnt that, at the human and population scale, natural systems are complex, dynamic and nonlinear in behaviour, and that they display emergent properties. Social scientists have laid bare the misplaced presumptions of objective observation and value-free interpretation in the empirical sciences. Increasingly, the human genome is being perceived as an interactive system, not as a set of discrete, independent, on–off genetic switches.

Societies have been taken by surprise in recent decades by the appearance of HIV/AIDS, by the seemingly bizarre origins of BSE (bovine spongiform encephalopathy, or "mad cow disease") and nvCJD, by the startling rise and spread of antibiotic resistance among many common infectious disease agents, by the depletion of stratospheric ozone, by the unexpected concentration (via environmental distillation processes and food-chain bioaccumulation) of semivolatile synthetic chemicals in high-latitude regions, by the collapse of a succession of ocean fisheries, and by the relatively rapid retreat of glaciers and sea-ice in a warming world.

Within this type of world, science can no longer strive for complete knowledge, precision and simplicity. Decisions cannot be postponed until high levels of "certainty" are attained. Nor are there objective "facts" or universal values: science is as much about perceiving and interpreting as it is about observing. Hence the need, today, for an idiom of scientific enquiry and communication, and of attendant social discourse and decision-making that recognizes the incompleteness and provisionality of our understanding of the present world and our capacity to foresee the future world. Decisions will need to be taken in a spirit of compromise and accommodation, and according to the tenets of prudence lest (seemingly) low-probability, high-damage events occur because of a reluctance to take preventive actions.

The recent maturation of Western scientific thinking has coincided, maybe fortuitously, with the advent of human-induced global environmental change. Our social–political response to the latter should be enhanced by the conceptual fruits of the former. There is, then, a chance to achieve simultaneously the following: an interdisciplinarity of science appropriate to the complex problems at hand; a social engagement of scientific research and evaluation; a better appreciation of the intrinsic uncertainties and complexities of the world around us; and, consequently, a more prudent and farsighted approach to managing the risk inherent in the changing world.

For the theme of global climate change and human health – taken here to be representative of the wider agenda of global environmental change and health – this analysis suggests four possible implications for research and policy.

The first is that there will be health effects arising from climate change that can be forecast, and hence prepared for, but not predicted. The issue is whether political judgements that carry high costs can be made well in advance of the health and other social outcomes that justify such expenditures. Or whether to await the inevitable "crisis" and respond belatedly, and often less effectively but more expensively. One example of this would be the spread of infectious disease due to warmer and more humid conditions, another the onset of "new" i.e. geographically displaced, diseases as climate conditions migrate. The work on malaria cited in this book is a good example of this.

The second implication is to recognize that climate-change-induced ill-health may adversely affect those already vulnerable to disease and chronic illnesses. This could increase the "harvesting" effect of premature illness and death amongst the most exposed and vulnerable. Handling this level of "predictable" uncertainty will involve a matter of equity and social justice in political analysis. Hence the need for careful pro-active and cooperative links between epidemiologists and policy analysts, set in a more open "civic" dialogue of the kind introduced here.

The third implication is the capacity for unusual combinations of pathogenic activity to combine into genuinely unpredictable health effects. Some of these combinations could become life threatening to ecosystems, and possibly so to humans, yet very expensive to handle. We do not know. What we will require is a more systematic monitoring of unusual health effects in general populations, and especially vulnerable groups in society. That will require much dedication and careful planning, again involving full transparency.

Finally, there is the wider issue of climate-change impacts and who pays for the costs of treatment and anticipatory behaviour and policy. This would appear to require revenue for greenhouse emission levies or some other fund supplied by a "polluter pays" principle. As we have seen from this text such a link between cause of illness and payment of treatment is difficult to justify politically. Climate-change politics may increasingly result in some sort of "carbon levy" to pay for impact mitigation. Health consequences should not be lost in the heat of battle. This is why "uncertainty science" warrants its own day in the sun.

References

Burke, D. (1998). The regulatory process and risk: a practitioner's view. In *Science, Policy and Risk*, pp. 67–72. London: The Royal Society.
Eiser, R. (1994). *Attitudes, Chaos and the Connectionist Mind*. Exeter, U.K.: University of Exeter Press.
Fishkin, J. (1992). *Democracy and Deliberation: New Directions for Democratic Reform*. New Haven, CT: Yale University Press.

Funtowicz, S. O. & Ravetz, J. R. (1993). Science for the post-normal age. *Futures*, **25**, 733–55.

Hexham, M. & Sumner, D. (eds.) (2000). *Science and Environmental Decision Making*. Harlow, U.K.: Prentice Hall.

Hulme, M. (2000). Climatic Research Unit, University of East Anglia. www.cru.uea.ac.uk/home/text.htm

Intergovernmental Panel on Climate Change (IPCC). (1990). *Climate Change: The IPCC Scientific Assessment*. Cambridge: Cambridge University Press.

Intergovernmental Panel on Climate Change (IPCC). (1996). *Climate Change 1995*. Cambridge: Cambridge University Press.

Intergovernmental Panel on Climate Change (IPCC). (2000). *Special Report on Emission Scenarios*. Cambridge: Cambridge University Press.

Jasanoff, S. & Wynne, B. (1998). Science and decisionmaking. *Human Choice and Climate Change: The Societal Framework*, ed. S. Rayner & E. Malone, pp. 1–88. Columbus, OH: Battelle Press.

Knight, F. (1921). *Risk, Uncertainty and Profit*. New York City, NY: Harper and Row.

Lafferty, W. & Meadowcroft, J. M. (eds.) (1996). *Democracy and the Environment: Problems and Prospects*. London: Edward Elgar.

Levins, R. (1995). Preparing for uncertainty. *Ecosystem Health*, **1**, 47–57.

Marris, C., Langford, I., Saunderson T. & O'Riordan, T. (1997). Exploring the "psychometric paradigm": comparisons between individual and aggregate analysis. *Risk Analysis*, **17**, 303–12.

McMichael, A. J. & Beaglehole, R. (2000). The changing global context of public health. *Lancet*, **356**, 495–9.

O'Riordan, T., Cameron, J. & Jordan, A. (eds.) (2001). *Reinterpreting the Precautionary Principle*. London: Cameron May.

Renn, O., Webler, T. & Wiedermann, P. (eds.) (1995). *Fairness and Competence in Citizen Participation: Evaluating Models for Environmental Discourse*. Dorderecht: Kluwer.

Royal Society (1983). *Risk Assessment: a Study Group Report*. London: The Royal Society.

Royal Society (1992). *Risk: Analysis, Perception and Management*. London: The Royal Society.

Royal Society (1997). *Science, Policy and Risk*. London: The Royal Society.

Schrader-Frechette, K. (1996). Methodological risks for four classes of scientific uncertainty. In *Scientific Uncertainty and Environmental Problem Solving*, ed. J. Lemons, pp. 12–39. Oxford: Blackwell.

Stewart, J. (1996). *Innovations in Public Participation*. London: Local Government Management Board.

Tansey, J. T. & O'Riordan, T. (2000). Cultural theory and risk: a review. *Health, Risk and Society*, **1**, 71–90.

Tarnas, R. (1991). *The Passion of the Western Mind*. London: Pimlico Press.

Thompson, M. & Rayner, S. (1998a). Risk and governance part 1: the discourses of climate change. *Government and Opposition*, **17**, 139–66.

Thompson, M. & Rayner, S. (1998b). Cultural discourses. In *Human Choice and Climate Change: the Societal Framework*, ed. S. Rayner & E. Malone, pp. 265–344. Columbus, OH: Battelle Press.

van Asselt, M. B. A. (2000). *Perspectives on Uncertainty and Risk: the PRIMA Approach to Decision Support*. Dordrecht: Kluwer Academic Publishers.

Weick, K. (1999). Sensemaking as an organizational dimension of global change. In *Organizational Dimensions of Global Change*, ed. D. Cooperrider & J. Dutton, pp. 39–56. London: Sage.

Weinberg, A. (1972). Science and trans-science. *Minerva*, **10**, 209–22.
Worcester, R. (2000). Science and society: what scientists and the public can learn from each other. MORI International, 32 Old Queen Street, London (www.mori.com).

Bibliography

Greenpeace (1998). *Genetic Engineering: To Good to go Wrong?* Canonbury Villas, London: Greenpeace.

Grove-White, R., Macnaughton, P., Meyer, S. & Wynne, B. (1997). *Uncertain World: Genetically Modified Organisms, Food and Public Attitudes in Britain.* University of Lancaster, Lancaster: Centre for the Study of Global Change.

Krebs, J. & Kacelnik, A. (1997). *Risk: a Scientific View*, pp. 31–44. London: The Royal Society.

Lee, K. (1993). *Compass and Gyroscope: Integrating Science and Policy for the Environment.* New York City, NY: Island Press.

Marris, C., Langford, I. & O'Riordan, T. (1996). Integrating sociological and psychological approaches to public perceptions of environmental risks: detailed results from a questionnaire survey. CSERGE Working Paper GEC 96-07, University of East Anglia, Norwich.

Ministry of Agriculture, Fisheries and Food (1998). *The Food Standards Agency: a Force for Change.* CM3830. London: The Stationery Office.

O'Riordan, T., Marris, C. & Langford, I. (1997). Images of science underlying public perceptions of risk. In Royal Society (1998), 13–30.

Renn, O. (1998). Three decades of risk research: accomplishments and new challenges. *Journal of Risk Research*, **1**, 49–72.

Slovic, R. (1992). Perceptions of risk: reflections on the psychometric paradigm. In *Social Theories of Risk*, ed. S. Krimsky & D. Golding, pp. 84–116. New York City, NY: Praeger.

Index

Page numbers in italic indicate a reference to an illustration.

334